Carp: Biology and Culture

Springer
London
Berlin
Heidelberg
New York
Barcelona
Hong Kong
Milan
Paris
Santa Clara
Singapore
Tokyo

R. Billard

Carp: Biology and Culture

Springer

Published in association with
Praxis Publishing
Chichester, UK

Editor: R. Billard, Museum of Natural History, Paris, France
Translator: Dr. Lindsay Laird, Department of Zoology, University of Aberdeen, UK
Translation editor: Dr. Jenny Watson, Department of Zoology, University of Aberdeen, UK

Original French edition, *les carpes: biologie et élevage*
Published by © INRA, Paris, 1995

This work has been published with the help of the French Ministère de la Culture—Centre national du livre

SPRINGER-PRAXIS SERIES IN AQUACULTURE & FISHERIES
SERIES EDITOR: Lindsay Laird, M.A., Ph.D., FIFM, University of Aberdeen, UK
CONSULTANT EDITOR: Selina Stead, B.Sc., M.Sc., Ph.D., Scottish Agricultural College, Aberdeen, UK

ISBN 1-85233-118-6 Springer-Verlag Berlin Heidelberg New York

British Library Cataloguing in Publication Data
Billard, R.
Carp : biology and culture. - (Springer-Praxis series in aquaculture and fisheries)
1. Carp
I. Title
639.3′7483

Library of Congress Cataloging-in-Publication Data
Carp : biology and culture / R. Billard [editor] : [translator, Lindsay Laird].
p. cm. — (Springer-Praxis series in aquaculture & fisheries)
"Original French edition, Les carpes: biologie et élevage, published by INRA, Paris, 1995"—T.p. verso.
Includes bibliographical references (p.) and indexes.
ISBN 1-85233-118-6 (alk. paper)
1. Carp. 2. Fish-culture. I. Billard, R. (Roland) II. Title.
III. Series.
SH167.C3C3713 1999
639.3′7483—dc21 98-50710 CIP

Printed by MPG Books Ltd, Bodmin, Cornwall, UK

Cover design: Jim Wilkie
Typesetting: Originator, Gt. Yarmouth, Norfolk, UK

Printed on acid-free paper supplied by Precision Publishing Papers Ltd, UK

Preface

The rearing of carp for consumption is an ancient practice and takes place in ponds where extensive production methods are being adapted in more and more countries. At more than 11.5 million tons, the present level of carp reared accounts for some 50% of total world freshwater production (and 75% of fish production). Although Asian countries in particular are concerned with this, especially China, development is taking place in richer countries, caused by increasing demand for fish for consumption. Consumers are attracted to aquatic products, fish in particular, because of their nutritive and dietary qualities. These have been widely publicized with stress being put on the role of polyunsaturated fatty acids in the prevention of cardiovascular illnesses.

With world fishing yields stabilizing, owing to diminishing stocks of sea-fish and the consequent reduction of fish quotas, demand for fish will become increasingly difficult to satisfy and it will be necessary for manufacturers of fish products to ask fish farmers to fill this gap. This will require regular and standard supplies to cover consumers' demands. Diverse and irregular supplies from fishing cannot cater entirely for these needs and manufacturers will welcome the availability of farmed produce.

Catches from the fishing industry will always be wanted for their originality, diversity and fresh condition. Traditional consumers will continue to choose them but new consumer categories require new products (cooked dishes, processed products). The consumption of aquatic, in particular fresh-water, products has remained constant for decades probably because of their traditional presentation. This satisfies a precise category of consumer, but does not respond to the needs of a wider clientele. With new, developing markets maintained by well-structured systems of farmed production and processing, fish and other aquatic products will achieve a status based on quality, quantity and predictability of production rarely known before, at least in industrialized countries.

Various scenarios of production exist to cope with the foreseeable increase in demand. The potential for production by fish-farming at sea is limited in many countries because of competition from other activities, pollution and high production costs. Farmed production is often restricted to 'top of the range' products in limited amounts exploiting a relatively large number of fresh-water species (perch, tench, zander, pike, etc.). Mollusc production (oysters, mussels) which is very important in Europe is struggling to develop further because of the difficulty of managing and maintaining the right quality of environment. Trout production in fresh-water is unlikely to increase because of the limited number of favourable sites (good quality water not exceeding 20–22°C is required) and problems of environmental impact not yet totally resolved. Pond fish farming, in its present unintensive form (barely exceeding 200 kg/ha per year), has the potential for further development, but remains very seasonal; present systems do not allow for continuous production throughout the year. By adjustment and intensification, this type of production is capable of supplying significant quantities in temperate zones. Farming is well suited for some species (common carp, Chinese carp, American catfish) which comprise most of the current world production.

Other types of fresh-water fish farming could be developed, whether intensive (catfish, sturgeon) or more extensive (pike, tench, perch, etc.); but production costs will remain high, at least until all stages of production have been mastered. There is, in fact, a tendency towards diversification; brief analysis of world aquacultural output shows that although demand for fish is strong in industrialized countries, individual production of certain farmed species (e.g., trout) is limited. Consequently, demand switches from these species to other species or products. The fishing industry has always offered a great variety of species and so the consumer similarly expects a large range from the fish-farming industry.

Current trends include the establishment of production–manufacture–distribution systems similar to that in the case of catfish in the USA and integrated with other agricultural activities. This integration can have several forms:

- —supply of cereals and vegetable proteins for the manufacture of granules for fish. At present, fishmeal represents a sizable fraction of meal and oil for fish which will soon be insufficient to satisfy the demand from fish farming. So, recourse to vegetable proteins will soon be necessary and their food value will have to be studied as well as ways of incorporating them into granules. This has happened in the case of catfish production in the USA, where granules destined for fish growth contain cereals and vegetable proteins (such as soya) produced by local agriculture.
- —use of farming effluents (pig and chicken droppings) as organic fertilizers.
- —of course, increases in fish consumption will reduce consumption of other agricultural food products. Therefore, cattle, pig and chicken farmers will have to make adjustments and possibly consider fish-farming as a substitute.

This agro–fish-farming arrangement could come about according to different scenarios, at least at a production level (see diagram).

1. Intensive monoculture in well-adapted ponds with food of exclusively exogenous origin made up of artificial nutrients.

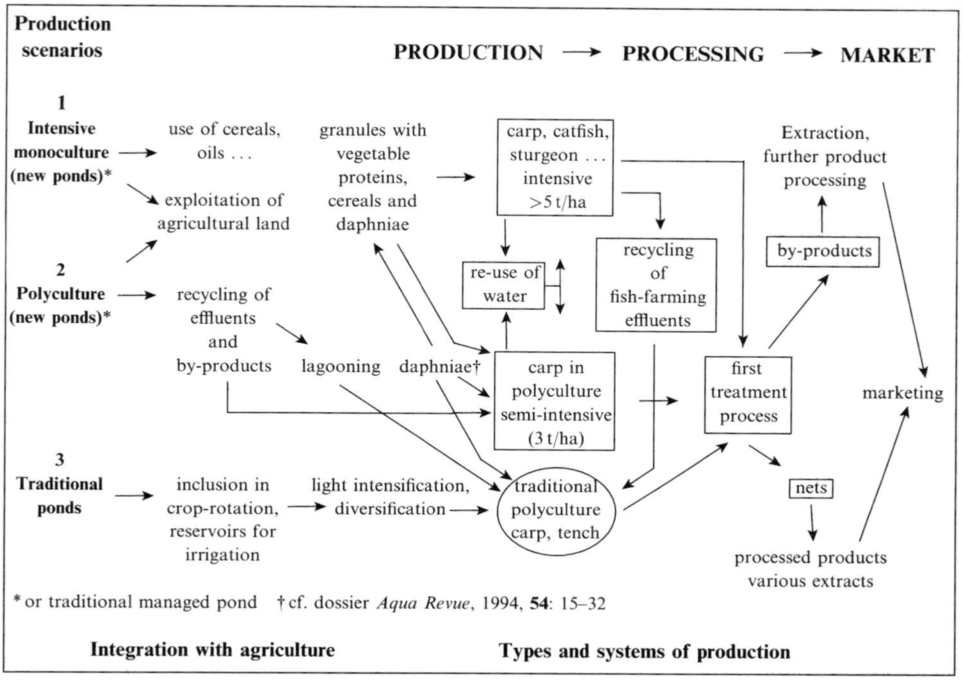

2. Polyculture also practised in well-adapted ponds with stimulation by fertilization, particularly organic, of the trophic structure, in order to produce plankton as food. Assuming intensification of effluent recycling, lagoons will produce living daphniae. Daphniae can also be provided in dried form in granules.
3. Production from traditional ponds suitable for diversification and slightly intensified by complementary feeding and organic fertilization, chiefly by farming effluents (of mammals/birds) but also from fish farming (scenarios 1 and 2). In fact the quality of water in scenarios 1 and 2 will be noticeably altered with regard to the physicochemical norms admissible for rivers, and the water could not be released in such condition into the natural environment; it could be recycled and purified in large ponds serving as lagoons. From this point of view a certain complementarity exists within the three scenarios. Simple re-use of water can be effected by pumping to other fish-farming ponds.

Processing and distribution are key parts of the system. At the processing level it is essential to evaluate all the products (fillets and by-products) and at the marketing level it is advisable to fully identify products which would appeal to the consumer.

The techniques used in different areas of this agro–fish-farming system will be described in this book; however, it is more concerned with scenarios 2 and 3 with an approach to the management of the trophic structure in ponds. Several, mainly European, specialists have contributed sections or chapters to this book. Some material is taken from data drawn from classical works on pond fish farming (marked * in the text, a list is given in the appendix) as well as further more recent or specific material taken from original papers in the specialist literature (the most significant references are inserted at the end of each chapter but are not generally mentioned in the text).

There is a great difference between theory and practical know-how as this book makes clear. Knowledge of reproduction, larval rearing and management of pond ecosystems is becoming consistent and such knowledge is being set down in the form of practical 'recipes'. Those relating to genetics, pathology and nutrition are important but are not all operational in pond fish farming, especially in the context of intensification. The inadequacies become even greater if the final parts of the system are considered: processing, distribution and economic analysis. Although there are clear opportunities in these areas, they are seldom made accessible in the literature.

The presentation of Chinese production methods shows how imprecise some of the available literature is. These very practical methods are essentially based on empirical know-how conveyed by word of mouth without ever being written down. This explains, at least in part, the difficulties of dissemination in other countries. Therefore, to the practitioner reading this volume it will appear that pond fish-farming techniques are somewhat vague with little standardized character. Refinement of these techniques will be for the fish farmer who will have to adapt to his/her own situation those recipes and descriptions taken in particular contexts. The importance of recording data of an experiential nature on all aspects of fish culture is emphasized throughout this book.

A great number of people have contributed to the making of this book and must be thanked. There are the authors of the principal contributions (p. 335) and those who provided edited elements, in particular R. Berka, V. Cinot, L. Horvath, J. Mullet i Cerda, J. Pokorny, O. Schlumberger, Zdenka Svobodova ... Various readers have been good enough to give critical advice at various stages of its creation: Christiane Ferra, Jacqueline Marcel, Catherine Mariojouls, Ghislaine Perches and Josette Sevrin-Reyssac, R. de Courson, L. Dabbadie, A. and J. Heyman, G. Maisse, C. de Montalambert, J.P. Pommereau, Y. Racapé ... Documentation was gathered by F. Nadot and M. Margout, the typing of the manuscript was by J. Barthelemy and J. Faurillon and proof reading by Françoise André. The drawing up of this volume has received financial support from the French Ministry of Agriculture and Fisheries (Aquacultural Office). The editor of this work would like to thank, in advance, readers and particularly professionals for comments they may wish to offer and which might contribute to improving a further edition of the book.

Contents

Processing, marketing and economics

Appendices

Contributing authors:

P. Bergot
R. Berka
R. Billard
Th. Boujard
A. Demaël
B. Fauconneau
P. Haffray
M.G. Hollebecq
M. Jacquinot
P. Kestémont
B. Lanoiselée
H. Le Louarn
C. Mariojouls
J. Marcel
P. Melotti
M. Natali
M. Morand
C. Salmomoni
J. Sevrin-Reyssac
M.A. Szumiec
L. Varadi

1

Biology of the cyprinids

1.1 SYSTEMATICS AND BIOGEOGRAPHY[1]

From the taxonomic point of view the cyprinids belong to the order Cypriniformes which is traditionally grouped with the Characiformes, Siluriformes and Gymnotiformes to make up the superorder Ostariophysi. These groups have certain common features such as being found predominantly in fresh water and their possession of Weberian ossicles, an anatomical structure originally made up of small pieces of bone formed from four or five of the first vertebrae; the most anterior bony pair is in contact with the extension of the labyrinth and the most posterior with the swim bladder. The function is poorly understood but it is presumed that this structure takes part in the transmission of vibrations from the swim bladder to the labyrinth and in the perception of sound, which explains why the Ostariophysi have such a great capacity for hearing. These structures are used for avoiding predators and play a part in social behaviour and in the formation of groups of individual species. It is essential to take into account this capacity for sound perception when devising methods for catching these species. Most cypriniformes have scales and teeth on the inferior pharyngeal bones which may be modified in relation to diet. *Tribolodon* is the only cyprinid genus which tolerates salt water: there are several species which move into brackish water but which return to fresh water to spawn (e.g. *Cyprinus acutiodorsalis*). All of the other cypriniformes live in continental waters and have a wide geographical range; they have colonised all continents with the exception of Australia.

The cyprinids constitute one of the largest families and best known of the cypriniformes. They are widely distributed in North America, Eurasia and Africa but are absent naturally from South America, Madagascar and Australia where man has introduced several species such as the common carp (*Cyprinus carpio*) (generally called carp in the remainder of this work). In these countries there are equivalent zoological and ecological neighbours: characins in South America, Melaenotaenids

---

[1] R. Billard.

in Australia and New Guinea. The diversity of cyprinid species is at its greatest in China and in South-East Asia but is lower in Africa and North America where there are nevertheless 36 genera and 280 species (118 belonging to a single genus *Nototropsis*). Fossil cyprinids go back to the Palaeocene in Europe, the Eocene in Asia and the Oligocene in North America. The cyprinids are dispersed to the east of the Wallace line, reaching Borneo, but have not crossed the Strait of Makassar. The carp originating from Western Asia is no longer represented in the wild form other than by the sub-species *Cyprinus carpio haematopterus* in Eastern Asia and *C. carpio carpio* in Eastern Europe.

From the point of view of karyology, cyprinids generally possess 48–50 pairs of chromosomes but some species such as the common carp and the Crucian carp have sometimes more. The goldfish of Northwest China, *Carassius auratus gibelio*, has as many as 75 pairs of chromosomes and a related species in Japan has 100: this is undoubtedly the result of the phenomenon of polyploidy. Hybridisation occurs frequently in cyprinids, resulting partly from environmental changes, in particular changes in their spawning habitat: these changes can be natural or due to man's activities. The cyprinids combine a great number of species and for most of these the number of individuals and most often the biomass is high; they are confined to restricted environments which submit them to strong social interactions and oblige them to occupy diverse ecological niches. Continental freshwaters support 41% of the species although the volume of water (126 000 km^3) is much less than that of marine waters (1.3 billion km^3) (the volume of water in lakes represents 1/1000 of the total and that of rivers 1/10 000). The mean number of individuals per species has been estimated at 117 billion (thousand million) in the sea and can vary between 83 million and 83 billion in fresh water. It can therefore be seen that the volume of water available to each individual is much reduced in fresh waters (between 1.5 and 1500 m^3). Figures given by Horn (1979) are only general indications with numerous exceptions but these suggest that in freshwater there is strong competition between individuals for space, food and for spawning. The cyprinids are especially important as they represent around 20% of the total number of species in continental waters (in Europe and Asia this can reach 50%), In a river, in undisturbed conditions, it is possible to correlate the total number of species including the cyprinids with the increase in the size of the watercourse and drainage basin. This is linked, among other things, to the increase and diversification of sources of food and habitats, in particular in the river and surrounding the side streams which are vital for spawning. Winfield and Nelson (1991) (see General bibliography) have dedicated a major work to the biology, systematics and exploitation of cyprinids.

1.2 SOME CHARACTERISTICS OF THE BIOLOGY OF THE CYPRINIDS

Cyprinids, in common with other members of the superorder Ostariophysi have several distinctive biological and morphological characteristics. Many cyprinids can tolerate a very wide temperature range: an adult common carp can survive temperatures from 1 to 35°C. The optimum temperature for rearing juveniles is above 30°C. They can also survive rapid temperature fluctuations. Many species of cyprinids can also tolerate major fluctuations in concentrations of dissolved oxygen, which occur frequently in summer in still waters due to photosynthetic activity. They can survive conditions of supersaturation or almost anoxia (<1 mg O_2/l) for a few hours in summer and for longer periods in winter under the ice. They are therefore capable of anaerobic metabolism. Carp can also survive a wide range of pH, from 5 to 9.

The dietary regime is extremely diverse: cyprinids feed at all trophic levels: higher plants, phytoplankton, zooplankton, zoobenthos, bacteria attached to detritus and even fish in the example of *Aspius* (asp) and *Ptychocheilus* (e.g. Colorado squawfish). Some species are extremely opportunistic and vary their diet according to season and environment (e.g. roach). Carp in ponds will begin by feeding on benthic organisms and when these are exhausted they will hunt larger zooplankton in open water. Many of their morphological and physiological characteristics are associated with this diversity and versatility in feeding regime and assist in the capture, break-up and digestion of prey. Cyprinids have no teeth in their jaws but some species possess pharyngeal teeth (e.g. *Ctenopharyngodon idella*, the grass carp) or a type of strainer at the base of the gills which allows the filtering of zooplankton to take place. Many species have barbels around the mouth. These are sense organs which, through touch and taste, help in the search for food. The length of the digestive tract also varies considerably in omnivorous species; it is 15 times the length of the body in herbivores which are feeding on plants but is considerably reduced (three to five times) when the diet is based on artificial pellets. Digestion of plants is always incomplete, for example that of algae by the silver carp (*Hypophthalmichthys molitrix*). In practice, few fish are completely vegetarian and in fresh water, the most spectacular of these are found among the Chinese carps.

In natural tropical or sub-tropical conditions herbivorous carps can gain several kilograms in weight every year. Growth is more modest in ponds in temperate regions with weight reaching 500 g at the end of the 2nd year and between 1 and 2 kg in the 3rd year for common, grass and silver carp species (Figure 1.1). Growth is faster in intensive culture because of the feeding rate which, in summer, can be up to

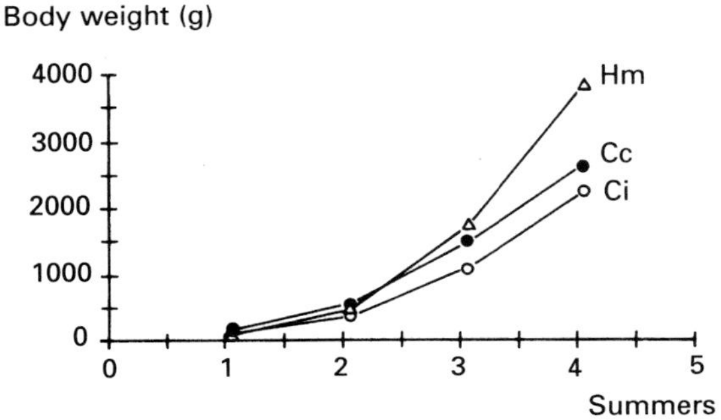

Fig. 1.1. Growth curves under natural conditions in Western Europe: common carp *Cyprinus carpio* (Cc) in Belgium (after Timmermans, 1989); silver carp *Hypophthalmichthys molitrix* (Hm) and herbivorous carp *Aristichthys* (or *Hypophthalmichthys*) *nobilis* (Ci) in Brittany, France (after Le Louarn, pers. comm.).

3% body weight per day for the common carp and even as much as 10% for grass carp. Cyprinids reach a wide range of sizes at maturity; some species measuring 10–11 mm (e.g. *Danionella translucida* Roberts, 1986), while other species are over a metre in length and can weigh over 100 kg; examples being *Barbus tor* in India and *Caltocarpio siamensis* in Indochina.

Modes of reproduction are extremely diversified in the Ostariophysi; in most species fertilisation is external but for some species there is internal fertilisation without viviparity (Breder and Rosen, 1966). In Natal, in South Africa, there is a species *Barbus viviparus* Weber 1987 whose name results falsely from observation of apparent viviparity. Parental care is very diverse; eggs may be released without any protection, spawned on various substrates, deposited inside molluscs, or in nests which may or may not be guarded or attached to the abdomen of the female (Bunocephalids). There are also species which incubate eggs in the buccal cavity. Egg size is very variable and ranges from less than 0.5 mm to several centimetres (up to 5 cm for the Ariids) in diameter.

The diversity of modes of reproduction is less within the cyprinid family although there is a wide variety of egg types. Some adhere to various substrates (plants, gravel, stones) some non-adhesive and floating above the bottom or in open water (demersal, semi-demersal or pelagic). The bitterling lays its eggs in mussels using an ovipositor which appears in females as a temporary sexual characteristic. In many species, at spawning time, the males have spawning tubercles on the head, operculae, fins (especially the anal fin) and the scales. In general, cyprinid spawning takes place in the spring in temperate regions; eggs are often small (around 1 mm diameter or even 0.5 mm as in tench), embryonic development is brief and on hatching the larvae must search for small food organisms such as ciliates, rotifers and then small crustaceans which are only found in sufficient quantities when water temperatures are increasing. This rise in temperature also determines the final maturation and spawning in addition to the presence of suitable substrates for spawning for species which deposit their eggs on plants.

1.3 CYPRINIDS AND AQUACULTURE

Over 11 million tonnes of cyprinids are produced by aquaculture worldwide each year: they are far and away the most exploited group of farmed fish. Annual production of salmonids is no greater than 0.6 million tonnes. Around 15 species of carp are cultured (Table 1.1) but most production is based on a few species (Table 1.2). The culture of cyprinids has ancient roots, particularly in China (a book by Fan Li was produced in the seventh–sixth century BC) but only the common carp, the Crucian carp (*Carassius carassius*) and the goldfish (*Carassius auratus*) have been truly domesticated with the management of all stages of the life cycle, particularly reproduction, for a long time. Controlled spawning of the Chinese carps is relatively recent dating back to the 1960s and relies on the technique of hypophysation; prior to this the juveniles were simply captured from rivers or lakes and on-grown in ponds. The characteristics of wild fish still remain in these species and these appear

Table 1.1. Principal species of farmed cyprinid, common and scientific names, main dietary regime in adults.

Common name	Scientific name	Diet
Common carp	*Cyprinus carpio* includes many different races	Benthos, zooplankton
Chinese carps		
Silver carp	*Hypophthalmichthys molitrix*	Phytoplankton, microzooplankton
Grass carp	*Ctenopharyngodon idella*	Plants
Mud carp	*Cirrhina molitorella*	Detritus
Bighead	*Aristichthys nobilis*	Zooplankton
Black carp	*Mylopharyngodon piceus*	Molluscs
Major species of Indian carp		
Catla	*Catla catla*	Zooplankton
Calbassu	*Labeo calbasu*	Detritus
Rohu	*Labeo rohita*	Detritus, plants
Mrigal	*Cirrhinus mrigala*	Detritus, plants
Bream	*Parabramis pekinensis*	Plants
Crucian carp	*Carassius carassius*	Detritus
Goldfish	*Carassius auratus*	Plankton
Tench	*Tinca tinca*	Detritus, benthos
Roach	*Rutilus rutilus*	Omnivore

during the capture of broodstock which are able to jump over 1 m in height. This is also true for the Indian carps whose juveniles are still taken from the wild. After several centuries of culture and managed spawning the common carp has been subjected to mass selection so that there are now several stocks which are not properly identified in fish culture units of Europe and Asia. Because of the number and diversity of fish species, it is likely that man has not engaged in a systematic creation of races which can be adapted to the production environment but rather that he has utilised the different species already in existence. The story is different for ornamental fish (such as the many different varieties of goldfish). The wide geographical range and number of cyprinids, as well as all of their physiological characteristics, particularly their resilience to wide temperature ranges and water qualities, their good growth rates at optimum temperature with acceptable food conversion coefficients, explain their success in aquaculture. Their ability to feed at different levels in the trophic web, particularly for Asiatic species, has been put to profitable use by fish farmers in developing the concept of polyculture. This is based on the association in a single water body of several species of fish attached to the main ecological niches in a pond which is treated with organic fertiliser to maximise the endogenous production of the system.

Table 1.2. Distribution and production of the principal species of farmed cyprinids throughout the world (FAO Fish Circular 815, Review 10, 1998).

	Species	Number of producing countries	Production (t/year) 1991 × 1.000 t	Main producer	
				Country	%
Hypophthalmichthys molitrix	Silver carp	35	2888	China	97
Ctenopharyngodon idella	Grass carp	25	2438	China	99
Cyprinus carpio	Common carp	2	1992	China	80
Aristichthys nobilis	Bighead	15	1418	China	99
Carassius carassius	Crucian carp	6	693	China	99
Labeo rohita	Rohu	3	493	India	86
Catla catla	Catla	3	420	India	99
Parabrama pekinensis	White amur bream	1	379	China	100
Cirrhina molitorella	Mud carp	2	130	China	100
Mylopharyngodon piceus	Black carp	2	120	China	97
divers cyprinids		16	41	Pakistan	44
Puntius gonionotus	Thai silver barb	3	32	Thailand	84
Puntius javanicus	Java barb	1	28	Indonesia	100
Osteochilus hasseltii	Nilem carp	1	12	Indonesia	100
Notemygonus crysoleucas	Golden shiner	1	9.5	USA	100
Carassius auratus	Goldfish	3	2.9	Italy	96
Rutilus rutilus	Roach	3	2.5	France	100
Tinca tinca	Tench	8	1.3	France	54
Misgurnus anguilicandatus	Pond loach	2	0.5	Korea	51

Several species:*Probarbus jullieni, Leptobarbus hoeveni, Misgurnus, Puntius* sp. whose production is only a few tens of tonnes are omitted.

One of the characteristics of the group is unfortunately a major handicap; this is the presence of small bones. It appears that global production is slowing down and even declining in the tropics (where there is a preference, for example, for tilapia in Israel, southern China and Taiwan) because these bones make eating the fish disagreeable, particularly to children. This characteristic is counteracted in European countries by filleting and by removing the bones.

1.4 INTRODUCTION OF CYPRINIDS BY MAN

A recent worldwide survey carried out by FAO has shown that over the last century there have been a total of 1354 primary introductions taking 237 species into 140 countries. Forty per cent of these introductions have been for farming purposes, the rest for sport fishing, commercial fishing and for the aquarium trade. Nineteen per cent of these introductions have been of salmonids, 16% tilapias, 10% cyprinids and 9% centrarchids.

The carp was probably one of the first species to be dispersed by man. From its origin in western Asia, the carp dispersed naturally to China and Siberia. It also reached the Danube basin where the Romans farmed it and transferred it to Western Europe. Later on, it was spread by monks during the 13th to 15th centuries (p. 101). Very detailed information on carp farming was given by Duhamel du Monceau in his *Traité général des pêches* (1769–1782) (see General bibliography). More recent developments have seen introductions for farming through all continents from the 1870s, followed later by Chinese species with a peak in the first decade of the 20th century and another in the 1960s which had as an objective the control of macrophytes and the elimination of micro-algae. Common carp have been introduced into 59 countries, Chinese carps into 113, goldfish into 26 and tench into 82.

These movements, while testifying to the interest in this group, have not been without damage to the environment: increase in water turbidity after the introduction of common carp and competition for food which has been serious in ponds and in rivers as in the case of *Pseudorasbora parva* (accidentally introduced into Europe at the same time as the Chinese carps and now actively spreading there). Added to the risks of genetic pollution to local stocks is the risk of population explosion for species that reproduce successfully in the new environment, often at the expense of competitors (e.g. common carp in Australia). There is also the danger of introducing diseases, in particular those caused by parasites.

BIBLIOGRAPHY

Berrat T.M., 1981. *An Atlas of Distribution of the Freshwater Fish Families of the World.* University of Nebraska Press, Lincoln and London.

Breder C.M., Rosen, D.E., 1966. *Modes of Reproduction in Fishes.* The Natural History Press, Garden City, New York, 941 pp.

Horn M.H., 1979. The amount of space available for marine and freshwater fishes. In: M.S. Love, G.M. Caillet (eds), *Reading in Ichthyology*. Goodyear Publ. Comp., Santa Monica, California, pp. 29–31.

Timmermans J.A., 1989. Données sur la croissance de quelques espèces de poissons dans les étangs de Campine. Travaux de la Station de Recherches forestières et hydrobiologiques, B-190. Groenendaal-Hœilart (Belgium). Série D. No. 56: 3–34.

Environment

2

The aquatic ecosystem and water quality

2.1 THE FOOD WEB IN PONDS USED FOR THE CULTURE OF CYPRINIDS[1]

The plants and animals present in ponds affect each other, directly or indirectly, and make up a complex trophic or food web which can be exploited by man for the production of fish. These organisms also have an effect on their environment: the concentration of oxygen, carbon dioxide, nutrients, transparency of the water and many other parameters depend largely on their abundance and species composition. In ponds used for fish culture, plankton development is greatly encouraged which differentiates them from natural ponds and from rivers. In practice the pond is fertilised or cultivated and the most abundant species are not necessarily those found in other equivalent fresh waters. The managed pond most frequently contains a high density of fish which by their presence alone have a profound effect on the environment.

This chapter outlines the characteristics of the plankton in carp ponds, the effect they have on fish production and how the farmer can change the composition and abundance of the plankton.

2.1.1 Principal constituents of the food web

Three main groups which have an important role in fish production can be distinguished among the plankton in ponds: bacterioplankton, phytoplankton and zooplankton. As well as those organisms which live in open water there are also those which are attached to the sediment, constituting the benthos (Figure 2.1).

Bacterioplankton

The bacterioplankton of lakes has been widely studied but there is little information available on that of ponds used for fish culture. In practice, most of the research is

[1] J. Sevrin-Reyssac.

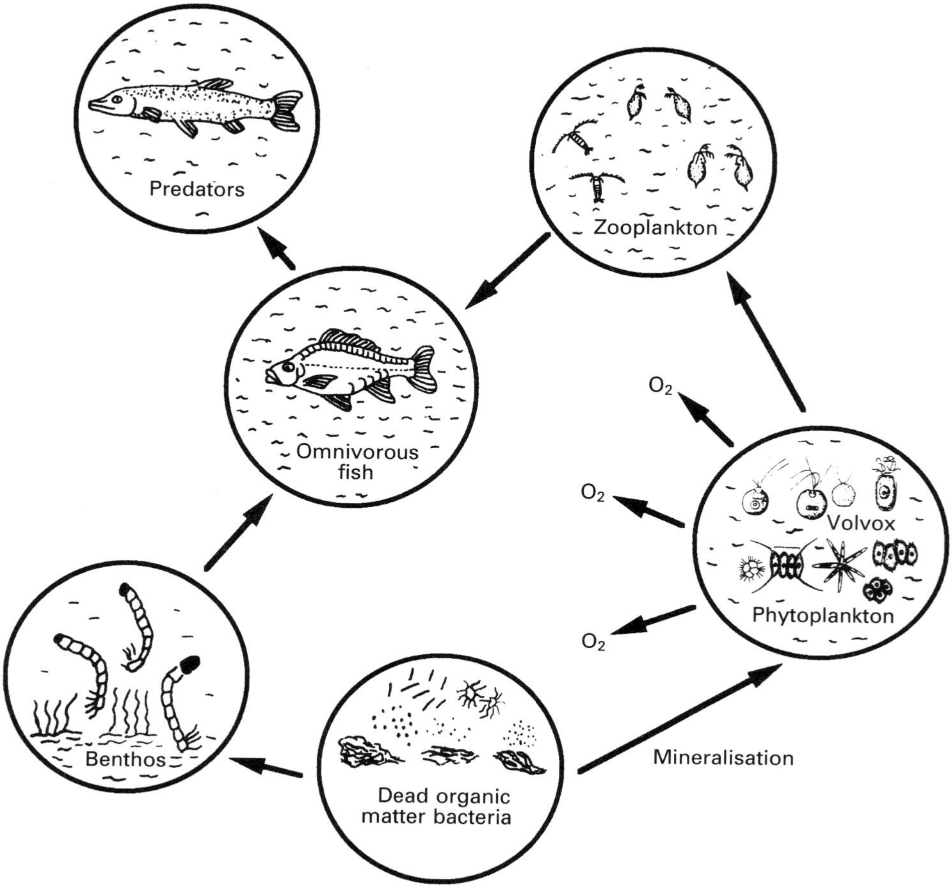

Fig. 2.1. The trophic web in a pond used for fish culture.

carried out on pathogenic bacteria. Although important, these are not the most active species in the biochemical processes at work in the pond.

As with those in the soil, bacteria living in water come from varied groups and have very different roles. There are those which fix molecular nitrogen, autotrophic nitrifiers, and various decomposers of organic matter. In ponds, bacteria have two important functions:

—they participate in the recycling of dead organic matter and convert it into nutrients which can be assimilated by plants;
—they provide food for many planktonic or benthic invertebrates.

Bacteria and the recycling of dead organic matter

In ponds there is considerable production of organic waste which is chiefly composed of nitrogen compounds. It is derived from the waste produced by fish and other aquatic animals such as zooplankton and benthic invertebrates. Planktonic animals and plants, which have not been eaten die and are deposited on the mud where they accumulate before being subjected to bacterial activity. Over a period of 24 h, the number of zooplanktonic organisms sedimenting on an area of $1\,m^2$ has been found to be from around 200 000 to over one million according to estimates made in 1989 in ponds in la Dombes, France. The spreading of wastes such as manure from agriculture as fertilisers further increases the organic matter in the pond.

The decomposition of organic matter produces nitrogen compounds, the most important of which is ammonia (NH_3) which is toxic. However, this is only stable in alkaline waters; where the pH exceeds 9 the proportion of ammonia is greatest. In neutral waters such as most treated ponds, ammonia is converted to the ammonium ion ($NH_3 + H^+ = NH_4^+$) which is not considered to be dangerous to aquatic animals.(2)

The nitrogen cycle is probably the most important biochemical process in a pond. It corresponds to the conversion of organic nitrogen (from wastes) into mineral (inorganic) nitrogen, especially nitrates, which can be assimilated by most plants. This conversion could not take place without the action of various species of bacteria.

From the ammoniacal stage this reaction proceeds as follows:

—nitrifying bacteria (1) (*Nitrosomonas*) convert ammonia into nitrite (NO_2^-) which is extremely toxic. This process has a major requirement for oxygen:

$$4NH_3 + 7O_2 \text{ (\textit{Nitrosomonas})} \longrightarrow 4NO_2^- + 6H_2O$$

—nitrifying bacteria (2) (*Nitrobacter*) convert nitrite to nitrate. This reaction also entails the consumption of oxygen:

$$2NO_2^- + O_2 \text{ (\textit{Nitrobacter})} \longrightarrow 2NO_3^-$$

Bacteria from other genera also take part in these processes but *Nitrosomonas* and *Nitrobacter* are the most important.

The production of nitrate from the ammoniacal or nitrite form thus depends on the activity of bacteria and the presence of oxygen. However, conversion does not take place instantaneously. The time required depends on temperature, pH and oxygen availability. An increase in temperature accelerates the process but even at 30°C with favourable pH and oxygen it takes around 13 h for nitrifying bacteria to turn ammonia to nitrite and a further 14 h for the conversion of nitrite to nitrate. This can give rise to major problems. The bacterial flora may not act rapidly enough to mineralise organic matter and prevent the appearance of toxic substances. Significant phytoplankton mortality (either natural or induced through the application of a herbicide for example) can be followed by a sudden increase in the

(2) The ammonium ion (NH_4^+) and the ammonia molecule (NH_3) are often wrongly grouped together as "ammonia".

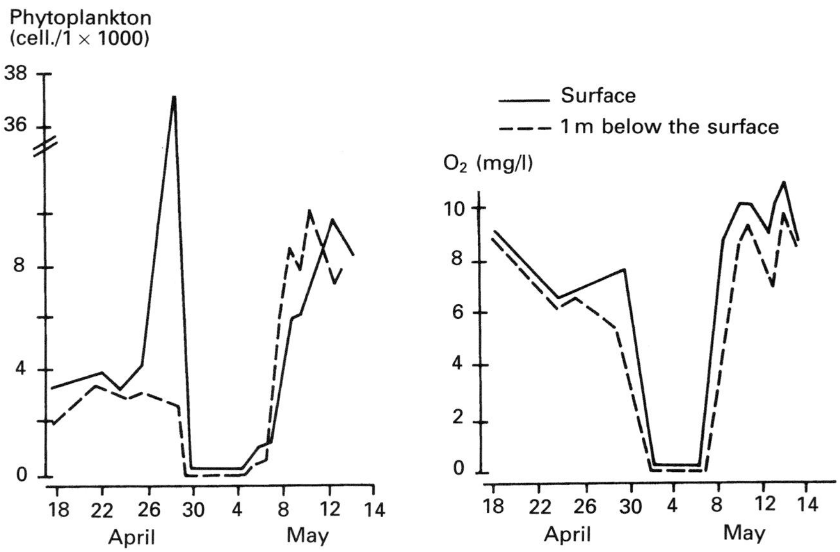

Fig. 2.2. Development of another algal bloom in a pond used for fish culture followed by its breakdown and the total disappearance of oxygen in the environment over several days (after Boyd *et al.*, 1975).

concentration of ammonia and at the same time a major oxygen deficit, the latter because of oxygen consumption by bacteria (Figure 2.2). This combination of two "nuisance" factors causes skin lesions or gill haemorrhages in carp.

As has been frequently observed, oxygen is used up rapidly during nitrification: this can block the action of the bacteria, preventing conversion of other organic wastes in the environment. The reaction can even be put into reverse, i.e. nitrate converted to nitrite. Where there is severe deoxygenation, as often occurs close to the bottom of a pond, inorganic nitrogen is converted back to organic nitrogen or free nitrogen, N_2; the latter is used by some species of cyanobacteria (also known as blue–green algae) which often cause problems in the summer. The rate of nitrification is also influenced by pH. Where sufficient oxygen is available even a slight increase in pH strongly influences the conversion of ammoniacal nitrogen to nitrite. For example, this process is three times faster at pH 7.8 than at pH 7.1. The reverse is true for the reaction which follows. Oxidation of nitrite to nitrate is more rapid at a lower pH. The maintenance of favourable conditions for aquatic life thus depends on the participation of bacteria in the nitrogen cycle and the progress of the biochemical pathway requires time. The conversion of organic matter is favoured and accelerated by an increase in the concentration of dissolved oxygen which can be achieved by artificial aeration.

Organic phosphorus derived from dead organisms and faeces is also converted to inorganic phosphorus through bacterial action: this is utilised rapidly by phytoplankton.

Bacteria as a source of food

Bacteria are very rich in protein and therefore constitute an excellent source of food for many aquatic organisms. These include ciliate protozoans (Paramecium) and all of the other filter feeding zooplankton which consume enormous quantities of them. A very dense zooplankton population can develop where bacteria are the food source. The bacteria are also consumed directly by several species of cyprinid. While ingesting particles of organic detritus several millimetres in diameter suspended in the water, carp also take in the abundant bacteria covering the detritus, boosting its nutritional value. This is especially true for the silver carp which, when filtering water, absorbs many bacteria attached to suspended particles.

Leaving aside pathogenic species, positive and negative aspects of the bacterioplankton can be summarised as follows:

—Positive aspects: some bacteria convert dead organic matter and toxic forms of nitrogen into inorganic salts which can be assimilated by the algae in the phytoplankton. All of the bacterioplankton provides food for filter-feeding zooplankton and for detritivorous organisms living in the silt (worms, chironomid larvae) which, in their turn, are consumed by most species of cyprinids. Finally, to a degree, bacteria are also consumed directly by fish.

—Negative aspects: the biochemical processes carried out by the bacteria require high oxygen consumption, which may lead to the asphyxiation of fish. The depletion of oxygen due to bacterial proliferation, notably following massive algal mortalities, is probably the chief cause of fish mortality in the summer (Figure 2.2). Finally, where there is deoxygenation, some bacteria convert nitrate to nitrite, thus contributing further to the degradation of the fishes' environment.

Phytoplankton

The observer on the bank of a pond can only see these microscopic algae when their mass proliferation results in a characteristic colour change to the water. What differentiates the plankton in ponds used for fish culture from those in a more natural environment is, above all, their abundance which is the result of the application of organic or mineral fertilisers and the presence of a high biomass of fish. The phytoplankton production can be 10 to 15 times higher in fertilised than unfertilised ponds.

Several species of phytoplankton are found in freshwater. Depending on the pigments they contain these microalgae can be green, red, yellow, blue or brown. Where they are found in the natural environment (lakes, rivers) a greater number of species occur together than in the ponds used for fish culture where the environment is characterised by a low depth of water, high level of organic matter and reduced transparency. These conditions lead to the disappearance of a large number of sensitive species generally found in clear water that is poor in mineral salts. Fish-culture ponds usually have neutral water as a result of liming, acidophil algae; such as those which are abundant in forest pools which are rich in humic acids, also disappear.

However, other species which have an affinity for water which is rich in organic matter proliferate in fish ponds. These mainly belong to three groups: brown algae or diatoms, green algae or chlorophycaea, blue–green algae or cyanobacteria (Figure 2.3).

Diatoms have two main seasonal peaks in abundance: spring and autumn. They are mainly small cells which, because of their size, are easily ingested by zooplankton. They can also be directly digested by fish. These algae float passively in the water. In some species the cells are attached end-to-end forming chains of varying length (Figure 2.3). Diatoms are enclosed in a thin silicon envelope which is pierced by many pores and sometimes has fine spines. Diatoms are dependent on dissolved silicon in the water for their growth and multiplication.

Green algae (chlorophycaea) and the cyanobacteria develop mainly in the summer, a period during which the two groups are in competition. The commonest green algae in ponds are very small in size. They multiply very fast which allows them to colonise the environment very rapidly when conditions are favourable. Many species float passively in the water in the same way as diatoms and also have a tendency to clump into colonies. Others are solitary and, using one or more flagellae, have a certain degree of motility (Figure 2.3).

These small green algae are looked on very favourably in fish ponds. Their size and thin cell membrane make them easy for zooplankton to assimilate.

Cyanobacteria are, on the contrary, regarded as undesirable for various reasons. Some species (*Oscillatoria*, *Aphanizomenon*, *Anabaena*) are arranged in long filaments, hence the name "filamentous algae" (Figure 2.3). This arrangement renders them of little value to filter-feeding zooplankton. Other species, for example those belonging to the genus *Microcystis*, are toxic. Finally, some species have a substance in their cells which gives a muddy taste to the water and to fish flesh.

The cyanobacteria have a tendency to develop where green algae are in decline. They occupy the niche left vacant and proliferate because they are able to adapt to all types of environment (Figure 2.4).

As with bacteria the algae in the phytoplankton affect the physico-chemical properties of the environment while providing a source of food for zooplankton and several species of cyprinids.

Phytoplankton and the physico-chemical properties of the environment

Phytoplankton play their part in the nitrogen cycle following that of bacteria. For growth and multiplication, algae utilise the nitrate which is formed by the biochemical transformations carried out by the bacteria. Green algae assimilate nitrogen in the ammoniacal form preferentially. Phytoplankton thus play an important role in the process of auto-purification. Chemical measurements carried out in ponds have shown that concentrations of nitrate in the water are very low or nil. This is partly because of absorption by silt and partly due to assimilation by

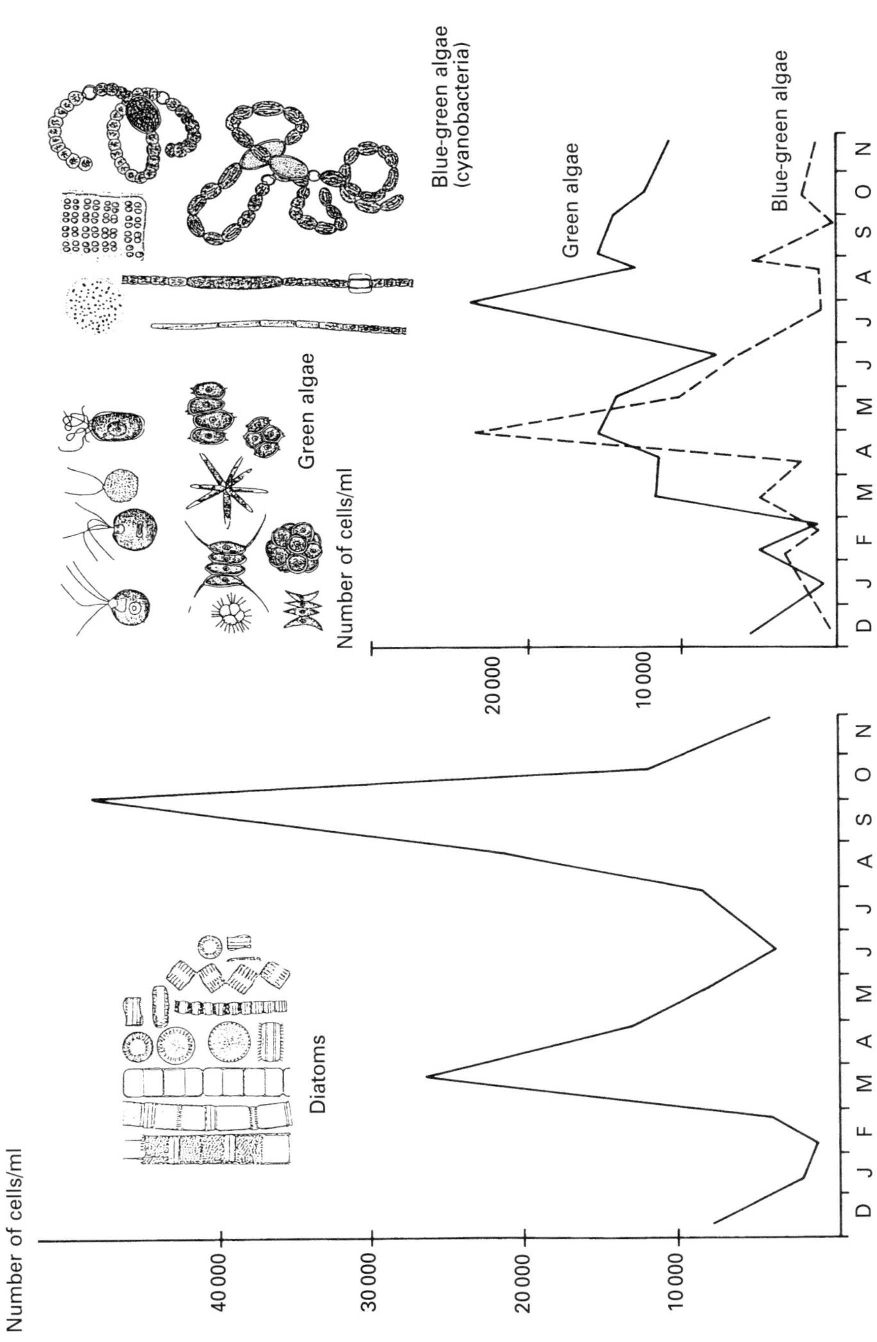

Fig. 2.3. Variations in the principal groups of phytoplankton in the Jourdineries pond (Indre) from December 1983 to November 1984. Fertilisation based on liquid phosphates (14/48) and dried poultry waste (after Sevrin-Reyssac and Gourmelen, 1985).

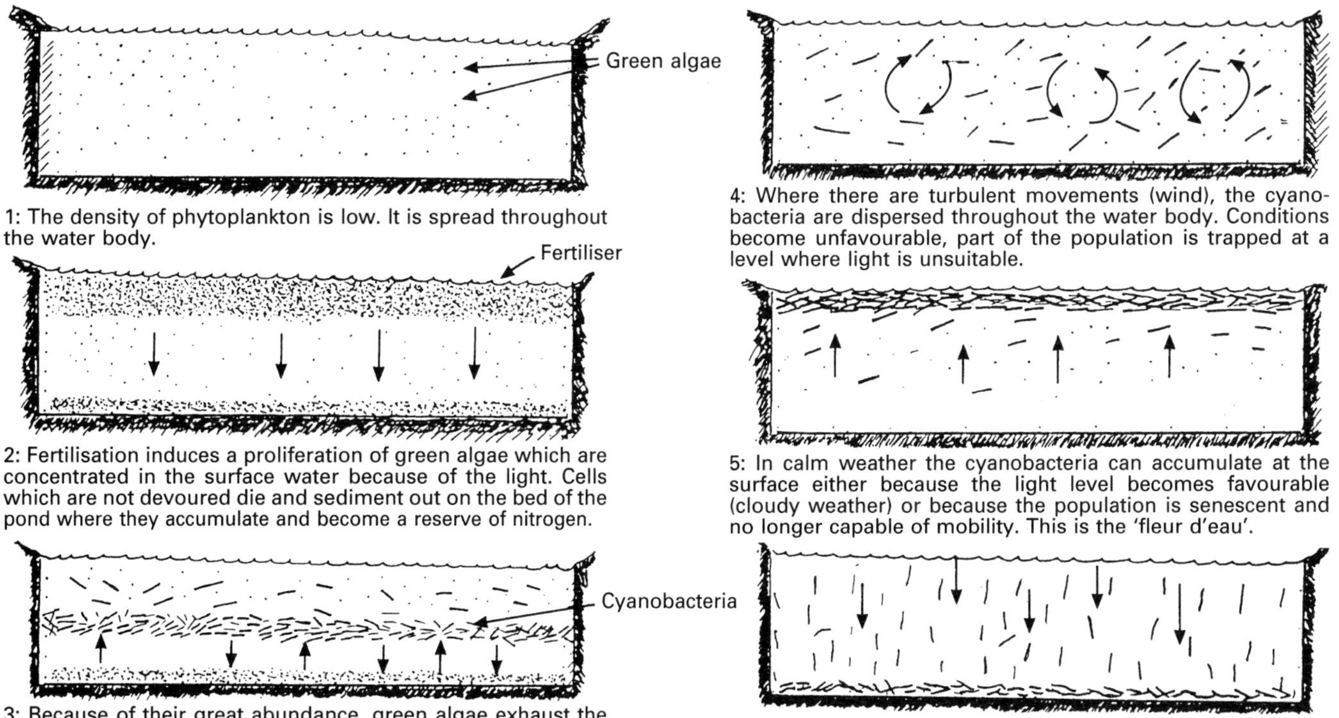

1: The density of phytoplankton is low. It is spread throughout the water body.

2: Fertilisation induces a proliferation of green algae which are concentrated in the surface water because of the light. Cells which are not devoured die and sediment out on the bed of the pond where they accumulate and become a reserve of nitrogen.

3: Because of their great abundance, green algae exhaust the nitrogen reserves. Without this element, the algae die *en masse* and sediment. The cyanobacteria then appear at the level in the pond where light conditions are most favourable to them. Mobile species can migrate vertically, thus utilising nitrogen resources which have accumulated on the silt.

4: Where there are turbulent movements (wind), the cyanobacteria are dispersed throughout the water body. Conditions become unfavourable, part of the population is trapped at a level where light is unsuitable.

5: In calm weather the cyanobacteria can accumulate at the surface either because the light level becomes favourable (cloudy weather) or because the population is senescent and no longer capable of mobility. This is the 'fleur d'eau'.

6: Following a natural or induced (herbicides) mortality, the 'fleur d'eau' disappears but the proliferation of bacteria which results can lead to significant oxygen deficits.

Fig. 2.4. Formation and development of a bloom of cyanobacteria.

phytoplankton. Planktonic algae also have a very important role as oxygenators in the pond.

Phytoplankton as a source of food

Phytoplankton form the essential link in the underwater conversion of mineral salts and dissolved carbon dioxide gas into living organic matter (the algae themselves). They thus form one of the principal food resources for the zooplankton.

Phytoplankton are also directly consumed by certain species of cyprinids. This is well known for the silver carp but is also true for roach. Eventually, following death and sedimentation, the algae are utilised by the detritivorous fauna which spreads out above or in the mud as well as by fish such as common carp which forage in the mud. Lumps of mud are ingested and the nutrients are extracted from these in the digestive tract.

Positive aspects of phytoplankton in fish ponds (mobilisation of nitrogen compounds, oxygenation of the water, acting as nutrients for various organisms) largely outweigh the negative ones; the latter are usually associated with blooms of cyanobacteria.

Problems associated with phytoplankton

The mass development of cyanobacteria during the summer is the main problem posed by phytoplankton to the fish farmer. This is generally associated with ponds stocked with fish which feed on zooplankton, as is the case with the majority of the cyprinids.

Favourable conditions for the appearance of cyanobacteria. Cyanobacteria generally bloom in summer as a result of a combination of conditions which often occur in fish ponds. The main factors controlling their abundance are: light intensity, temperature, chemical composition and dissolved oxygen concentration of the water.

Cyanobacteria are inhibited by the strong light intensity which prevails at the water surface during summer. They remain at a depth where illumination is more favourable to them. When the sky is cloudy, as before a storm, light intensity at the surface decreases and the algae move upwards in the water column. This phenomenon can be clearly observed as they form a very distinctive bluish-green film on the surface of the water. When the rain arrives and the surface of the water is disturbed this film disappears but sometimes forms again on the following day and even for several days after if the weather remains suitable. Some species (*Oscillatoria*) have a great capacity for movements within the water column and stabilize where light conditions are most favourable. This gives them a great advantage over other types of algae (for example, green algae) which are not so mobile. The non-appearance of the blue–green skin at the surface does not therefore mean that cyanobacteria are absent from the pond. They may form very dense populations in the underlying layer (Figure 2.4).

Most species prefer relatively warm water (25°C). The main blooms occur in summer when the temperature of the water is above 20°C. However, the effect of this parameter is primarily indirect. An increase in temperature causes the appearance of conditions which favour cyanobacteria in comparison with green algae; penetration of light is attenuated because of the high density of algae (Figure 2.4), the water column becomes vertically stratified, there is a lack of nitrogen in the upper layer of water and the concentration of oxygen near to the sediment is very low.

A deficit of nitrogen favours the cyanobacteria in competition with the green algae as they can utilise either atmospheric nitrogen (N_2) which can be fixed by some species (*Anabaena*) using special cells (heterocysts) or nitrogen which accumulates on the surface of the sediment (Figure 2.4) which they are able to reach because of their mobility (genus *Oscillatoria*): green algae are unable to do this. Growth of cyanobacteria is also favoured by low oxygen concentrations. Some species are actually inhibited when concentrations are high. It has been demonstrated that the development of cyanobacteria depends on the maintenance of low oxygen concentrations for a sufficiently long period of time.

The most spectacular manifestation of cyanobacteria proliferations in the pond is the formation of a full scale algal bloom. The huge accumulations of algae at the surface can cover the whole of the pond and can be from a few millimetres to several centimetres deep. The phenomenon may be ephemeral or last for several days. Wind action may disperse the cells temporarily but the bloom reappears when turbulence ceases (Figure 2.4).

Problems caused by cyanobacteria. Some species are known to be toxic (*Anabaena flos-aquae*, *Aphanizomenon flos aquae*, *Caelosphaerium kuetzingianum* and, most of all, *Microcystis aeruginosa*) and have caused mass mortalities of fish, particularly carp. Terrestrial animals which have drunk water containing cyanobacteria have also been poisoned. These phytotoxins have an effect at very low concentrations; for some the effect on the organism is similar to that of curare. However, most fish-kills linked with cyanobacteria are attributable to the deoxygenation which occurs when the bloom breaks down.

Cyanobacteria release substances (geosmin) which give an unpleasant taste to fish flesh. This problem also occurs in winter with the development of a species (*Phormidium*) which lives on the surface of the mud. It can be identified by the characteristic bluish film which is left when the pond is emptied. To get rid of this smell and muddy taste the fish must be kept in clean water for several days.

Zooplankton

Composition

From the point of view of fish culture it is useful to distinguish two types of zooplankton (Figure 2.5).

Large sized (2.5–5 mm). These are mainly crustaceans belonging to the cladoceran family and the genus *Daphnia*. The two most common species in temperate regions are *Daphnia magna* and *Daphnia pulex*. They are highly visible and therefore favoured prey for fish which hunt by sight. Their slow swimming is also an aid to predators.

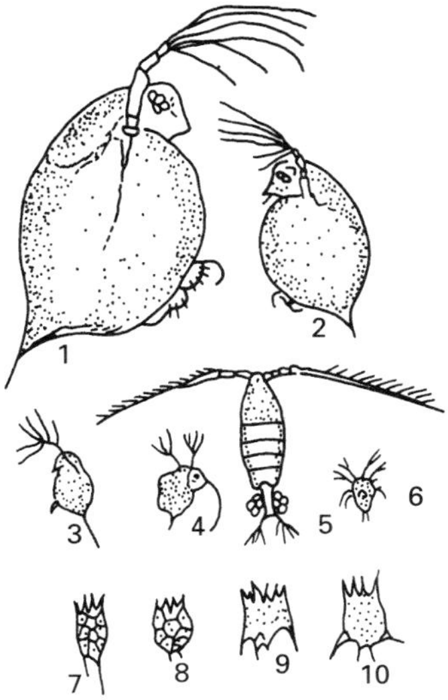

1—*Daphnia magna* (large daphnia)
2—*Daphnia pulex*
3—*Daphnia longispina*
4—*Bosmina longirostris*
5—cyclopoid copepod
6—nauplius larva of a copepod
7 to 10—rotifers.

Fig. 2.5. Large (2.5–5 mm) and small (< 1 mm) forms of zooplankton.

Small sized (<1.5 mm). These are other species of cladocerans; small daphnids such as *Daphnia longispina* and, particularly, *Bosmina longirostris* which proliferate during the summer. Small forms of other crustaceans, the copepods, also come into this category although adult females can be extremely visible when they are carrying egg sacs. Their rapid, jerky swimming allows them to escape predation more effectively than cladocerans.

The small forms also include many species of rotifers. There are some planktonic species which swim freely in the water while others attach themselves to submerged plants using an appendage termed a "foot". Their small size and transparent body makes them difficult to locate for fish which hunt by sight. Their value in the food chain comes, essentially, from the role they play in larval feeding. Fry consume them for a few days towards the end of the absorption of the yolk sac.

Zooplankton feeding

The great majority of the zooplankton are filter feeders. Cladocerans and rotifers are able to retain particles which are suspended in the water column using their filtering

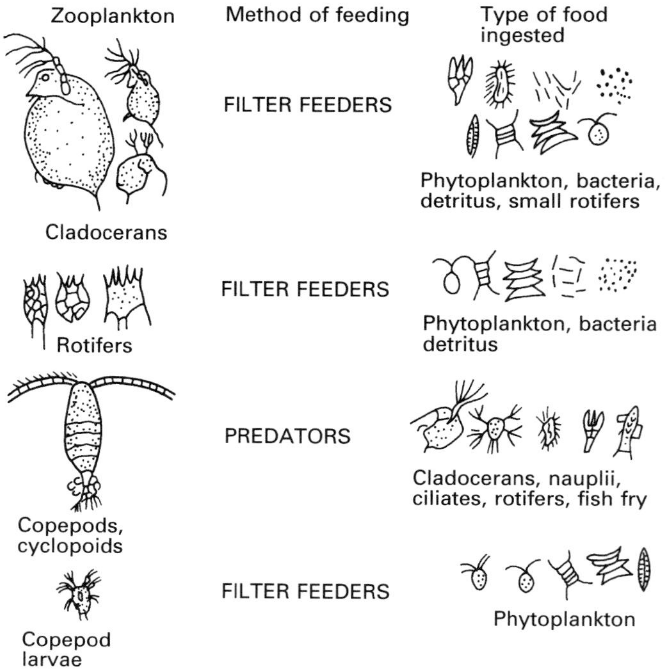

Fig. 2.6. Principal components of the zooplankton and their prey.

apparatus. These particles include phytoplankton, detritus, bacteria, and small ciliates (Figure 2.6).

The size of particles ingested depends on the size of the animal. Large types (*Daphnia magna*) are more competitive than small ones as far as filtration is concerned. Their dietary range is more varied and they can retain larger particles. *Daphnia magna*, which is 5 mm long at the adult stage, can ingest both small organisms such as bacteria or larger types of plant material (up to 120 μm in diameter).

The copepods which adapt best to shallow ponds and to waters which are rich in organic substances belong to the cyclops group. These are carnivorous and very aggressive predators. Only their larval forms feed by filtration. The adults attack rotifers and small cladocerans and also bigger organisms such as large cladocerans and even fish eggs, alevins and fry. Their aggression may even be directed towards larvae of their own species (Figure 2.6).

In fish ponds it is important for zooplankton to have abundant food available so that they in turn can fulfil their role as nutrients for the fish. It is not enough for the numbers of individuals to be sufficient: they must be of suitable nutritional quality. If poor-quality rotifers (*Brachionus*) are used to feed larval fish, a reduction in weight (not length) of the fish will result. A reduction in the lipid, protein and carbohydrate content is noticeable.

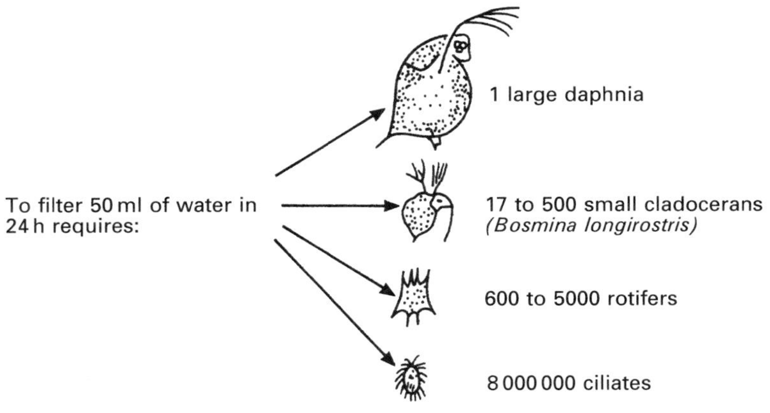

Fig. 2.7. Filtration capacity of the main constituants of the zooplankton.

The effect of zooplankton on the environment

The impact of different zooplanktonic organisms on the environment depends mainly on their filtration capacity. In their adult stage large daphnia filter between 50 to 80 ml of water every 24 h. The volume depends on several factors such as temperature, amount of food available and the developmental stage of the animals. Twelve to 20 individuals filter 1 litre of water every day on average (Figure 2.7). The quantity of food ingested is considerable. In a period of 1 h, tens of thousands of green algae are absorbed by a single individual. Because of such predation this species plays a very important role within the environment as it considerably reduces the number of algal phytoplankton available. The water therefore becomes more transparent and light penetrates further which favours the development of submerged higher plants (Figure 2.8). The frequent presence of submerged macrophytes in ponds which only contain *Silurus glanis* (which do not eat zooplankton) is explained by the presence of these large types of zooplankton.

Other filter-feeding organisms in the zooplankton are much less efficient (Figure 2.7). They thus have little effect on the phytoplankton. Despite their huge numbers during summer (several thousand individuals per litre), they are unable to restrict the proliferation of algae. Transparency therefore becomes very low, light penetration is limited to the superficial layer which leads to the reduction and even disappearance of submerged macrophytes (Figure 2.8).

In a carp pond, zooplankton have a net positive effect, the only negative aspect being the presence of carnivorous copepods which may be eliminated using insecticides during the larval rearing stages of fish (p. 132).

Benthos

Benthos is the term applied to plants or animals which live in or on the bottom of a body of water. Some are microscopic (diatoms, bacteria) while others are clearly

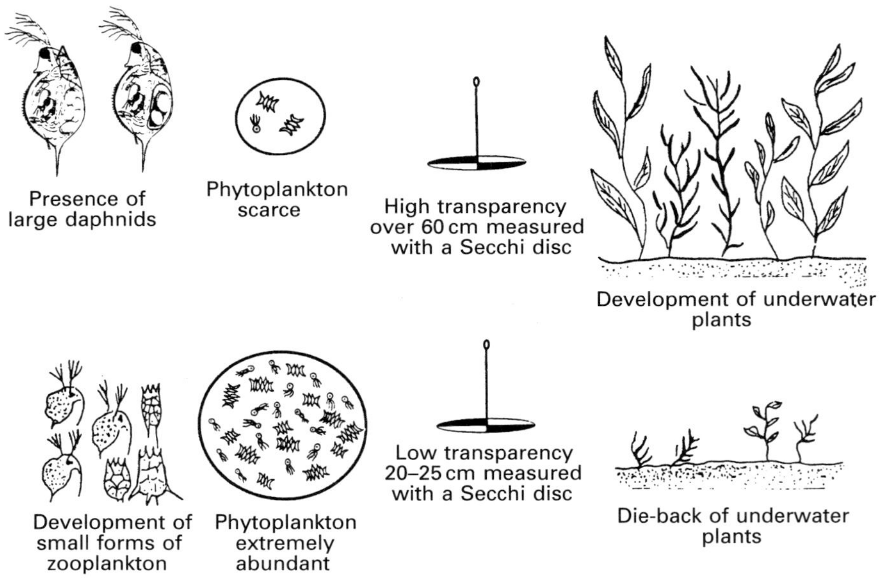

Fig. 2.8. Change of the environment and the plankton in relation to fish stocking. If the large daphnia are eliminated (predation by fish), the small forms of zooplankton proliferate but, in spite of their large number, they are not sufficiently efficient to prevent the mass development of phytoplankton.

visible to the naked eye (worms, insect larvae). They are strongly associated with environments which are rich in organic matter. In a pond which is not stocked with fish the worms and insect larvae develop to the point where they constitute a living carpet on the surface of the substrate. This is particularly marked for worms of the genus *Tubifex*. These are 3–4 cm in length and reddish in colour. Only the posterior part of the body is visible, the anterior part being buried in the mud as the animal searches for its food. Insect larvae, particularly chironomids, also colonise the superficial layer of the mud. In some lakes which are rich in organic matter, these represent up to 75% of the benthic biomass.

Benthos as a source of food for fish

Macro-invertebrates (worms, insect larvae) do not represent a significant source of food even if the fish density is low. In practice, their production is highly seasonal and the generation time too long to maintain a sufficient biomass. For example, for chironomids there are 7 or 8 cycles each year. Rearing systems developed in Israel have given a production of 200 g/m^2 (wet weight of animals) every 2 or 3 days in ponds enriched with poultry waste and fish food. Ten grams wet weight of such animals converts to 1 g of fish flesh.

Where carp are present the number of macro-invertebrates is considerably reduced. In experimental ponds in Brenne, France stocked with 15 g carp fingerlings it was observed that the mud no longer contained such organisms at the end of the culture season while in similar ponds which were not stocked the mean density of insect larvae and worms was 200 individuals/m^2.

Microscopic organisms are also eaten by fish. Benthic microalgae, diatoms and other unicellular or filamentous algae have been found in the digestive tracts of roach. The presence of lumps of mud in the stomach of carp or other species of benthic-feeding fish shows that the surface of the sediment plays a role in their feeding, probably because of the accumulation of dead phytoplankton and zooplankton.

The effect of benthic organisms on the environment

Mud has the ability to capture a major proportion of the nutrients (phosphorus, nitrate) coming from manure. The amount of phosphorus fixed by the silt will, depending on the pond, be between 70% to 90% of that spread on the water. The same is true for calcium. It is estimated that the top 5 cm of sediment are important for the exchanges which take place within the water; there is a progressive decrease to a depth of 20 cm.

The microscopic algae which develop on the surface of the sediment utilise the nutrients which are found there. The production of algae at the interface between the sediment and the water depends on the quantity of nutrients in the sediments (particularly nitrogen). These algae release oxygen and thus promote the recycling of organic matter by bacteria. Bacterial activity is also favoured by macro-invertebrates and by bottom-feeding fish which shift the mud as they search for food. This mixing improves the contact with the water and promotes the turnover of substances.

Large benthic invertebrates aid the exchanges between the sediment and water because they consume silt and detritus directly, thus increasing the porosity of the sediment and the mixing of particles. Worms especially act as conveyors by transporting sub-surface material to the surface. This action increases the speed of recycling of nutrients from the silt back into the aquatic environment. Laboratory experiments have shown that worms belonging to the same group as the tubificids turn over the whole of the silt layer (5 cm deep) every 2 weeks. As well as moving the sediment, worms and insect larvae introduce huge quantities of water into the substrate which helps the exchanges between the mud and the water. It can therefore be seen that the maintenance of a benthic community improves the utilisation of the nutrients in a pond.

2.1.2 The relationship between plankton and fish

Fish that consume plankton directly influence not only the composition and abundance of the zooplankton population but also numerous physico-chemical environmental parameters.

Influence of fish stocking on the composition of the zooplankton

The fry of all species of cyprinids are planktonivores. After eating rotifers following absorption of the yolk sac, they move on to cladocerans; the dietary preference for these is particularly strong in carp less than 1 month old.

Adult cyprinids have a varied diet, part of which includes zooplankton, but predation on these is very selective. They prey only on large individuals, particularly on large cladocerans which are highly visible. These are therefore depleted very rapidly. A 3-year-old carp can consume 80 000 daphnia in 24 h. In experimental ponds which were rich in large daphnia (around 100 individuals/l) 300 g carp (stocked at a rate of 1 individual/m^3) totally eliminated them in around 9 days. Under identical experimental conditions 12 g carp fingerlings (1 individual/m^3) removed them totally in around 19 days. Observations made recently in the centre Region, France and in Dombes, France confirmed results obtained previously by Polish workers: large zooplankton disappear very rapidly when a pond is stocked with fish and small forms proliferate. The latter are far harder to see and, because of their small size, are not retained by the gill rakers of the carp.

Roach are omnivores. They eat phytoplankton, decomposing plant debris, insect larvae living in the mud (chironomids) and in mid-water (mayfly) as well as some of the zooplankton. Examination of roach faeces collected in May showed a high number of benthic diatoms and other species of diatoms which live attached by a peduncle to the surface of large plants. Green filamentous algae were frequent components of the diet, particularly in spring (*Cladophora*). The latter constitutes a good source of food (12.07% protein dry weight, 5.16% wet weight). There were also lumps of mud. If there is not enough food in mid-water the roach is able to feed on the benthos. Previous studies have found the digestive tract to be full of the debris of decomposing leaves from lime trees. roach also consume the cyanobacteria which develop on the surface of the mud (*Phormidium*). These give a muddy taste to their flesh. The effect of roach on the composition of zooplankton is very similar to that of carp.

The feeding behaviour of the silver carp on zooplankton is very different. The fish is an open-water filter feeder which passively gathers suspended particles which are in its path: phytoplankton, detritus and small zooplankton. Although these fish are regarded as phytoplanktonivores they also ingest, in large quantities, small forms of zooplankton which are trapped on the filter-feeding apparatus at the same time as the phytoplankton. At high fish densities this species has been shown to reduce the density of zooplankton by a factor of 16. Analyses of stomach contents have demonstrated the presence of many rotifers and the small cladoceran, *Bosmina longirostris*.

During the summer of 1989, experiments carried out in Sologne, France confirmed the effect of silver carp on small forms of zooplankton. In an example of a mixed culture of silver carp and common carp, small forms were less abundant than in a similar system where common carp were reared alone. However, silver carp do not reduce the density of the large zooplankton in the same way as common carp.

Silver carp will not eat artificial diets. No differences were observed between the growth of individuals fed on pellets and those which remained unfed. From the aspect of the utilisation of the food web in a pond, silver carp have several attractions. Besides the fact that they consume phytoplankton (which are poorly utilised by other filter-feeders) they ingest the small zooplankton. These forms proliferate in summer and because they are in open water are utilised little or not

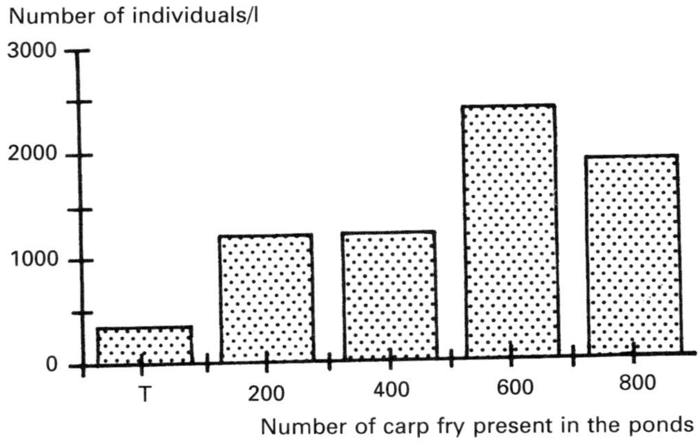

Fig. 2.9. Variations in the number of individuals in the "small form" category of zooplankton in relation to the density of carp fry. T = neighbouring pond with no fish (after Boulon, 1989).

at all by the other fish. In addition to this, the silver carp has no impact on the large cladocerans which are actively preyed on by other cyprinids.

The influence of cyprinids on the abundance of plankton: consequences for the environment

The abundance of phytoplankton and zooplankton is largely dependent on the nature and significance of the fish stocking which has taken place. All of the observations which have been made in ponds show that fish are visual predators on the large species of cladocerans; phytoplankton and small zooplankton reach their greatest densities when the biomass of fish is high. In fact, the greater the biomass of fish, the more rapidly the larger zooplankton are eliminated and the more the smaller types increase (Figure 2.9). These smaller forms are no longer in competition with the large cladocerans for food. The elimination of these cladocerans thus favours them and allows them to proliferate. Their small size and the cloudiness of the water due to the abundance of algae allows them to escape the attentions of visual predators. Samples of zooplankton taken several weeks after stocking with fish are, consequently, very rich in small types (rotifers, small cladocerans, copepod larvae) numbers of which may reach several thousand individuals per litre. Where cladocerans disappear phytoplanktonic algae also proliferate as they are no longer being actively consumed. In trials, the numbers of phytoplankton did not exceed 10 000 cells/ml in ponds without fish while they reached 80 000 to 100 000 cells/ml in ponds where large cladocerans had been eliminated by carp fingerlings (individual weight 10 to 12 g).

Fish have a profound indirect effect on the environment through their predation on large daphnia. The proliferation of phytoplankton can bring about a nitrogen deficit and create favourable conditions for cyanobacteria. Most of the plant matter which is not eaten dies and sediments. The cells accumulate on the mud and, as they decompose, there is a risk that anoxic conditions will develop. In contrast, in a pond which is not stocked with fish or one which is stocked with fish which do not consume the large cladocerans, phytoplankton levels are low even where fertiliser is spread because the large cladocerans prevent the proliferation of algae, even when they are present at low numbers (<10 individuals/litre).

The effect of grass carp and silver carp on the environment is very different from that of other cyprinids because of the differences in their feeding behaviour. Grass carp feed mainly on higher plants and act as a scythe and can be used in bodies of water which cannot be drained as a means of preventing the encroachment of macrophytes. grass carp have a stimulating effect on the phytoplankton because of the fertilising action of the wastes produced. This species converts higher plants in the pond which are otherwise trophic "dead ends" into a natural fertiliser which increases the growth of algae. However, there is some doubt, particularly in open water, as to the effectiveness of this species in eliminating higher plants. They will not eat tough, coarse plants; these will therefore still tend to invade the environment. For a long time it was believed that grass carp did not compete with any other species of fish. However, recent studies have shown that, in the absence of their favoured types of plants, they will prey on large cladocerans (*Daphnia magna*).

Silver carp do not consume large cladocerans; the reduction in phytoplankton which has been observed in ponds containing this species results much more from the persistence of large cladocerans than from the actions of the fish themselves. The silver carp thus has an indirect effect on the quantity of phytoplankton. However, this effect is usually reduced in ponds because the silver carp is normally raised with other species of cyprinids which actively consume large daphnids.

It has been suggested that silver carp are able selectively to retain cyanobacteria because of their relatively large size. They are said to be able to reduce the likelihood of the formation of an algal bloom; if this were so it would be of considerable benefit in terms of improving the quality of the environment. In practice this role in the control of cyanobacteria has not been definitively established. Studies carried out in Brenne, France have shown that these algae can be highly abundant in spite of the presence of silver carp (Figure 2.10).

2.1.3 Influence of the type of fish culture on the quality and quantity of plankton

The seasonal pattern of phytoplankton development in ponds is generally characterised by the lowest levels occurring in winter, intense development in spring followed, during summer by a less active phase. A further burst, usually less pronounced than that in spring, occurs in the autumn. As with terrestrial plants, growth picks up in March/April and declines in November. These seasonal changes depend on temperature (an increase accelerates cell division), daylength, and, particularly, the application of fertilisers.

Such applications determine the development cycle of the phytoplankton (Figure 2.11) but the response of the algae to fertilisers only takes place if the pH is not too low (it should be as close as possible to neutral) and if the turbidity (presence of suspended particles in the water) is not too great. Excess detritus in the water reduces

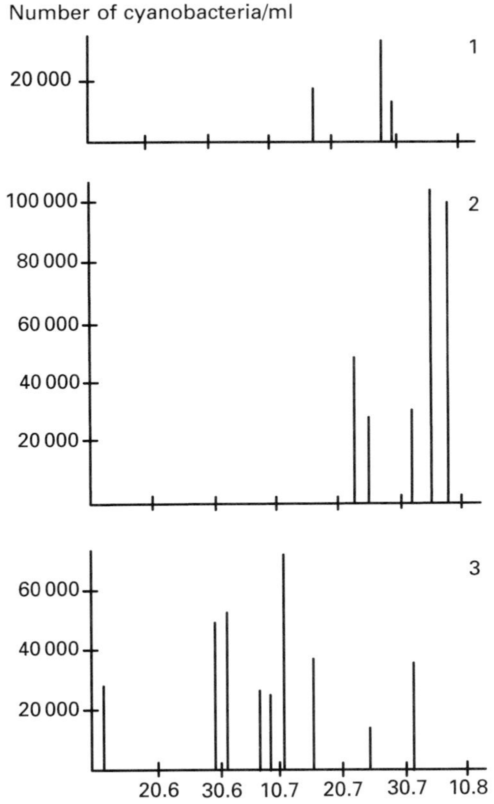

Fig. 2.10. Development of cyanobacteria in the presence of silver carp in polyculture. The three ponds, 1, 2, and 3 have the same surface area and the same biomass of fish. From June to August 1986, they were given organic fertiliser (poultry waste) at a rate of 150 kg dry weight/ha/day (pond no. 1), 100 kg/ha/day (pond no. 2) and 50 kg/ha/day (pond no. 3). The abundance of cyanobacteria appears to depend more on the amount of organic matter spread (it decreases with increase in supply) than on the biomass of silver carp (after Sevrin-Reyssac *et al.*, 1990).

the density of algae so much that there is a strong decline in the chemical quality of the water which is expressed by the accumulation of nitrogenous wastes (notably ammonia) which are not being recycled. In good conditions primary production can be 10 to 15 times higher in fertilised than in unfertilised ponds.

Several methods are available for optimising the environment for fish.

Prevention of the winter decline in numbers of organisms

The decline which occurs in winter is not totally linked to atmospheric conditions. It is also associated with the absence of fertilisers. The winter decline in zooplankton numbers results from that of the phytoplankton, its food source. Results from ponds in Brenne, France show that numbers are four to ten times lower than in summer. In

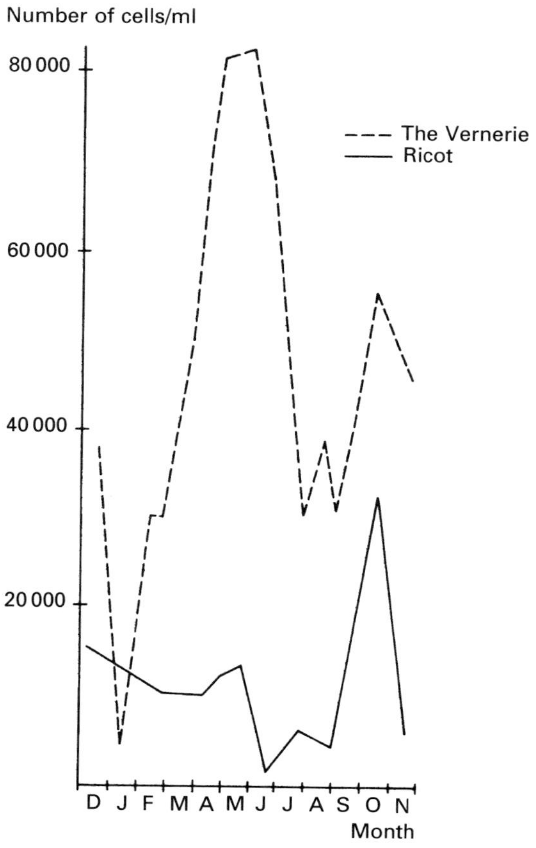

Fig. 2.11. Effect of fertilisation on the number of phytoplankton cells from December 1983 to November 1984. The Vernerie pond (Indre) was given organic fertiliser (34 kg of manure/ha spread in two batches) and mineral fertiliser (200 litres of 14/48 spread in 17 batches. The Ricot pond (Indre) received no fertiliser (after Sevrin-Reyssac and Gourmelen, 1985).

addition, generation time increases with the drop in temperatures. The time taken from hatching for an individual to reproduce is shown below for three major groups of zooplankton:

- rotifers: 1.5 days at 20°C; 6 days at 10°C;
- cladocerans: 8 days at 20°C; 20 days at 10°C;
- copepods: 23 days at 20°C; 45 days at 10°C.

It therefore takes longer for plankton to colonise the environment when water temperatures are low.

Traditionally, fertilisation of ponds is halted in winter as it is felt to have no useful purpose. It is believed that fish, particularly carp (fry and adults) do not feed during

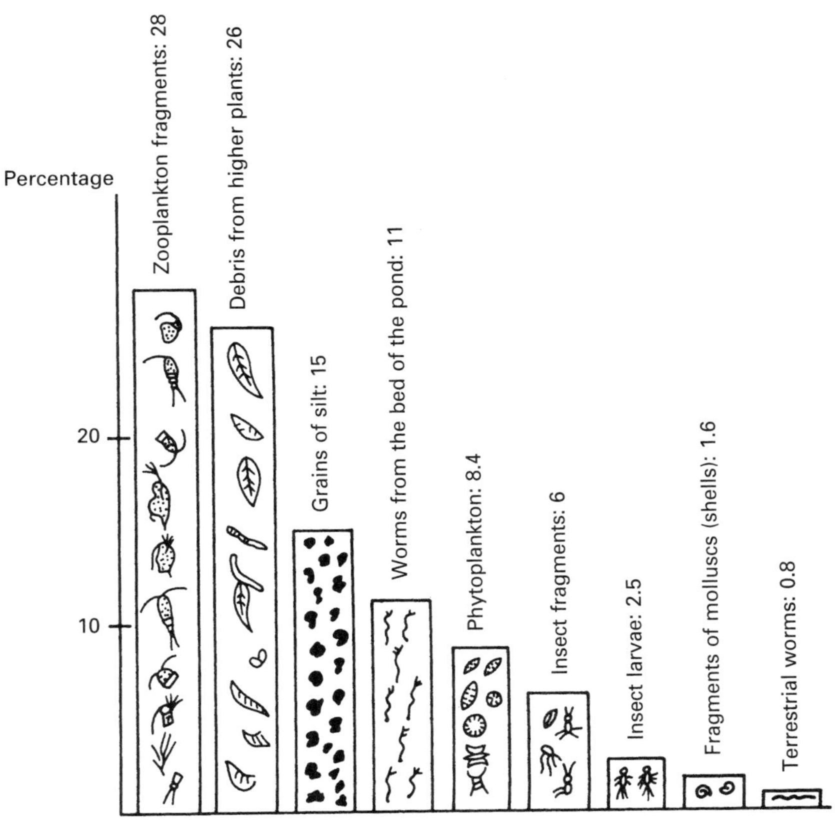

Fig. 2.12. Food of carp fry during the first winter. Results from an inventory of intestine contents from information given by Taran (1936). Percentages are calculated from the frequency with which the author found different items.

the cold season but remain dormant on the mud. The belief that the fish are in a state equivalent to hibernation has brought about this lack of attention to their feeding. However, studies of the contents of the stomachs of 148 carp fingerlings (less than 1 year) during their first winter have shown that they have been feeding even at low temperatures (3 to 8°C). More than 70% of the stomachs examined were full of animal and particularly plant remains (Figure 2.12). Unable to find enough food in mid-water the carp have turned to the sediment and fed on plant debris and small dead animals. In addition to this it is thought that the food is not only ingested but also digested. In the anterior part of the intestine the food organisms are intact or only partially broken down while in the posterior part, only chitinous debris such as crustacean carapaces or the cephalic apparatus of chironomids remains. This intake of food, even though quantities are small, prevents the fish from exhausting their reserves and gives better resistance to the diseases which occur in the spring.

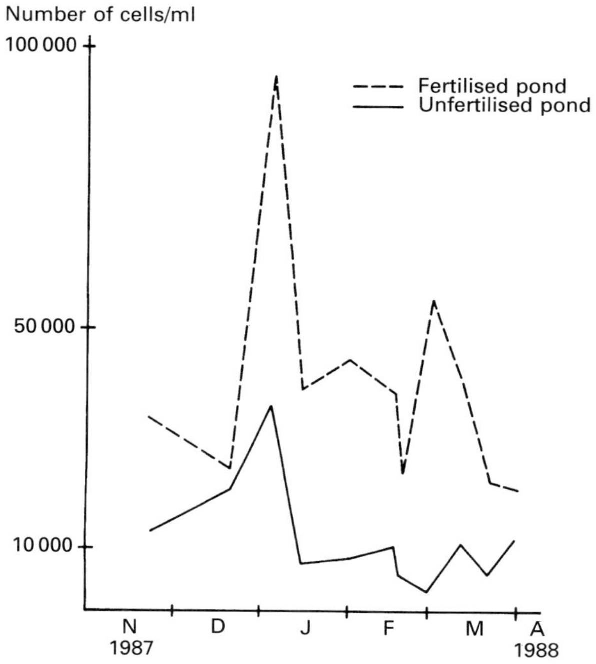

Fig. 2.13. Quantitative variations in the number of phytoplanktonic cells in a fertilised and unfertilised pond in Brenne over winter (after Zimmermann, 1990).

Many trials have been carried out both in low volume experimental environments and in ponds in Brenne and Sologne, France to see if, in spite of poor weather conditions, plankton production can be stimulated through the application of fertilisers (organic matter including poultry wastes and manure and also mineral fertilisers) in the same way as in summer (pp. 160 and 163).

When the temperature is below 10°C the activity of the bacterial flora is low so that the response of the microorganisms to organic fertiliser is weak. However, phytoplankton are very responsive to the application of organic fertiliser in winter (Figure 2.13). Many species of green algae which are common in waters rich in organic substances can use dissolved organic matter directly before it is converted to minerals. These algae are "mixotrophs", that is they have two types of feeding behaviour: autotrophic (using minerals and dissolved carbon dioxide through photosynthesis) and heterotrophic (using dissolved organic substances directly). This characteristic puts them at a great advantage in winter when light levels are unsuitable for photosynthesis. Algae belonging to the genus *Scenedesmus* which are very common in ponds can also use ammoniacal nitrogen directly, incorporating it by photosynthesis. The response of the phytoplankton to the application of organic fertiliser (poultry waste) in winter is therefore better than when mineral fertiliser is added (Figure 2.14).

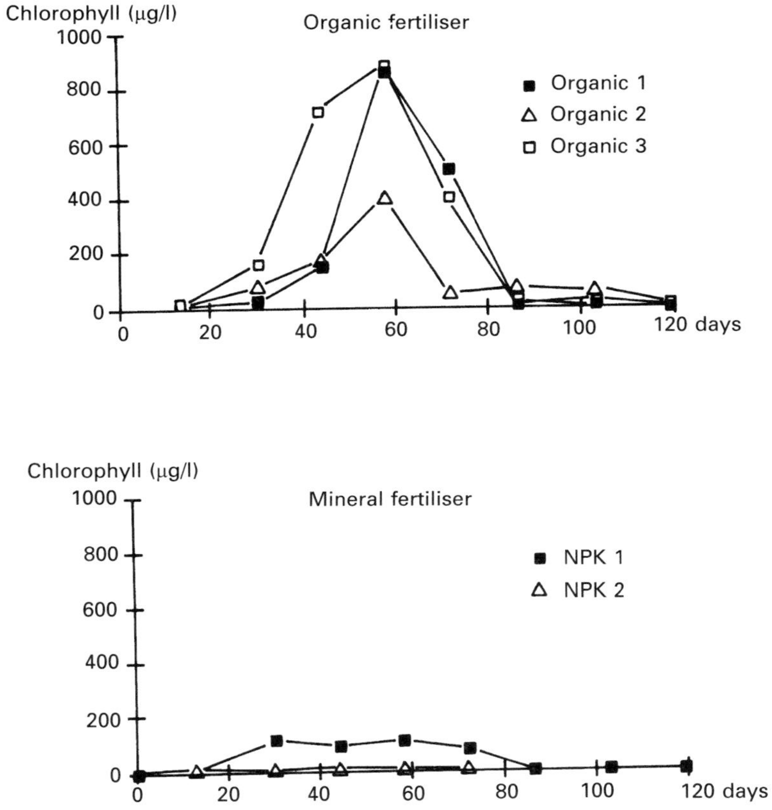

Fig. 2.14. Comparison of the effects of mineral fertiliser (NPK mix) and organic fertiliser (poultry waste) on phytoplankton over winter (from chlorophyll sampling). Measurements from 700-l experimental tanks.

Organic fertiliser also stimulates the development of zooplankton during winter (Figure 2.15). Results obtained in Brenne, France in small 400 m^2 ponds show that the most favoured organisms are the filter feeders which are major consumers of green algae. Thus the mean number of cladocerans increases by a factor of 5 and the number of rotifers doubles in relation to that of unfertilised ponds. Ciliate protozoans (paramecium) which feed on dissolved organic substances also benefit from this fertiliser. Only carnivorous cyclopoid copepods appear to show little benefit from fertilisation of the ponds.

Carp fingerlings in their first winter in experimental containers of 1 and 2 m^3 capacity showed a survival rate of 50% where fertiliser had been applied (bimonthly applications of manure at a rate of 80 to 100 g dry matter (DM)/m^3) while the mortality was 100% in unfertilised ponds. In the winter of 1987–1988 the mortality of fingerings was 65% in unfertilised containers and only 27% in those which were treated with organic matter (the same levels as before). In addition, a weight gain of 12% was obtained in the fertilised ponds while the surviving fingerlings in the

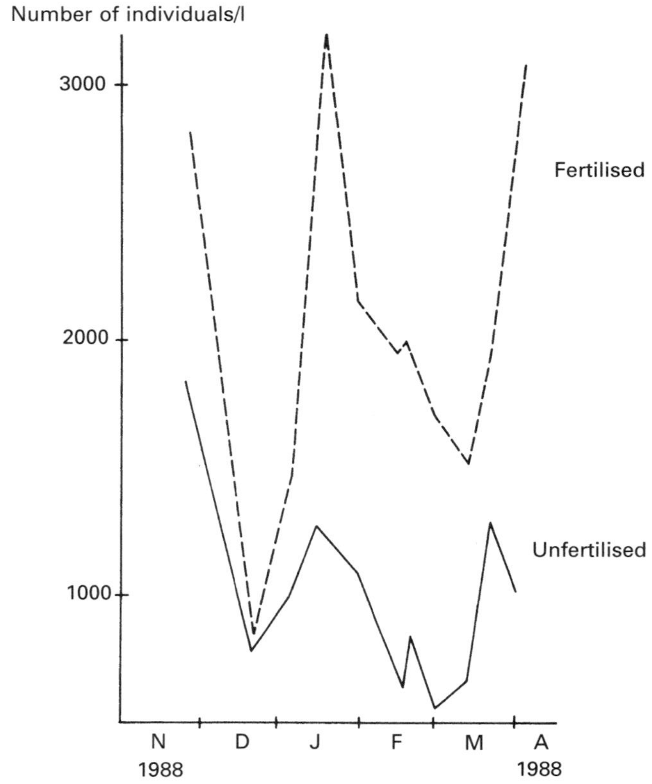

Fig. 2.15. Variations in zooplankton in two ponds in Brenne over winter; one is fertilised with poultry waste (80 g DM/m^3) twice each month, the other was unfertilised (after Zimmermann, 1990).

unfertilised ponds showed a weight loss of 9%. It appears that for part of the winter, fish find their food on the surface of the mud in the form of dead plankton which have fallen to the bottom. The autumn temperature decline causes a major mortality of these small organisms. Bacterial activity is markedly reduced in cold waters, decomposition is therefore very slow at the bottom of the pond which can act as a food reservoir, fulfilling the same function as zooplankton in summer.

Results available to date have shown that it is necessary to manage the nutrition of fish, particularly 1-year-old carp, in summer and throughout the cold season. Organic fertiliser is one way of maintaining trophic conditions and responding to the fishes' requirements.

Prevention of algal blooms in summer: how to reduce the numbers of cyanobacteria

During summer, the phytoplankton flora of ponds is very rich (Figure 2.11) because of the spreading of fertilisers and the disappearance of the large cladocerans which are rapidly eliminated by predatory fish. At first, the proliferation is usually composed of green algae. These deplete the nitrogen in the environment and create the conditions which favour the blooming of cyanobacteria. As yet no satisfactory method has been found of combatting this, either by preventing the appearance of the bloom or eliminating it without further damage to the environment.

It is possible to resort to the use of herbicides, artificial oxygenation and the addition of nitrogen. The most frequently used herbicides are copper sulphate and simazine. Other herbicides (Diquat®, Roundup®) are mainly used for the control of higher plants. By blocking photosynthesis, herbicides bring about the mass mortality of algae but their effectiveness depends largely on the physical and chemical conditions of the environment. This is particularly true for copper sulphate which is more active in acid than in alkaline waters. Treatment of a pond shortly after liming will, consequently, not have the same effect as where liming has not taken place. Recommended doses are generally around $1.0\,g/m^3$ in acid waters and $1.5\,g/m^3$ in alkaline ones. Copper toxicity increases markedly with temperature. An ineffective concentration which is harmless at a relatively low temperature may become toxic in warm water. By eliminating algae, this treatment also causes the destruction of part of the microfauna, particularly cladocerans.

Simazine is very effective in the elimination of algae and has the advantage of being innocuous as far as the microfauna is concerned at the levels used to control algal proliferation. Quantities recommended for the elimination of cyanobacteria are generally around 1 to $2\,gm^3$. As with copper sulphate, effectiveness decreases sharply as temperature falls. However Simazine is a persistant substance and its use has now been prohibited.

The mass destruction of plants results in an increase in dead organic matter and the consequent appearance of huge numbers of bacteria which all consume oxygen (cf. Figure 2.2). It is preferable for remedial action to be taken before the bloom develops. In ponds where there is a risk of blooms occurring, a preventative treatment may consist of placing jute sacks filled with copper sulphate at several points in the pond such as where current is caused by an aerator or, where possible, in the water inflow (see Figure 7.9).

Oxygenation through mechanical aeration

Artificial oxygenation creates conditions which are unfavourable for cyanobacteria. The agitation of the water layers breaks up the vertical stratification which is one of the conditions favouring their development; the increase in oxygen content is also detrimental to these organisms.

Nitrogen availability favours green algae in competition with cyanobacteria. The latter are eliminated by supplying nitrogen in the form of ammonia at a rate of 7 to $14\,g/m^3$/week. The applications should be made in small amounts over a period of time rather than all at once. The use of high doses can lead to the proliferation of green algae (notably *Hydrodictyon reticulatum*). However, when nitrogen supplies are again exhausted, these algae die and are likely to cause the recurrence of oxygen depletion. The only difference is the nature of the algal production—the cyanobacteria are replaced by green algae. However, the effects on the environment of the mass mortality of these cells are identical. Where nitrogen is spread regularly, a qualitative and quantitative stability develops in the phytoplankton. This maintains favourable conditions for green algae without the formation of intense blooms. Such stable conditions were found in Brenne, France in a 2.5 ha pond which was used throughout the summer to rear ducks. The wastes produced by the ducks gave a daily injection of nitrogen fertiliser which favoured the development of small green algae to the detriment of the cyanobacteria, the proportion of which remained below 5% of the total algal population.

2.1.4 Influence of artificial fish diets on the plankton

Observations made in Sologne, France in the summer of 1989 showed that the phytoplankton is just as abundant in ponds where the fish are fed on cereals or pellets as where organic fertiliser (manure) is spread; the response of zooplankton to artificial diets is also positive. In winter experiments (1989–1990) the effect of artificial feeding was even more marked. Phytoplankton and zooplankton were more abundant where feeding was taking place than where manure was spread (Figure 2.16). Carp fingerlings ingest less food in winter than in summer; it is therefore possible that more of the food acts directly as a fertiliser.

2.1.5 Conclusion

A thorough knowledge of the mechanisms which exist in a pond at the level of microorganisms and the interactions which exist between them and the environment should allow fish farmers to avoid errors of management which can have serious consequences, resulting at worst in fish mortalities. In most examples, the mortalities are due to a lack of oxygen which is largely the result of its consumption by bacteria. Operations performed in the hopes of improving environmental conditions (scything, applying herbicides) have the effect of considerably increasing the rotting organic matter and, in consequence, the biomass of bacteria. The risk of deoxygenation is greatly increased if the temperature rises following such action. Fish ponds are usually not very deep and therefore react rapidly to changes in air temperature.

Deoxygenation is also a risk when organic fertilisers are applied in spring. In Brenne, France, fish mortalities followed the spreading of poultry manure during April 1984; the temperature rose sharply by 10°C after the manure had been spread. Such accidents can be prevented by dividing up the fertiliser and spreading it in several applications rather than all at once.

An interesting aspect of the relationship between plankton and fish is shown by the effect of silver carp on phytoplankton. This species does not completely reduce the algal biomass. It does not clear the water nor does it limit the proliferation of cyanobacteria during summer. It is not yet clear whether it allows better utilisation of the first link in the food chain which is poorly exploited by other aquatic organisms.

No "miracle" solution for eliminating cyanobacteria has been found which does not harm the environment, but it appears that artificial oxygenation creates con-

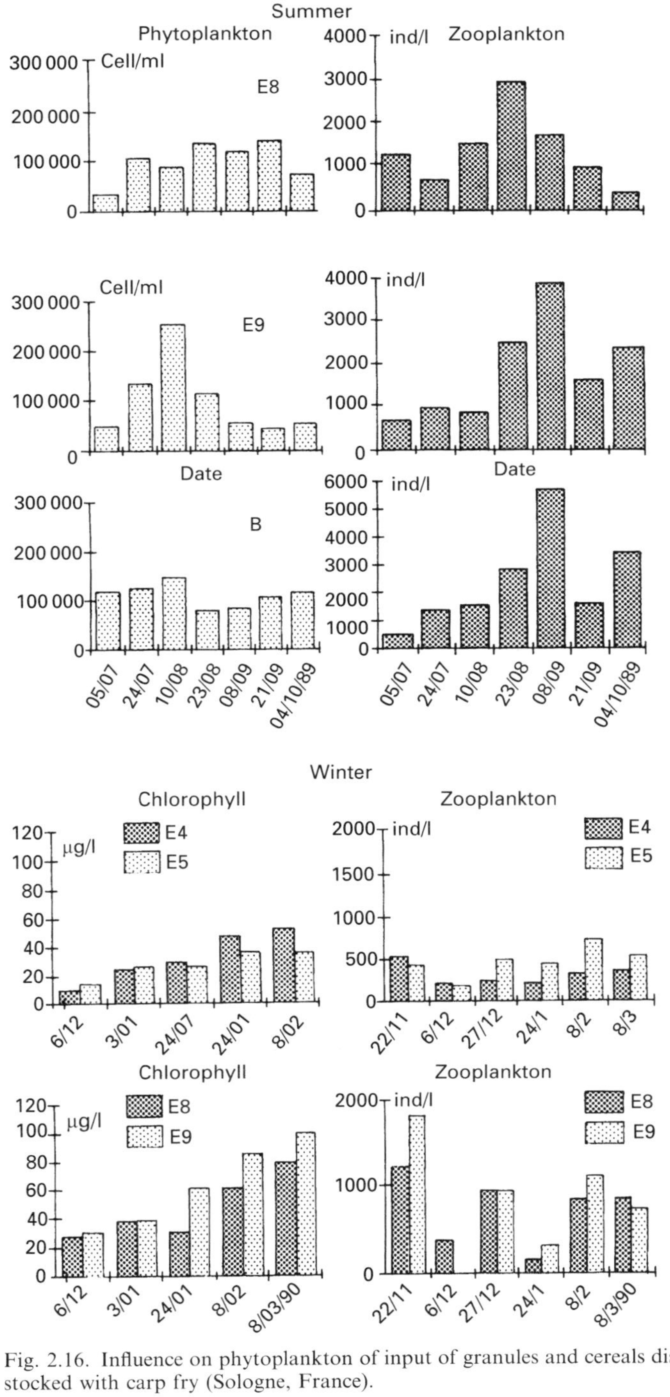

Summer

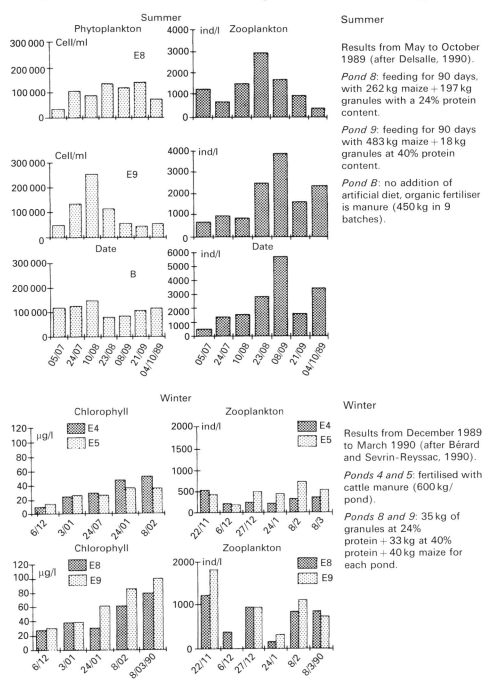

Results from May to October 1989 (after Delsalle, 1990).

Pond 8: feeding for 90 days, with 262 kg maize + 197 kg granules with a 24% protein content.

Pond 9: feeding for 90 days with 483 kg maize + 18 kg granules at 40% protein content.

Pond B: no addition of artificial diet, organic fertiliser is manure (450 kg in 9 batches).

Winter

Results from December 1989 to March 1990 (after Bérard and Sevrin-Reyssac, 1990).

Ponds 4 and 5: fertilised with cattle manure (600 kg/pond).

Ponds 8 and 9: 35 kg of granules at 24% protein + 33 kg at 40% protein + 40 kg maize for each pond.

Fig. 2.16. Influence on phytoplankton of input of granules and cereals distributed to fish in 0.5-ha ponds stocked with carp fry (Sologne, France).

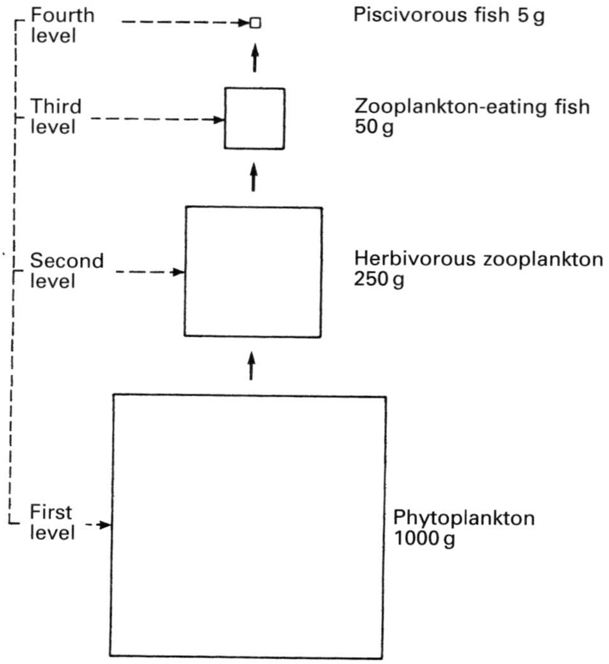

Fig. 2.17. Diagram of the trophic pyramid in the aquatic environment.

ditions which are unfavourable to their development. When a herbicide has been used to destroy them it is recommended that artificial oxygenation should be carried out, not only to avoid asphyxiating the fish but also to speed up the process of recycling the dead algae by bacteria.

It should also be noted that the passage from one level of the food chain to a higher level is accompanied by a considerable loss of energy (Figure 2.17). This also occurs when artificial feeding is carried out as part of the cereals or granules is retained by the microbial detritus chain and other planktonic organisms.

Some questions remain unanswered although many studies have been carried out on environmental conditions in ponds, particularly those stocked with cyprinids:

—What is the role of the small-sized zooplankton (rotifers, small cladocerans) in the feeding of fish?
—Do these constitute a source of food in winter when they have fallen to the bottom of the pond?
—What exactly is the trophic value of the water–sediment interface?
—How can cyanobacteria be eliminated without damaging the environment?

It also remains to be established what role artificial feed plays as food for fish and as part of the detritus chain. This is influenced by many different factors: environment (temperature), the type of the stocking and the composition and solubility of the feed.

2.2 WATER QUALITY AND ITS CONTROL[3]

2.2.1 The main aspects of water quality affecting the production of fish in ponds

The production of fish in ponds is largely dependent on the chemical and physical properties of the water. These can be characterised by a number of parameters: only the most important will be analysed here. Understanding these parameters allows the determination of the potential of the pond and the planning of improvements in water quality in order to provide the fish with a parasite-free environment which is more favourable for survival, growth and, eventually, reproduction. Providing fish with a good environment is the best way of preventing diseases.

There are some generally agreed standards for water quality. These are classified with several users in mind, not just fish farmers. Generally, the first class corresponds to water of excellent quality, free from pollutants, and suitable for all use including bathing and as a drinking water supply (Class 1A) or, if slight treatment is required, Class 1B. Class 2 represents water which is of good enough quality for irrigation and for industrial use in which most fish species can survive but that is not clean enough for bathing. Class 3 is mediocre-quality water which can only be used for irrigation, industrial cooling or for navigable canals. Fish survival is uncertain in Class 3 water. For cyprinids, Class 2 waters are considered to be satisfactory for pond rearing, as is Class 1. Other characteristics are given in Table 2.1 and the most important are listed below.

Temperature

Temperature has a profound influence on biological activity which, generally within the range which supports life, doubles for each 10°C rise (e.g. oxygen consumption or general metabolism will be two times higher at 30°C than at 20°C. All changes to the pond such as fertilisation, feeding and fishing must take temperature into account. Optimum growth in cyprinids occurs between 20 and 30°C. Ponds are heated through the surface and thermal stratification becomes established either during the day–night succession in shallow ponds, especially in the tropics, or during longer periods in the summer. Temperatures of 33–35°C, lethal for several species, may occur at the surface but fish seek out deeper water at temperatures where they can survive (as long as the oxygen levels are satisfactory) (see below and Figure 2.18). When temperatures in a pond become too high intervention should be limited. All handling of the fish and temperature shocks should be avoided. While fish are able to tolerate gradual temperature changes without ill-effect, they are unable to withstand sudden changes, particularly those of high amplitude. It is known that fish are better able to withstand extremes of temperature if they are adapted to them progressively.

[3] R. Billard.

Table 2.1. General characteristics of Class 1B and 2 waters. These can be used as a reference for suitability of water quality for the production of cyprinids in ponds. An EEC directive (16.17.9.1975) has defined several aspects of water quality for cyprinids. BOD: Biological Oxygen Demand, COD: Chemical Oxygen Demand.

	Class 1B	Class 2	Cyprinid waters
Temperature (°C)	20–22	22–25	10–25
Dissolved oxygen mg/l (% saturation)	5–7 (70–90)	3–5 (50–70)	5.00
pH	6.5–8.5	6.5–8.5	6–9
pH during active photosynthesis		6.5–9	
Conductivity (µSiemens/cm at 20°C)	400–750	750–1500	
Suspended solids (SS mg/l)	≤30	30.00	25.00
Biochemical oxygen demand (BOD_5 mg/O_2/l)	3–5	5–10	6.00
Oxidation potential (mg/l)	3–5	5–8	
Chemical oxygen demand (COD mg/l)	20–25	25–40	
Total nitrogen (Kjeldhal)	1–2	2–3	
NH_4^+ (mg/l)*	0.1–0.5	0.5–2	0.20
NO_3^- (mg/l)	0–1	<44	6.00
Fe total (mg/l)	0.5–1	1–1.5	
Mn (mg/l)	0.1–0.25	0.25–0.5	
Cu (mg/l)	0.02–0.05	0.05–1	
Zn (mg/l)	0.5–1	1–5	0.10
As (mg/l)	≤0.01	0.01–0.05	
Cd (mg/l)	≤0.001	≤0.001	
Cr (mg/l)	≤0.05	≤0.05	
HCN (mg/l)	≤0.05	≤0.05	
Pb (mg/l)	≤0.05	≤0.05	
Se (mg/l)	≤0.01	≤0.01	
Hg (mg/l)	≤0.0005	≤0.0005	
Phenols (mg/l)	≤0.001	0.001–0.05	0.005–0.02
Detergents (mg/l)	≤0.2	0.2–0.5	
Oils–fats	None	Traces	
Coliforms (n/100 ml)	<5000	5000–50 000	
Escherichia coli (n/100 ml)	<2000	2000–20 000	
Faecal streptococcus (n/100 ml)	20–1000	1000–10 000	
Colour (mg Pt/l)	10–20	20–40	
Radioactivity (SCPRI category)	I	II	

* The toxicity to fish depends on pH and the temperature of the water (cf. Table 2.7), for the equations for ammoniacal nitrogen (see p. 13).

Dissolved oxygen

Although the oxygen requirements of pond-dwelling cyprinids (>3 mg/l) are lower than those of cold-water species such as salmon (>5–6 mg/l), this parameter is still of fundamental importance. The solubility of oxygen in water depends mainly on

Table 2.2. Balance of gains and losses of dissolved oxygen (DO), in mg/l, due to various processes in ponds used for the culture of catfish (after Boyd and Lichtkoppler, 1979 and Boyd, 1982*).

Gains during the day due to	Extremes (mg/l)	
– Photosynthesis by the phytoplankton	5–20	
– Diffusion	1–5	
	6–25	
Losses* during the night due to:	Mean (mg/l)	%
– Respiration by plankton	4.32	52.4
– Respiration by the fish	1.08	13.1
– Respiration by benthic organisms	0.73	8.9
– Diffusion	2.11	25.6
	8.24	

* The losses are calculated for a 1 ha pond with 3.2 tonnes of fish/ha; temperature 28°C, chemical oxygen demand COD (chemical oxygen demand): 75 mg/l, DO at dawn: 15 mg/l.

temperature but also on atmospheric pressure and salinity. Oxygen enters the water by diffusion from the air but this route is of much less importance in ponds than the input from photosynthesis by phytoplankton (Table 2.2). Losses occur through the respiration of a range of organisms and diffusion from water to air. Oxygen is only produced during the daytime. Levels of oxygen are at their highest in late afternoon at the surface where photosynthesis is most active (Figures 2.18 and 2.19). This phenomenon is all the more marked when the plankton bloom is at its highest because of the reduction in the penetration of light. The decline in oxygen shown in Figure 2.18 is of little consequence as the amplitude is limited; in general it should be borne in mind that the period during the disappearance of the phytoplankton bloom

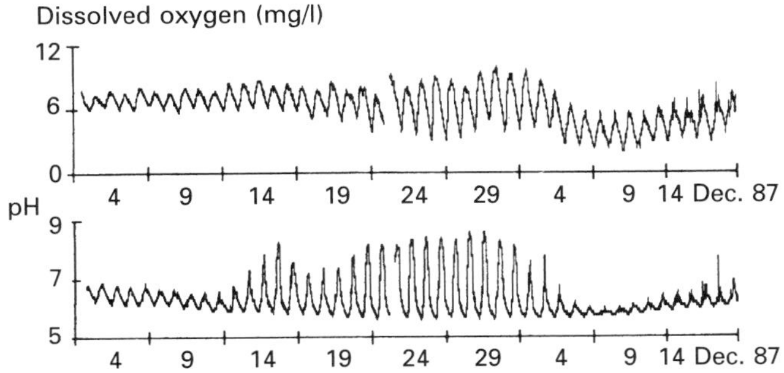

Fig. 2.18. Effect of a phytoplankton bloom on daily fluctuations in dissolved oxygen and pH. This information was obtained in French Guyana in a 1000 m^2 pond, following daily organic fertiliser applications beginning on 10 November 1987. This water has a tendency to acidity and is poorly buffered, meaning that the pH response is stronger than that of oxygen. The algal bloom, assessed daily by measuring the transparency of the water began on 16 November, reached a maximum between the 26 and 30 and then regressed slowly under the pressure and grazing by zooplankton which tripled in numbers between the 3 and 7 December. After this the levels of oxygen decreased (after Ginot, 1990).

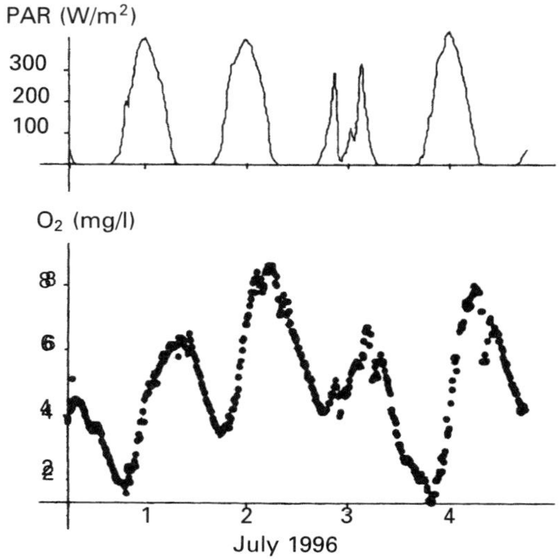

Fig. 2.19. Effect of the passage of clouds on the level of dissolved oxygen in the pond. The Photosynthetically active radiation (PAR) represents the solar radiation between 350 and 700 nm used for photosynthesis (W/m^2) (after Ginot, 1990).

(either through grazing by zooplankton as shown here or through mass mortality of phytoplankton) is often critical with regard to levels of dissolved oxygen. In cloudy conditions, photosynthetic activity will be lower and after a few sunless days the level of dissolved oxygen in the water can be dangerously low. The information shown in Figure 2.19 was collected in July from a 2 ha pond in la Dombe, France which had previously been treated with organic fertiliser. The oxygen probe was 40cm below the surface and the water was not stratified; the rise in levels during the two fine days was followed by a significant drop on the 3rd July when respiration, at about 0.5 mg O_2/l/h was not balanced by the production through photosynthesis. Such an occurrence, although of concern to the fish farmer, is happily infrequent as enough oxygen is usually produced, even in cloudy weather, because the algae are adapted to use low light levels (Ginot, 1990).

When major blooms develop, algae concentrated at the surface (most frequently cyanobacteria, p. 18) die and their decomposition causes the total depletion of oxygen through bacterial activity (Figure 2.2). Oxygen is only restored when the population of algae is re-established.

The oxygen concentrations over the course of the day are shown in Figure 2.20. These values have been obtained by applying a diurnal development model for dissolved oxygen to data from an organic fertilisation experiment in French Guyana. Similar results have been obtained in France and in Guyana over 100 days in 10 ponds (surface area between 200 m^2 and 3 ha) where stratification had not developed. The total respiration budget calculated for catfish ponds in America included the respiration of planktonic and benthic microorganisms as well as that of fish (Table 2.2). In general, losses of oxygen due to fish represent slightly more than 10%

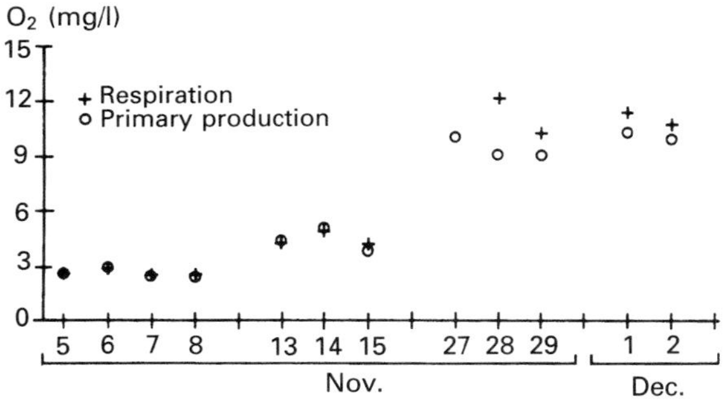

Fig. 2.20. Oxygen balance over a day and change over 1 month in French Guyana with an indication of the quantities of oxygen produced and consumed (after Ginot, 1990).

of the total. It was noted that the oxygen balance is either remarkably close to equilibrium or with a slight oxygen deficit. This latter situation does not occur, unless accidentally, except in very fertile waters or when an aerator is in operation. The gap between daily values for oxygen production/respiration is between 2 to 3 mg O_2/l/h for oligotrophic ponds or in winter and around 6-10 mg O_2/l/h for ponds managed normally (fertilised or with artificial diet given). It can reach 15 to 25 mg O_2/l/h with a general budget deficit in ponds which are highly fertilised. According to Ginot (1990), it seems that for equal biomasses of fish, a pond receiving organic fertiliser consumes more oxygen than one where the fish are fed artificial diets alone. This balance does not apply where clear stratification exists. Here, oxygen production is confined to the surface layer where there is supersaturation and a release of gas to the atmosphere (Figure 2.21).

The minimum O_2 requirements for most pond fish are low and the lower lethal limit may even be below 1 mg/l. Several species can survive a short exposure (a few h) to levels between 0 and 0.3 mg/l oxygen. The optimum range for rearing is above 5 mg/l, but fish exposed continuously to levels between 1 and 5 mg/l will survive but be more sensitive to diseases, have a lower growth rate and a poorer food conversion efficiency. In general there are many interactions between the levels of dissolved oxygen in the environment and feeding (p. 35), the oxygen consumption of fish per unit weight tends to increase with the quantity of feed distributed.

It is sometimes possible to improve the level of dissolved oxygen in fish culture units. Risks of a deficit occurring become serious when the water gets too warm or when there is a major algal bloom (Secchi disc reading <25 cm). The situation can become critical towards the end of the night. Knowing that oxygen concentration decreases regularly during the night, Boyd and his co-workers in Alabama, USA, measured concentrations at dusk and then 2 or 3 h later and extended the line obtained (Figure 2.22). An even more critical situation occurs when an algal bloom collapses and decomposes (see Figure 2.2) or when there is insufficient turnover of water such as in a pond covered with ice for a long period of time.

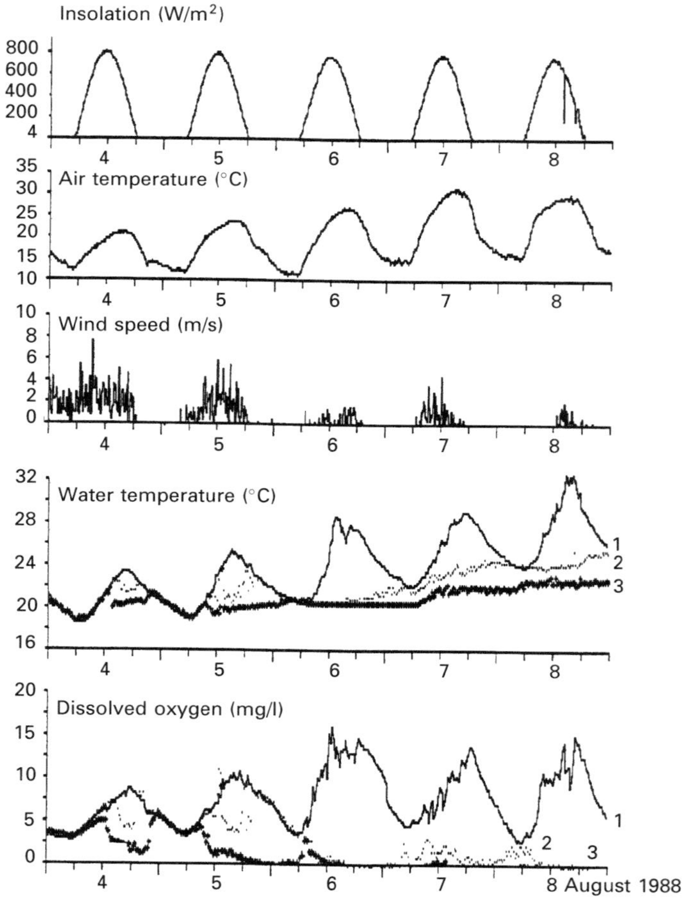

Fig. 2.21. Example of stratification in a pond. During the warm season, when insolation (W/m^2) increases, measured in the morning, with a light wind, and in a pond which is deep enough where the water is not very transparent a stratification of temperature and oxygen begins. The example here is a 3 ha pond, strongly fertilised (therefore very turbid) with a Secchi disc transparency of 30 cm. Depth is 1.2 m and temperature and oxygen probes are placed at a depth of 20 cm (1), 60 cm (2) and 100 cm (3) in relation to the surface. After 2 or 3 days of stratification, the major part of the water column is completely deoxygenated: this is dangerous for the whole pond community.

The most effective actions are based on the mechanical agitation of the water with various aerators placed at suitable locations around the pond (p. 240). In an emergency, water can be churned up with a motor boat or with aerators powered by a tractor. However, the low efficiency of the transfer of oxygen through air–water diffusion should be remembered (Table 2.3). In practice, it is difficult to get atmospheric oxygen to dissolve in water, particularly when temperatures are high. The effect of agitation is to increase the surface area of the air–water interface and

Table 2.3. Overall coefficient for O_2 transfer between air and water *Kla* for ponds of 1 m depth at a temperature of 20°C.

Condition of the pond	*Kla*
Pond in calm weather	<0.01
Pond in medium wind, without stratification	0.01–0.02
Pond in high wind	0.03–0.05
Pond vigorously aerated*	0.1–0.15

These values were obtained by adjusting a model of daily pattern of dissolved oxygen over 100 days in ten ponds with surface areas ranging from 200 m^2 to 3 ha. The differences result from the varying morphology of the ponds: the lower end of the range is a small, badly ventilated pond and the upper, large exposed ponds. Calculation of the effective transfer of oxygen is by multiplying the coefficient of oxygen deficit observed by its duration.

Example: during the course of a 10 h night, in calm weather and a mean deficit of 5 mg/l the pond will gain $0.01 \times 10 \times 5 = 0.5$ mg oxygen per litre. Natural aeration is therefore ineffective in compensating an oxygen deficit in calm weather. For temperatures (T in °C) other than 20°C, it is necessary to correct the coefficient of transfer using the equation $Kla(T) = Kla\,20 \times 1.024\,\exp(T - 20)$ Ginot, 1990).

* Aerator 10 hp/ha or winter storm.

thus the diffusion of oxygen into the water. In addition, agitation breaks up the vertical stratification which mixes the upper, oxygen-rich layers with the less-well oxygenated water below.

It is beneficial to prevent thermal and oxygen stratification by using devices such as paddle wheel aerators which help to keep the mass of water in the pond in motion, particularly during heatwaves or in the absence of wind. In the United States, a submerged pump (water mixer) which pushes the water horizontally was used by Boyd (*Aquaculture Magazine*, 1922b, 26–30) to mix surface waters with those from nearer the bottom effectively. This allowed a reduction in normal aeration in catfish ponds. Aeration applied selectively using little energy will often be sufficient. Other methods have been suggested to accelerate the decomposition of organic matter (spreading potassium permanganate or calcium hydroxide) but these compounds are rarely used and do not have a rapid effect.

pH

pH depends on the alkalinity and the hardness of the water. The alkalinity of water corresponds to the level of carbonates and bicarbonates and the hardness is an indication of the level of divalent metallic ions (mainly Ca^{++} and Mg^{++}). The level of calcium bicarbonate, $Ca(HCO_3)_2$, which depends on the concentration of calcium carbonate, $CaCO_3$, magnesium bicarbonate and of CO_2 in the water has a buffering effect, preventing sudden changes in pH. Low alkalinity leads to poor buffering capacity and generally a low pH. A high buffering capacity prevents fluctuations in the acid-base equilibrium caused by strong photosynthetic activity: the carbonate bases allow the maintenance of a large reservoir of CO_2 in the water which is essential for photosynthetic activity.

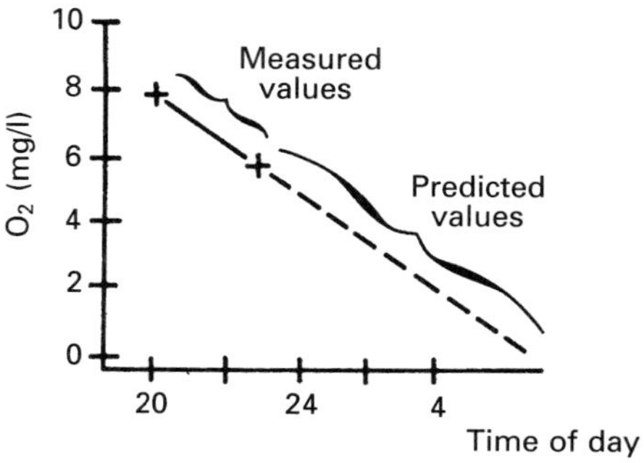

Fig. 2.22. Graphical method for prediction of the development of dissolved oxygen levels in water during the night (after Romaire and Johnston, 1978).

The pH of the water depends on the local geology and the calcium brought into the pond and also, above all, on the carbon dioxide level. During the daytime, phytoplankton and higher plants consume CO_2 during photosynthesis which leads to a rise in pH. The reverse effect occurs during the night which brings about a daily fluctuation in pH (Figure 2.18). In low-alkalinity water, the range can be from 6 to10 or even more over the course of a day. In alkaline waters pH variations are lower, between 7.5–8 and 9–10.

Where hardness is low, pH can exceed 11 during intense photosynthesis; pH 4 and 11 are recognised as the lethal limits for fish (Figure 2.23). When values (measured in the early morning) are outside the range 6–9, growth and production are poor. In fact, the most favourable pH for the culture of fish in ponds is between 6 and 9. Total alkalinity and total hardness, expressed as calcium carbonate equivalent are most favourable at around 150 mg/l. Minimum levels must be of the order of 30 mg/l to obtain good production. The fish can withstand concentrations exceeding 250 mg/l. The calcium composition is expressed in different units depending on the country (Table 2.4) and may be altered by spreading calcareous compounds (Figure 2.23).

The decision to change the pH of a pond should be based on chemical analysis, particularly that of total alkalinity. This should be of the order of 10 to 20 mg/l $CaCO_3$ equivalent for the maximum response to the spreading of fertiliser (Table

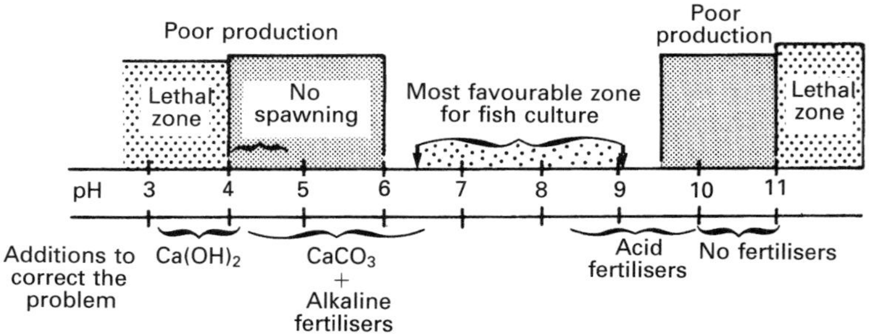

Fig. 2.23. Relationship between the pH of pond water and performance of fish-culture units.

Table 2.4. The units for expressing hardness and alkalinity of water and their correspondence* with different forms of calcium (after Barbe *et al.*, 1991, *Echo-Système*, **20**: 1–20).

	1 SBV	1 degree English	1 degree German (dH)	1 degree French (TH)	1 meq/l (milliequivalent/l)
Ca^{++} mg/l*	20	5.7	2.24	4	20
CaO mg/l*	28	8	10	5.6	28
$CaCO_3$ mg/l*	50	14.3	17.8	10	50
meq/l	1	0.286	0.35	0.2	1
Degree French (TH)	5	1.43	1.78	1	5
Degree German (dH)	2.8	0.8	1	0.56	2.8
Degree English	3.49	1	1.25	0.7	3.49
SBV	1	0.286	0.35	0.2	1

* Correspondence: 1 mg Ca^{++} = 1.8 mg CaO = 3.5 mg $CaCO_3$.
1 mg CaO = 0.55 mg Ca^{++} = 1.94 mg $CaCO_3$.

2.5). Changes are only justified where the growth of fish is mainly dependent on natural production and where the level of $CaCO_3$ is less than 10–15 mg/l. It is clearly only profitable in ponds where fertiliser is also being spread; however, modifications to pH can be justified in unfertilised ponds in very acid waters (pH <5), where the growth of fish is poor. It is best in this case to use slaked lime, $Ca(OH)_2$ or quicklime, CaO. The quantity to be distributed can be determined in the laboratory from the chemical analysis of mud samples (Box 2.1), but empirically it is possible to proceed with an application of 2 t/ha of $CaCO_3$ and to alter the alkalinity one or two months later if no improvements have been seen. However, beyond a certain tonnage, the profitability of this operation cannot be guaranteed. A rise in the level of $CaCO_3$ of 15 mg/l in the water increases fish production by 25%. Applications should be carried out in the autumn or at the beginning of winter before fertiliser is spread because if the applications are carried out simultaneously phosphorus can pre-

Table 2.5. Quality of ponds according to their levels of $CaCO_3$ and free CO_2 and the possibility for correction by addition of other compounds (information from Dr Pokorny).

Type of pond (level of Ca)	$CaCO_3$ mg/l	Free CO_2 mg/l	Applications of calcium* (t/ha)	
			Winter	Spring
Poor	0–25	0–0.1	1–2	0.5–1
Medium	25–75	0.1–1.2	0.3–0.6	0.1–0.4
Rich	>75	>12	0.1–0.2	0

* Crushed limestone, slaked lime.

Box 2.1
Estimating the quantity of amendments to be used in ponds for the removal of acidity in muds
(according to Boyd, 1979* and 1986 in Billard and Marcel, 1986*)

This technique, established in the south-east of USA, applies to acid muds free from sulphurs and only requires a pH meter and a solution of P-nitrophenol made up to pH 8 (obtained by dissolving of 7.5 g of boric acid (H_3BO_3), 37 g of potassium chloride (KCl) and 5.25 g of potash (KOH), adding 10 ml of the completed P-nitrophenol to 1 l of distilled water). The mud must be sampled from its entire loose depth and assessed randomly (25 samples from a pond of 1 to 4 ha). Samples of equal volume are air dried and passed through a sieve (mesh size 0.85 mm); 20 g of the sieved amount is mixed with 40 ml of the solution in a 100 ml beaker and stirred at intervals with a glass rod during a 1-h period. The pH is then measured and the difference from the initial pH of 8 is multiplied by 5.6 to obtain the dose of $CaCO_3$ to be spread in tonnes per hectare. For example, if the reduction is 0.5 units of pH, the amount of $CaCO_3$ to be used will be 2.8 t/ha.

A variation of this technique referred to by Valdeyron (1993, *Aqua-revue*, **50**: 16–17) consists of mixing first, for 1 h, 20 g of sediments in 20 ml of distilled water and taking the pH (value A), then adding 20 ml double concentration of the made-up solution mentioned above and retaking the pH after shaking for 20 min (value B). The quantities of $CaCO_3$ to apply in kg per hectare in order to arrive at water with 20 mg of Ca^{++} per litre are given in the following table.

pH of the mud	Value B									
(value A)	7.9	7.8	7.7	7.6	7.5	7.4	7.3	7.2	7.1	7
5.7	91	182	272	363	454	544	625	726	817	908
5.6	126	252	378	504	630	756	882	1008	1134	1260
5.5	202	404	604	806	1008	1210	1400	1612	1814	2016
5.4	290	580	869	1160	1449	1738	2029	2318	2608	2898
5.3	340	680	1021	1360	1701	2042	2381	2712	3062	3402
5.2	391	782	1172	1562	1948	2344	2734	3124	3515	3900
5.1	441	882	1323	1765	2205	2646	3087	3528	3969	4410
5.0	504	1008	1512	2016	2520	3024	3528	4032	4536	5040
4.9	656	1310	1966	2620	3276	3932	4586	5342	5980	6652
4.8	672	1344	2016	2688	3360	4032	4704	5390	6048	6720
4.7	706	1412	2116	2822	3528	4234	4940	5644	6350	7056

Which amendments to use?

The ground limestone or calcium carbonate ($CaCO_3$) in powder form (80% <80 μm) containing about 50% of CaO costs 60 €/ton; it dissolves slowly and can be carried dry in bulk. It will not greatly vary the pH.

Quicklime (burnt and crushed limestone) containing 90 to 95% CaO costs 136 €/ton; it is very soluble and can greatly vary the pH so that when scattered in open water it must be apportioned in fractions.

Hydrated lime (or slaked) has the same characteristics of solubility as quicklime but only contains 70% CaO.

* See General bibliography.

cipitate. The effect of treatment generally lasts from 3 to 5 years but a "maintenance dose" of 20–25% of the initial dose can be given each year.

Normal enrichments are not always enough to counteract the wide fluctuations in pH which may occur during the day. The use of agricultural gypsum, $CaSO_4$, to increase total hardness is advocated. The dose level is given by the following formula:

$$\text{mg/l gypsum} = (\text{total alkalinity} - \text{hardness} \times 2.2 \times 2)^{(4)}$$

Another approach is to manage the pond and its fertilisation in order to avoid excessive algal development (p. 34)

Levels of other essential minerals

As in agriculture, nitrogen, phosphorus and potassium are the most important elements. These are present in water and in the sediments and can vary over the course of the year. Phytoplankton and floating macrophytes (duckweed) remove them directly from the water and macrophytes use their roots to draw them from the sediments (at least nitrogen and phosphorus). Optimum levels, per litre of water are around 0.2 to 0.3 mg of potassium and 1.5 to 2 mg for nitrogen. Appropriate levels of potassium are lower and it is not generally added to ponds. The N/P ratio (dissolved nitrogen/dissolved phosphorus) should remain around 4/8. There is a risk of the development of a phytoplankton bloom when N/P falls below 2. Fertilisation is described on p. 186.

Salinity

The salinity of water is the total concentration of dissolved ions and is an important parameter because it influences the osmotic status of the fish. It is particularly important in lagoon-rearing systems. Fish, particularly fry and juveniles, are very sensitive to sudden changes of salinity but many species can withstand gradual changes and tolerate a range of salinities. Most species of cyprinids, which live only in freshwater (p. 1) can withstand salinities up to 10 ppt but for satisfactory growth they should not exceed 5 ppt.

Turbidity and colour

The turbidity and, to a lesser degree, the colour of ponds limits the penetration of light and hence photosynthesis. In marshy areas, the water is often brownish in colour. This is caused by humus originating from plants. These are generally acid waters of low alkalinity and are low in production. Application of calcareous compounds is likely to eliminate the humates and improve the quality of the water. Turbidity is caused by the presence of suspended particles, plankton, dead organic matter or silt. It is beneficial to eliminate these and treatment by flocculation

---

(4) 2.2 is the conversion factor for agricultural gypsum (80% $CaSO_4$) and $CaCO_3$, 2 is a multiplying factor taking account of the capture of minerals by the sediment (Marcel 1991).

using aluminium sulphate (15 to 25 mg/l) has been suggested and found to be effective. These treatments are only appropriate in alkaline waters (otherwise they should be modified) and they do not remove the causes of turbidity. Dead organic matter and suspended minerals cause mechanical damage to fish gills (effects are particularly pronounced when levels reach 40 mg/l) and block the filtering apparatus and the digestive tube of cladocerans and copepods at concentrations of 100 to 500 mg silt/litre. The effects on phytoplankton of turbidity due to suspended silt can be especially catastrophic. Microalgae are dragged to the pond bottom as the particles settle. They di.e.there and are therefore unable to either fulfil their oxygenating function or remove nitrogenous substances and phosphorus.

Bacteria

Ordinary heterotrophic bacteria are a major component of the trophic web of a pond (p. 12) but in the tropics there is a risk of introducing the bacteria associated with salmonella or even cholera when organic manure is spread. It is essential to ensure that manure and wastes are free from these diseases. Microbiological standards are being developed for waters used for fish culture. In Sweden, the number of thermostable coliforms must be below 1000 per litre, the total number of coliforms below 10 000 with tolerance for occasional overshoots (Ackefors *et al.*, 1990). Similar norms are in place in Japan. These standards are much stricter than those allowed for class 1B or 2 (see Table 2.1).

Levels of pesticides and heavy metals in the water and in fish flesh[5]

A wide range of pesticides is found in water and tends to concentrate in fish flesh. Normal standards remain to be defined; for example, in Japan allowable concentrations (in mg/l) in water are less than 0.025 for DDT, 2 for Dieldrin, 1 for DEP, 0.01 for Aldrin and Pentachlorophenol, 0.03 for BHC and Heptachlorine and 0.08 for Endrin.

Concentrations found in the fish themselves have not been thoroughly established for cyprinids. It is known that the concentrating power of fish muscle is high (between 1000 and 10 000 times that of the water). In general, levels of DDT of a few hundred μg/kg wet weight in muscle and 1 mg/kg in liver have been found while permissible levels are between 1 and 5 mg depending on country and species. The allowable limits for human health are 0.5 mg PCB/kg of flesh in (the former) Czechoslovakia and 2 mg in Germany and Sweden.

Levels of heavy metals in fish ponds are generally lower than those in reservoirs, rivers and estuaries and concentrations in fish flesh are likely to be low although few measurements are available. The level of mercury (Hg) has been measured at 0.03–0.16 μg/l in running water, 0.02–0.06 mg/kg in zoobenthos, 0.1–0.3 mg/kg in non-predatory fish and 0.5 mg/kg in piscivorous fish (wet weight). Levels of 10 mg Hg/kg wet weight flesh were found in pike in Swedish lakes which received the effluent from paper mills. Levels of mercury of 0.1 mg/kg were allowed in

---

(5) Information supplied by Dr Svobodova.

Czechoslovakia and 1.5 mg in Norway. Levels of between 0.5 and 1 mg are allowed in most European countries. These limits can vary according to species: 1 mg Hg/kg live weight for eels, pike and zander and 0.5 mg/kg for other species in Germany.

2.2.2 Water quality in the hatchery–nursery system

In a combined hatchery–nursery requirements are stricter than in the grow-out ponds described previously because eggs and larvae are more delicate than the later stages of fish. Water for the hatchery–nursery can have several origins: ponds, lakes, rivers, springs or drinking water sources. Whatever the origin, the chemical composition should be checked before the hatchery is established. Measurements should be made at regular intervals over at least one year; these should include total nitrogen, ammoniacal nitrogen, nitrite, nitrate, phosphate, BOD (biochemical oxygen demand), pH and hardness. Certain "biological" tests should also be performed: these could involve test incubation and fry rearing on-site (or in a nearby hatchery, using water which has been brought from the test site) to see if survival and growth are satisfactory compared with the norms for the region. In existing systems the water should be analysed from time to time, particularly when anything unusual or an accident occurs.

Water quality in a conventional hatchery (Table 2.6)[6]

Fish hatcheries require water of the highest quality. As well as being needed for the incubation of eggs and larvae it will be heated, pumped and recycled. Although used for incubating pond fish species, water should be above class 2 (Table 2.1).

Levels of dissolved gases

Of all the dissolved gases, oxygen usually attracts most attention from fish farmers: the concentration in the water decreases as temperature, salinity and altitude increase. When the temperature in a hatchery for pond fish species is raised from 18 to 30°C and the stocking density of eggs, alevins or fry is relatively high, it is generally advised that oxygen concentration should be close to saturation (>70 to 80%). For example, in fresh water at 30°C, at sea level, saturation is achieved at 7.5 mgO_2/l which corresponds to the requirements of larvae of herbivorous fish. However, too high a level of oxygen is harmful; carp alevins cannot withstand levels of 20 mg O_2/l. Potential sources of water for the hatchery may be oxygen deficient (underground or spring water) or have an excess of oxygen (pond water during the day) and may therefore need oxygenation or degassing before being fed into the rearing system (see below).

Other dissolved gases such as nitrogen (N_2) are toxic above 100% saturation, carbon dioxide gas (CO_2) at over 10 mg/l, chlorine (most frequently encountered in drinking water sources) at above 3 μg/l, hydrogen sulphide (H_2S) at over 3 μg/l and

---

[6] J. Marcel.

Table 2.6. Optimum and extreme values for several water quality parameters in the hatchery (several sources including EEC directive, 1975, EPA Redbook 1979 and Meyer *et al.*, 1983).

			Limits	
		Optimum	Minimum	Maximum
Ions (total)	mg/l	600–1700		4000
Conductivity	μS	1000–2700		6000
Total alkalinity (CO_3Ca)	mg/l	50–200	20	400
NH_3 (cf. Table 2.7)	mg/l	0		0.02
Ions NH_4^+	mg/l	1.5–2.0		
NO_2^- nitrites	mg/l	0.05–0.2		0.5
NO_3^- nitrates	mg/l	1–10		15
H_2S (hydrogen sulphide)	μg/l	<3		
pH	units	7.0–8.0	6.5	9
O_2 dissolved	mg/l	5–12	4	
O_2 dissolved	% saturation	>70	50	150
N_2 (partial pressure)	% saturation	<100		103
CO_2	mg/l	<5		15
HCN (hydrocyanic acid)	μg/l	<10		
Cl (chlorine)	μg/l	<3		10
SS	mg/l	<30		
BOD	mg/l	<10		
COD	mg/l	<20		
Ca (calcium)	mg/l	50–160	10	
Cd (cadmium)	μg/l	0.4*–1.2†		4
Cu (copper)	μg/l	<6*–30†		
Fe (iron)	mg/l	<0.1		
Fl (fluorine)	mg/l	0.5*–5†		0.5
Hg (mercury)	μg/l	<0.05		0.5
K (potassium)	mg/l	<5		
Mg (magnesium)	mg/l	<15		
Mn (manganese)	μg/l	10		
Na (sodium)	mg/l	<75		
Ni (nickel)	mg/l	<0.01		
P (phosphorus)	mg/l	0.1		3
Pb (lead)	μg/l	25*–50†		
PCB	μg/l	2		
S (sulphur)	mg/l	<1		
Sulphates	mg/l	<50		
U (uranium)	Mg/l	0.1		
V (vanadium)	mg/l	<0.1		
Zn (zinc)	μg/l	30*–100†		

* Where alkalinity <100 mg $CaCO_3$/l.
† Where alkalinity >100 mg $CaCO_3$/l.

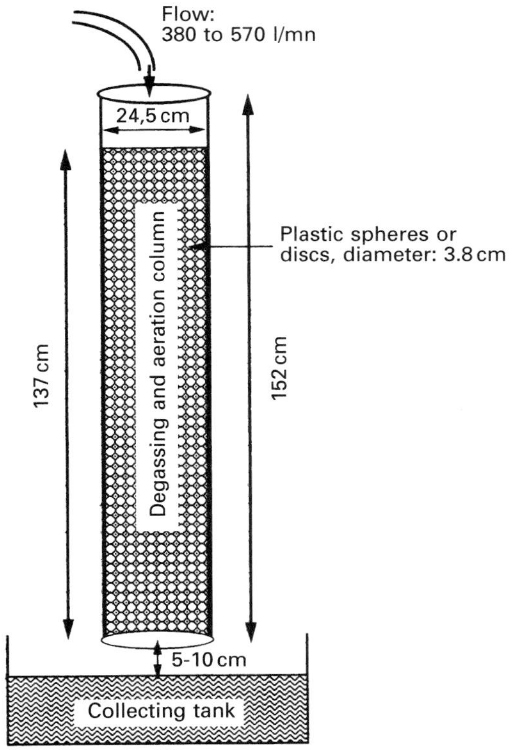

Fig. 2.24. Diagram of a simple degassing/aeration column installed at the water supply to a hatchery. This reduces the nitrogen supersaturation from 130 to 100% and increases the oxygen concentration by around 10% (after Owsley, 1979).

hydrocyanic acid at above 10 μg/l. Toxicity increases when the oxygen level in the water nears its limits for alevins and juveniles; CO_2 is more toxic when oxygen is deficient. Supersaturation with nitrogen and carbon dioxide gas often occurs when air is drawn into pipes as may happen at water intakes coming from a header tank which is not completely full. Chronic mortalities whose cause is not readily apparent may sometimes be caused by slight gas supersaturation. It is therefore important to take periodic measurements using a saturometer.

Installing a simple degassing–oxygenation column (Figure 2.24) upstream of the water distribution system gives excellent results in the elimination of dissolved gases (N_2, Cl_2) and oxygenation of the water. This column can be coupled with a storage (header) tank 1–1.5 m above the system so that water is distributed to tanks by gravity rather than by pumping.

Minerals and organic matter

Water taken from rivers or ponds is generally rich in suspended solids (SS). The total level should never exceed 30 mg/l for larval rearing. As most waters have higher levels than this, at least periodically, a means of filtration is required. This may consist of a simple gravel and sand filter through which water trickles and which must be cleaned regularly.

Total salinity, a measure of the substances dissolved in water, should be below 4 g/l for larval cyprinids. pH values between 6.5 and 9 are allowable; higher values are likely to be toxic for fry. When water is taken from a pond there will be a buffering effect; the alkalinity (carbonate–bicarbonate complex) limits pH fluctuations (p. 46). Optimum alkalinity and hardness (calcium and magnesium) is between 50–200 mg/l ($CaCO_3$ equivalents) (Table 2.6).

The nature of the nitrogen compounds present depends on the origin of the water. Water from boreholes, rivers or reservoirs used to supply drinking water can contain nitrate, which has a low toxicity; the allowable level is 15 mg/l. However, some boreholes in agricultural areas may have levels in excess of 30 mg/l.

Nitrite (NO_2^-) is more dangerous and its toxicity, greatest in waters of low alkalinity, varies between 0.1 and 0.2 mg/l.

Ammoniacal nitrogen is present in low quantities in waters supplying a hatchery, except when this comes from rivers or ponds. However, nitrogen is produced during juvenile rearing, particularly in the nursery where the fish are being fed. Ammoniacal nitrogen is very toxic in the unionized form (NH_3) and this toxicity increases with increase in temperature and pH (Table 2.7). Toxic effects have been reported for trout alevins from 0.025 mg/l for NH_3 and it is extremely likely that ammonia is also highly toxic to carp alevins. Because of this, hatcheries operating open water systems ensure that wastes are removed or use a system for water recycling via a biological filter. Methods of analysis do not give the proportion of NH_3 in relation to total ammonium (NH_3, NH_4^+). Calculations are necessary; tables give the percentage of NH_3 in relation to pH and to water temperature (Table 2.7).

Heavy metals are particularly toxic and should be systematically measured in new sites or if any anomalies occur in existing ones. The LD_{50} 96 h (mortality of 50% of the fish after 96 h continuous exposure) corresponds to concentrations of several tens of micrograms per litre. Only minute traces (maximum a few tens of μg/l) of zinc, copper, lead, cadmium, mercury and chromium can be tolerated. In a hatchery the use of galvanised and copper pipes must be avoided, particularly in waters of low alkalinity.

Pesticides are much more toxic for eggs and larvae than for adult fish; it is not possible to use the standards established for adults to define the concentrations which can be allowed in a hatchery. Toxicity levels are between 5 and 100 μg/l for brief exposures. The allowable level for prolonged exposure will be lower. The risks of contamination must not be overlooked and accidental spillages, including those from treatments to ponds, guarded against.

Water quality in fry ponds[7]

The water quality in ponds used for the rearing of fry corresponds broadly to that

---

[7] J. Marcel.

Table 2.7. Percentage of free ammonia (NH_3) at different pH and temperature (reported from Boyd, 1982)*.

	Temperature (°C)						
pH	8	12	16	20	24	28	32
7.0	0.2	0.2	0.3	0.4	0.5	0.7	1.0
8.0	1.6	2.1	2.9	3.8	5.0	6.6	8.8
8.2	2.5	3.3	4.5	5.9	7.7	10.0	13.2
8.4	3.9	5.2	6.9	9.1	11.6	15.0	19.5
8.6	6.0	7.9	10.6	13.7	17.3	21.8	27.7
8.8	9.2	12.0	15.8	20.1	24.9	30.7	37.8
9.0	13.8	17.8	22.9	28.5	34.4	41.2	49,0
9.2	20.4	25.8	32.0	38.7	45.4	52.6	60.4
9.4	30.0	35.5	42.7	50.0	56.9	63.8	70.7
9.6	39.2	46.5	54.1	61.3	67.5	73.6	79.3
9.8	50.5	58.1	65.2	71.5	76.8	81.6	85.8
10.0	61.7	68.5	74.8	79.9	84.0	87.5	90.6
10.2	71.9	77.5	82.4	86.3	89.3	91.8	93.8

* The ionised (NH_4^+) and un-ionised (NH_3) forms of ammonia are present in the water in variable proportions depending on pH and water temperature. The figures given indicate the overall levels (NH_4^+ and NH_3) although it is important to know the concentrations of NH_3 which is a toxic molecule and enters the blood via the gills.

described above. However, because of the size of fish at the start of rearing (1 and 13 mg respectively for first feeding carp and pike) and the degree of intensification (around 3 t/ha) some aspects have to be given particular consideration.

Temperature

Larvae which are only a few days old have a strict temperature range preference. In general, the thermal conditions should be similar to those in the hatchery. A temperature difference of more than 5°C at the time of transfer gives a thermal shock which adversely affects the survival of the larvae.

pH

For the main species (carnivores and cyprinids), the fry stage extends from April to August when light intensity and temperature allow active photosynthesis to take place. pH fluctuates widely during the day (up to 10 at the end of the day). It appears that some species such as pike are very sensitive to this. The buffering capacity of fry ponds should therefore be checked with particular care and any necessary corrections made at the water inflow (p. 45).

Dissolved oxygen

In fry ponds, dissolved oxygen is replenished by photosynthesis which leads to the daily fluctuations already described. In general, at biomasses of up to 3 t/ha, the dissolved oxygen balance remains positive except perhaps for a few hours before dawn. The installation of aerators (air hydro-injectors, 1 horsepower/ha) compensates for this deficit for a short while (p. 234). The oxygen saturation measured at the end of the afternoon (up to 130–150% saturation) has no effect on the survival and growth of fry. However, aeration prevents problems of stratification which occur in warm, calm weather (p. 40).

Nitrogen compounds

Production of ammoniacal nitrogen in fry ponds remains limited for most of the rearing cycle because of the small amount of food distributed. However, the elimination of NH_4^+ by phytoplankton is particularly effective during the period of intense photoperiodic activity. In contrast, in autumn, there is an accumulation of NH_4^+ resulting from low levels of phytoplankton activity. Exposure of 75 g carp juveniles to a level of 3 to 3.5 mg NH_4^+/l for a week at the start of October caused mortalities associated with lesions in the gill lamellae. In practice, reducing the amount of food given can prevent this accumulation from occurring.

The ecosystem of the pond can exert a buffering effect on a number of parameters or toxic elements such as nitrite, ammoniacal nitrogen (NH_3) but this does not apply to pesticides and to heavy metals which, if incorporated in the food chain, accumulate in the fry (p. 50). However, heavy metals can accumulate in the mud where they often form complexes (chelates) with organic substances: the chelates are not toxic. Thus only a proportion of the heavy metals introduced into an ecosystem finds its way into the trophic web.

2.2.3 Measurement of water quality

Overall analysis

The estimation of the density of phytoplankton is an alternative source of information on the potential of the pond. In ponds where there are few silt particles, the measurement of transparency using a Secchi disc (Figure 2.8) gives a good indication of plankton density. With a visibility of 30–60 cm, the quantity present will allow good production of fish and limit the penetration of light to a depth less than that favourable to the growth of macrophytes. If the visibility exceeds 60 cm, there will be significant growth of macrophytes, reducing the mineral salts available for the growth of phytoplankton. Where the visibility is less than 30 cm there will be a risk of an oxygen deficit as described previously. Regular measurements of clarity (once or twice a week) give the fish farmer information on the development of planktonic communities and, thus, the availability of food for fish. It may also be a good idea to count benthic organisms or both the large and small types of zooplankton. The apparent disappearance of benthic organisms and the total absence of large zooplankton (daphnids) suggest that predation is heavy. Algal production in the pond is therefore no longer or insufficiently exploited by the consumers, the larger forms of zooplankton (p. 28). It is therefore necessary to reduce the biomass of fish and/or to feed them (p. 174).

Classification of pond types

The natural productivity of ponds depends on several factors, of which the most important is probably the geology of the pond itself and of the surrounding basin; the exposure of the pond to sun, its orientation in relation to prevailing winds and the type of agriculture on the surrounding land are also significant. Combining these local physical factors it is possible to establish a system for classifying different types of ponds and to relate these to natural production (Table 2.5). One of the main traditional distinctions is between ponds in forests and those in cultivated land. The forest pond does not receive as much sunshine and is filled by water which is rich in organic acids, poor in minerals and hence the pond is not very productive. In comparison, a pond in a cultivated area receives fertiliser-rich runoff from the land as the result of the spreading of mineral fertilisers or organic wastes from animals. In the two types the necessary fertilisation and enrichments required are different. Ponds can also be classified by parameters determining production such as health status and production itself.

Measurement of the most important variables

Soil analysis

The fish farmer or technician can use various simple techniques to measure the most important parameters: N, P, pH, O_2, CO_2, total alkalinity and total hardness. There is a range of equipment and techniques from comparative colorimetry which gives sufficient accuracy in the measurement of a single parameter to kits which allow the measurement of a large number of parameters (e.g. Tetra, Merck, Hydrocure, Hach).

Measurement of oxygen

Two principal methods are used:

- —The *Winckler method*: this uses reagents which can be obtained from specialist suppliers and aquarist shops.
- —*Measurement using an oxygen electrode*: readings are taken using an oxygen meter. This device uses an electrode which reacts according to the concentration of dissolved oxygen in the water. The electrical signal is transmitted from the probe, through an amplifier, to a visual display which gives a value in mg/l O_2 or percentage saturation.

These meters are, for the most part, easy to use and sufficiently robust for work in the field. Some features require particular attention.

The probe itself. The two electrodes which receive a current proportional to the O_2 concentration are bathed in an electrolyte and are separated from the environment by a very thin membrane: this is the most fragile part of the oxygen meter.

Even a slight scratch on the membrane will lead to inaccurate readings. The electrodes can be repaired fairly easily, even in the field, but only with specialised equipment and with practice. If a little care is taken with the handling of the equipment membranes can last for several months.

Calibration of the probe. Here also there are two methods: the first by comparison with readings obtained by the Winckler method on the same sample and the second, which is the most frequently used, and most practical, is calibration in air. The wet probe is exposed to air and, knowing the temperature (and for increased precision the atmospheric pressure) the saturation values are obtained from a table supplied by the manufacturers. More sophisticated models have this calibration table stored in memory. Oxygen meters filled with built-in recording devices are very useful in the field.

For measurements in ponds it is preferable for the probe to have a mixing device attached to give reliable measurements; in practice all oxygen probes are designed to give accurate results for a particular current speed over the membrane. There are many types of oxygen meter on the market; it is advisable to compare several models in the field over a period of at least 24 h in order to determine any drift.

Portable oxygen meters (sometimes combined with pH meters) allow rapid measurements over a series of ponds and are useful in an emergency as they can be used by dropping the probe into the water from a moving vehicle in which the readings can be collected.

Removal and preservation of water samples for measurement

Simple equipment, such as weighted bottles, should be used for taking water samples. These are used from the bank or a boat. Another method uses bottles attached to a support which can be lowered to the desired depth. The bottle is closed while it is being dropped and, once in place, the stopper is withdrawn by pulling a string.

When the samples have been taken the most accurate results will come from immediate analysis, *in situ*, of the principal parameters such as dissolved gases, pH and temperature. For the dissolved gases, any time longer than 4 h will lead to errors being made and the sample cannot be considered to be representative even if the temperature has been lowered. Ammonia should be measured immediately, as should the main parameters (phosphate, nitrite, nitrate, etc.). If the measurements show a deficit or an excess for one particular element, repeat analyses should be carried out, at least twice.

BIBLIOGRAPHY

Ackefors H., Hilge V., Linden O., 1990. Contaminants in fish and shellfish products, 305–344. In: N. de Pauw, R. Billard (eds), *Aquaculture Europe 89 Business Joins Science*. EAS Special Publication 12. EAS, Bredene, Belgium.

Arce R.G., Boyd C.E., 1975. Effects of agricultural limestone on water chemistry, phytoplankton productivity and fish production in soft water ponds. *Trans. Am. Fish. Soc.*, **104**: 308–312.

Bérard A., Sevrin-Reyssac J., 1990. Production de bactéries, de phytoplancton et de zooplancton pendant l'hiver dans des étangs de pisciculture. Résultats des experiences menées en Sologne de novembre 1989 à mars 1990. Rapp. contret "Aliment demain," Minist. Agriculture et Minist. Recherche, Paris: 1–11.

Boulon R., 1989. Recherche de l'empoissonnement maximum de carpes miroir (*Cyprinus carpio*) en conditions de fertilisation organique; effets sur la biocénose. Mém. ENSAA, Dijon: 1–45.

Boyd C.E., Lichtkoppler F., 1979. Water quality management in pond fish culture. *Research and Development*, **22**.

Boyd C.E., Prather E.E., Parks R.W., 1975. Sudden mortality of a massive phytoplankton bloom. *Weed. Sci.*, **23**: 61–67.

Delsalle F., 1989. Le cladocère planctonique *Bosmina longirostris* en milieu piscicole: particularités morphologiques et biologiques, variations saisonnières, rôle dans le réseau trophique. Dipl. Agro. Approf., ENSA, Montpellier: 1–58.

Gallemard, P., 1986. Croissance de poissons d'étang en fonction de la fumure organique apportée. BTS, IUT Tours Fondettes: 1–17.

Ginot V., 1990. Modélisation nycthémérale de l'oxygène dissous en étang. Thèse, Univ. Cl. Bernard, Lyon I, 154 pp.

Jamet J., Lagouin Y., 1974. *Manuel des pêches maritimes tropicales*. Soc. centr. Equip. Territ. intern., Ministère Coopération, Paris, Tome I: 1–447.

Marcel J., 1991. Problème des pH élevés en étang de pisciculture. *Echo-Système, ITAVI, Paris*, **17**: 9-10.

Marcel J., Legouvello R., 1990. Deuxième alevinage. *Echo-Système, ITAVI, Paris*, **15**: 7–10.

Meyer, F.P., Warren J.W., Carey T.G. (eds). 1983. A guide to integrated fish health management in the Great Lakes basin. *Special Publication, Great Lakes Fish Commission*, 83-2: 262 pp.

Miquelis A., 1989. Particularités du phytoplancton dans les étangs de pisciculture. DEA, Univ. Aix Marseille, 3: 1–35.

Owsley D.E., 1979. Nitrogen gas removal using packed columns, 71–82. In: L.J. Allen, E.C. Kinney (eds), *Proceedings of the Bioengineering Symposium for Fish Culture*. American Fish Society, Fish Culture Section, Bethesda, Maryland, Publ. 1.

Poxton M.G., 1990. A review of water quality for intensive fish culture: 285–303. In: N. de Pauw, R. Billard (eds), *Aquaculture Europe Business Joins Science*. EAS Special Publication 12, EAS, Bredene, Belgium.

Romaire R.P., Johnston E., 1978. Predicting early morning dissolved oxygen concentrations in channel catfish ponds. *Trans. Am. Fish. Soc.*, **107**: 484–492.

Sevrin-Reyssac J., Giraud J.P., Delsalle F., 1990. La carpe argentée peut-elle limiter la quantité de phytoplancton? *Echo-Système, ITAVI, Paris*, **16**: 11–14.

Sevrin-Reyssac J., Gourmelen J.L., 1985. Le biotope «étang à roselières» en Brenne. Qualité des eaux et évolution saisonnière du plancton de quelques étangs de Brenne et de leur roselière. *Protoc. Mus. Hist. nat. Paris-Minist. Environnement*, conventions 82188 et 83079: 1–72.

Taran W.K., 1936. L'alimentation des carpillons d'un été pendant l'hiver. *Bull. Fr. Pisc.*, **8**: 217–222.

Zimmerman P., 1990. Aspects des populations planctoniques en hiver dans les étangs de pisciculture. Impacts de la fertilisation organique. Mém. DES, Paris VI: 1–56.

Farming

3

Reproduction[1]

3.1 REPRODUCTIVE BIOLOGY OF CYPRINIDS

3.1.1 Gametogenesis and release of gametes (ovulation and spermiation)

As with most species of fish, at the time of hatching the gonads of larval cyprinids are not morphologically differentiated. Differentiation is only apparent when the young fish are between 50–100 days old, depending on the species. In most cyprinids the female is the homogametic sex (XX). The age at first maturity depends mainly on the rearing temperature; for carp in the natural environment this is 4–5 years old in the Volga, 3–4 years in Poland, 2 years in France (Camargue) and 1 in Israel. When reared at a constant temperature of 15°C, female carp reach sexual maturity at 15 months. Males generally mature a year earlier than females in temperate regions; in France males can mature at 500 g body weight and females at 800–1000 g. In colder regions such as central Europe male carp usually mature at 2–3 years old (1 kg) and females at 4 years old (3–5 kg). The Chinese carps reach first maturity when they are older and heavier (Table 3.1). First maturity in the Indian carps occurs at 1–2 years old. The age and weight at first maturity depend also on the genetic stock, the temperature of the rearing water, feeding and growth rate.

There are several distinctive phases in female gametogenesis (oogenesis).

- —Multiplication of oogonia and their transformation into oocytes after the first meiotic division;
- —growth in size (from 30 to 300 μm diameter) of oocytes which at this stage are previtellogenic and are accumulating RNA;
- —accumulation of yolk, termed vitellogenesis, growth of oocytes from 300 to 800 or 1000 μ;
- —final oocyte maturation and ovulation.

The number of oocytes produced per cycle and per kilogram body weight (relative

[1] R. Billard.

Table 3.1. Information on the spawning of several cyprinid species under central European climatic conditions (taken in part from Horvath *et al.*, 1986, cf. General bibliography).

	Common carp	Grass carp	Silver carp	Bighead carp	Tench
Age and size at first maturation*					
Males age (years)	2–3	4–6	4–6	5–6	2–3
weight (kg)	1–2	4–6	3–4	5–7	0.4–2.5
length (cm)	25–30	50–70	40–60	50–90	25–30
Females age (years)	4–5	7	5–6	6–7	3–4
weight (kg)	3–5	6–8	3–6	5–10	0.75–3
length (cm)	30.40	50.70	40.60	60.100	25.30
Reproduction					
—spawning season†	April–June	May–July	May–July	June–July	May–June
—temperature °C (natural)	16–22	21–24	21–23	23–25	22–24
—temperature in the hatchery	20–24	22–24	22–24	22–26	22–24
—% ovulation after hypophysation	60–90	60–80	60–80	80–90	60–70
—eggs collected/kg (hatchery) × 1000	100–200	60–80	60–80	40–50	80–120
—ova					
• diameter (mm)	1.0–1.5	0.8–1.2	0.7–1.0	1.0–1.3	0.4–0.5
• number per kg × 1000	700–1000	800–900	900–1 100	600–700	2 000
—fertilised–hydrated eggs					
• diameter (mm)	2.0–2.5	3.7–5.3	3.7–5.3	4.0–6.0	0.6–0.7
• number per kg × 1000	80–120	16–18	18–22	15–20	600–700
Embryos and larvae					
—embryogenesis (duration in degree-days)	60–70	24–30	24–30	30–50	60–70
—yolk resorption (in degree-day)	60–70	60–70	60–70	60–70	100–110
—size of the larvae at hatching (mm)	4.8–5.0	5.0–5.2	5.0–5.2	5.0–5.2	3.5–3.6
—size at first feeding (mm)	6.0–7.0	6.0–7.0	6.0–6.5	7.0–8.0	4.5–5.5

—size of first prey (μm)	100–300	50–100	50–250	50–300	50–100
—size of larvae at 1 month (mm)	25–30	25–30	25–30	25–30	15–20
Incubation conditions (Zoug bottles)					
Volume of eggs/7 l Zoug jar	1–2.5	2–3	2–3	2–3	1–1.5
Water flow rate (l/mn)	0.5–2.0	0.1–0.5	0.1–0.5	0.1–0.5	0.5–2.0
Treatments					
—eggs: products (MG = Malachite Green)	MG	Formalin	Formalin	Formalin	MG
dose in mg or ml/l water	5 mg/l	0.1 ml/l	0.1 ml/l	0.1 ml/l	5 mg/l
duration/frequency	5 min/day	5–10 min: 4 h	5–10 min: 4 h	5–10 min: 4 h	5 min/day
—larvae: product	Formalin	Formalin	Formalin	Formalin	–
dose (ml/l)	0.1	0.1	0.1	0.1	—
duration/frequency	5 mn/day	5 mn/4 h	5 mn/4 h	5 mn/4 h	—
Expected performance in the hatchery					
% fertilisation	80–95	70–90	70–90	70–95	80–90
% hatch	90–100	75–85	75–85	75–85	90–100
% feeding larvae/kg ova × 1 000	500–700	400–600	500–600	400–500	1600–1800

* Sexual maturation occurs later and at a smaller size in western Europe: it is advanced in heated water.
† Spawning period of broodfish reared in ponds.

Table 3.2. Relative fecundity (number of ova per kg live weight female) for several species of cyprinids (after Jhingran and Pullin, 1985).

Bighead carp	67 000–130 000
Common carp	90 000–300 000
Crucian carp	110 000–150 000
Grass carp	80 000–120 000
Silver carp	100 000–150 000
Black carp	60 000–100 000
Mrigal	90 000–420 000
Rohu	100 000–400 000
Catla	100 000–250 000

fecundity) varies between species (Tables 3.1 and 3.2). Relative fecundity also varies with age and size (Figure 3.1).

Oogenesis is highly temperature dependent; a complete cycle requires at least 1000 degree-days (or 60 days at 20°C). When temperature conditions are optimal it is possible to obtain several spawnings a year, provided that the fish have been suitably fed. Final maturation requires temperatures above 18–20°C but if ovulation does not occur (for example because females and males have been kept in separate ponds or hypophysation has been performed too late) in the following 2–3 weeks, the oocyte breaks down and ovulation will not take place, or if it does the ova will be infertile. If the spring temperature rises happen early (March–April) and extend over 2–3 weeks or more, oocyte atresia is likely to occur. It would probably be too early to put the larvae into outdoor ponds this early so the females should be kept as broodfish and ovulation induced 2–3 months later when they have completed a new vitellogenic cycle. At the end of vitellogenesis oocytes can maintain their integrity for long

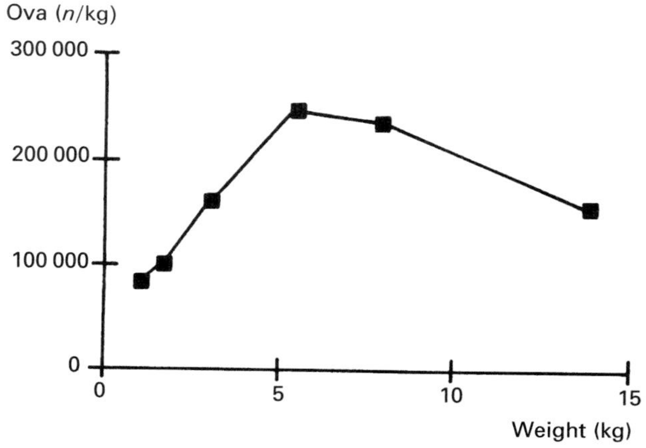

Fig. 3.1. Change in relative fecundity of female common carp in relation to increased body weight (after Alikunhi, 1966, *FAO Fish Synopsis*, **31**, 73 pp.).

periods of time (up to 9 months when the temperature is below 16°C). During this phase, transfer to water at a temperature of 20–25°C will bring on final oocyte maturation even without hormonal stimulation. The process of oocyte maturation and ovulation is also dependent on photoperiod and ovulation of many species of cyprinids is induced by the presence of spawning substrates (plants in the Dubisch ponds in Central Europe, kakaban in Asian countries) (p. 78).

Spermatogenesis comprises a phase of multiplication of spermatogonia followed by meiosis and spermatogenesis, which results in the accumulation of spermatozoa in testicular lobules and their subsequent emission (spermiation). Huge numbers are produced: 2000 billion spermatozoa per cycle (for a 1 kg male). Spermatozoa are formed in the testis from the autumn onwards and remain there until the breeding season. Whenever water temperature is above 5–8°C it is possible to extract small quantities of spermatozoa from the autumn onwards and this provides a means of sexing the fish. The spermatozoa are not all released by the end of the breeding season, even after hormonal stimulation, and remain in the testis during the following cycle. Their quality deteriorates as they age.

The endocrine determination of gametogenesis has been particularly well studied in goldfish and carp. Factors from the external environment, together with internal ones, induce the secretion of various neurohormones by the hypothalamus (in the brain) one of which, Gonadotropin Releasing Hormone (GnRH), causes the secretion of the gonadotropic hormones by the hypophysis; these are transported in the blood and stimulate the gonads. In cyprinids dopamine acts as an inhibitor superimposed on the stimulating mechanism. One of the methods of inducing ovulation is to remove this inhibition at the same time as providing stimulation (LINPE method, p. 76). At present, two gonadotropic hormones (GTH) have been identified in several cyprinids:

—GTH_1 which controls the first stages of gametogenesis including vitellogenesis in females;
—GTH_2 which primarily influences the final stages of gonad maturation (oocyte maturation, ovulation and spermiation).

Their action on the gonads takes place via steroids: testosterone, 11 keto-testosterone and 17α-hydroxy,20β-dihydroprogesterone (17–20P) in the male and oestradiol, oestrone, 17–20P in the female. A phenomenon resulting from social interaction has been observed in male goldfish and carp. The gonadotropin stimulation which induces spermiation is due at least in part to pheromones released by the female which are simply conjugated forms of progestins secreted during oocyte maturation and released into the water.

3.1.2 The biology of the gametes

Spermatozoa

The morphology of the carp spermatozoa is similar to that of most cyprinids; the head is roughly spherical (2–2.5 × 3.3 μm), the chromatin of the nucleus is not very

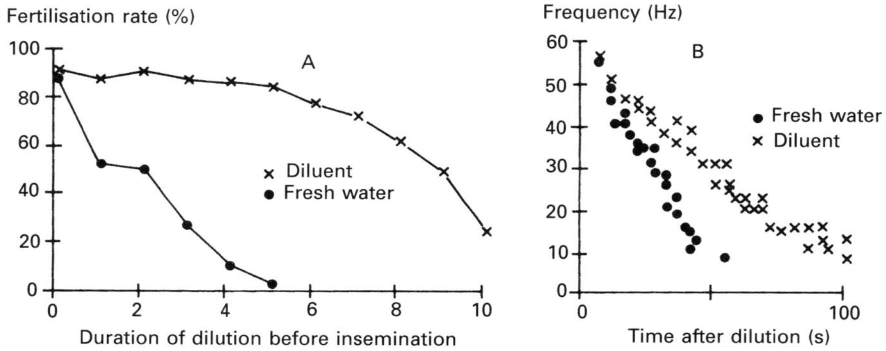

Fig. 3.2. Duration of survival of carp gametes diluted in fresh water or in diluent: A, change in fertilisation rate of ova; B, change in flagellum beat frequency.

condensed and the flagellum (40–50 μm) is inserted tangentially onto the head of the spermatozoa. There is no identifiable acrosome and the intermediate part is reduced to a few mitochondria. The spermatozoa of carp and other cyprinids are, as for most teleosts, immobile in the testis and in the seminal fluid. This immobilisation is due to the osmotic pressure (around 300 milliOsmoles/kg) and its decrease, which occurs on dilution in water, allows movement to begin immediately. The seminal fluid of carp is rich in potassium (around 80 mM), sodium (minimum 60 mM) and in amino acids. Sperm are only motile for a short time. Motility decreases steadily between the moment of activation and the cessation of movement which occurs 60–80 seconds after dilution in a physiological solution and between 40–60 seconds in fresh water (Figure 3.2).

Ova

Ovulated oocytes (ova) of cyprinids are extremely variable in size, from 0.4 to 0.5 mm in diameter in tench, from 0.6 to over 1 mm in Indian and Chinese carps and 1 to 1.5 mm in common carp (Table 3.1). The internal structure is similar to those of other teleosts with significant yolk reserves, cortical alveoli and granules and a small mass of cytoplasm localised at the animal pole beneath the micropyle and including the germinal vesicle (female pronucleus). The external envelope of the ovum (or vitelline membrane) undergoes major transformation after immersion in freshwater. There follows the development of an adhesive external layer (in carp) and a considerable increase in size through the process of hydration of the whole envelope for ova of Chinese and Indian carp which results in floating eggs (semi-demersal); these increase from 0.6–1.3 mm in diameter to 4–7 mm (Table 3.1) or from 3–8 to 30–80 mm^2 total external surface area.

During natural or artificial spawning the ova are accompanied by a small amount of ovarian fluid which is rich in proteins (4.1 g/l for carp and 2.7 g/l for silver carp) and in amino acids (10.6 μM/ml for carp) and has a pH varying between 8.5 for carp and 9.1 for silver carp. The osmotic pressure is 306 milliOsmoles/kg for carp and 218 for silver carp (spermatozoa are not activated at these values).

One of the characteristics of cyprinid ova in hatcheries is their rapid loss of fertility after ovulation. Whether *in vivo* in the ovary or *in vitro* (after stripping), the loss of fertility is almost complete within 4–10 hours for carp and in only a few hours for Chinese and Indian carps at normal spawning temperatures (50% of silver carp ova are not capable of being fertilised 30–40 minutes after ovulation). Keeping them in the refrigerator below 10°C does not improve things: it actually makes them worse. The reasons for this are not yet well known. It is not associated with the start of cortical activation, which does not occur spontaneously in the ovarian fluid. One hypothesis for carp is the loss of endogenous stores of oocyte ATP that decreases regularly and is exhausted 6 hours after ovulation and storage *in vivo* or *in vitro*. It is thus essential that fertilisation takes place very soon after ovulation: this explains the development of methods of "natural" spawning *in situ* in tanks or small ponds where males and females are put together and fertilisation occurs immediately after ovulation (p. 96).

After dilution the fertility of the carp ova decreases very rapidly in the water, dropping to 20% in 3 min and more slowly in a saline solution (diluent) where it decreases to 20% in 10 min (Figure 3.2). This is due to the phenomenon of cortical activation (discharge of material from the cortical granules into the perivitelline space) which causes the separation of the plasma membrane and the yolk envelope so that the spermatozoa can no longer contact and reach the ovum. In other cyprinids the drop in fertility is even more rapid: 30–40 s in water and 50–60 s in diluent for silver carp. At present there are no solutions available which prevent the ageing which occurs after ovulation: ova must be fertilised as rapidly as possible after ovulation and especially after immersion in water.

A number of factors influence the quality of ova and their fertilisation. Some, already mentioned, are the history and physiological state of the broodstock holding conditions, feeding, oxygenation and temperature (particularly during and immediately preceding the final phase of maturation). There are others such as the stage of oogenesis reached when ovulation is induced and the composition of ovarian fluid: the percentage fertilisation is negatively correlated with the levels of K^+ for carp and positively correlated with the levels of Ca^{2+} for silver carp.

3.2 METHODS FOR THE INDUCTION OF OVULATION AND SPERMIATION

Traditionally, broodfish spawn naturally in the rearing ponds but the success of spawning depends very much on climatic conditions and, in some years, no juveniles are available for restocking. Because of this fish farmers have sought to control spawning and have developed various approaches, some based on hormonal stimulation of captive broodstock in conditions where temperature is controlled, others based on the environmental factors which fluctuate naturally during spawning in the wild. These methods are used in combination with hatcheries and include not

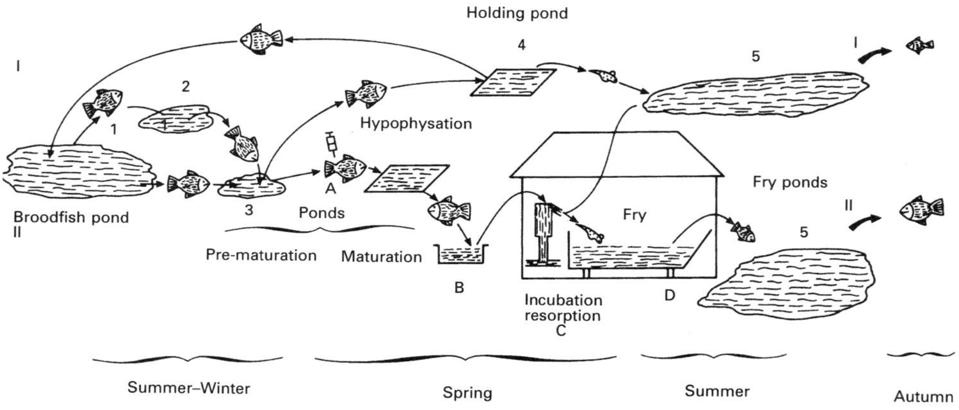

Fig. 3.3. Diagram of the number of operations associated with the management of spawning and the first-fry stage in (I) natural, controlled spawning and (II) hatcheries. ABCD, respectively hypophysation operations, artificial fertilisation, incubation and resorption of yolk/first-fry stage (see Figure 3.6 for details), (1) summer–autumn pond, (2) overwintering pond in cold countries, (3) pre-maturation pond, (4) spawning or holding pond, (5) fry pond (after Billard and Marcel, 1980).

only the management of spawning but also that of larval rearing which also suffers from climatic uncertainties (Figure 3.3).

3.2.1 Management of broodstock

Broodstock are usually kept in ponds (Table 3.3) but may sometimes be kept permanently in a hatchery and used several times a year for spawning with the resulting juveniles reared in small tanks (a few m^3).

In pond culture the density of broodstock varies according to species and to the available food. The density can vary from a few hundred kg/ha in Europe for carp to 1–2 t/ha in China for herbivorous species. In all cases a supplementary diet of protein-rich pellets (trout diet) must be given. In cold regions the broodstock are removed from their ponds in the autumn and held at a density of 100 to 450 5–6 kg individuals in 1000 m^2 ponds which are quite deep (1–2 m) and fed with water and/or aerated so as not to become frozen over. Early in spring the sexes are separated and placed into small ponds (1–2.5 ha) which are shallower (40 cm–140 cm) and sheltered from the wind.

These ponds warm up quite rapidly which promotes gametogenesis and the number of degree-days needed to complete sexual maturation can be completed at the appropriate time. Some characteristics of this method, as practised in Poland, are given in Table 3.3. In a milder Atlantic-type climate it is important to ensure that water temperatures exceeding 20°C do not occur too early in the season (see above). In all cases, supplementary feeding is required in these "pre-maturation" ponds and should begin when the temperature rises above 6°C. Protein-rich (30–40%) pellets

Table 3.3. Characteristics of different types of broodstock pond for overwintering, spawning and their holding capacity (e.g. carp culture in Poland, information from J. Wieniavski).

Type of pond	Overwintering	Breeding: Pre-maturation	Breeding: Reproduction pond
Duration of residence	Winter	Spring	12 to 48 h
Temperature (°C)	0.2–2.5	Up to 20°C	20–25
Number of ponds needed	1–2	*	*
Area (ha)	0.02–2	0.02–0.1	0.01–0.03
Depth (m)	1.8–2.5	1.0–1.5	0.3–1.0
Flow rate	†	As overwintering	‡
Stocking rate			
—number of individuals	Up to 1000/ha	Up to 500/ha	3 to 10/pond (2–5 males for 1–3 females)
—weight of individuals (kg)	1–8	4–8	4–8
—total weight (kg/ha)	4000–8000		

* Varies according to spawning techniques (cf. Figure 3.3).
† The water supply to overwintering ponds has the effect of oxygenating the environment and stabilising temperature.
‡ Compensation for losses of water through evaporation and infiltration.

should be provided with essential fatty acids; carbohydrates and saturated lipids should be avoided as these lead to the accumulation of fat. When broodstock are brought from outdoors into the hatchery it is usually necessary to raise the temperature from 15–16°C to 20–22°C. This warming up should be progressive (1 to 4°C/day) when hormonal treatment is being applied or in the form of temperature shocks (1°/h). (p. 79)

Oxygen requirements for broodfish are around 100 mg O_2/kg live weight/hour in water which is saturated with oxygen. This is particularly important during oocyte maturation and ovulation. During the rest of the rearing period, minimum levels should be around 6 mg/l. During handling in the hatchery (transfer, injection, holding prior to ovulation) 10–12 females or males of 5–6 kg can be held in a 3–4 m^3 tank, which has a water through flow of 1 l/min/kg live weight broodstock.

In the tropics, broodstock are usually kept with the sexes separated after spawning to avoid any natural breeding which might occur spontaneously in the following two months.

Males and females should be monitored separately for the selection of broodstock (see above). If allowed by legislation, they should be heat-branded on the side (mirror or leather carps) or on the head (scaled carps). These marks must be renewed periodically. All events affecting individuals should be recorded: performance during spawning (delay in response after treatment for induction of spawning), fecundity, volume of sperm, fertilisation and hatching rates, larval survival, growth, etc. These records will be very useful for selection purposes (p. 105) and decisions for improving broodstock.

In Central Europe, an average 5–6 kg female carp produces 700 000 eggs from which 400 000 larvae are obtained. Annual losses of broodstock, including those which are rejected, are between 10 and 15%; this should be made up every year by recruitment of new stock.

3.2.2 Control of spawning using hormonal treatments

Even when kept in satisfactory temperature conditions captive female broodfish will not ovulate spontaneously in the absence of the appropriate environmental stimulus. The traditional method of overcoming this problem is the use of hormonal stimulation. Spermiation in males is low when they are separated from females (p. 68) and a hormonal injection is used to increase the volume of milt but at a lower dose than is used for the females. The success of the use of hormones depends largely on the state of maturation of the gonads, i.e. the receptiveness to hormonal treatment. It is therefore necessary to measure and evaluate this state of readiness, particularly in females.

Maturity stage determination for ovaries

The amount of hormone to be injected depends on the state of maturity reached by the females which itself depends, among other things, on the temperature of the water in the rearing pond at the time of injection, the time during the breeding season and the accumulated degree-days from the start of the year in temperate zones (1000 to 1100 for common carp and 1300 to 1500 for Chinese carps). Several parameters are taken into account with a greater or lesser degree of objectivity: these include the suppleness, elasticity and roundness of the abdomen and the prominence and turgidity of the genital papilla. However, it is preferable to check on the position of the nucleus (germinal vesicle, GV) in the oocyte as this is a good sign of the stage reached in the female maturation cycle (Figure 3.4) and acts as a reference point for carp. To do this, a catheter is inserted into the genital orifice and some follicles

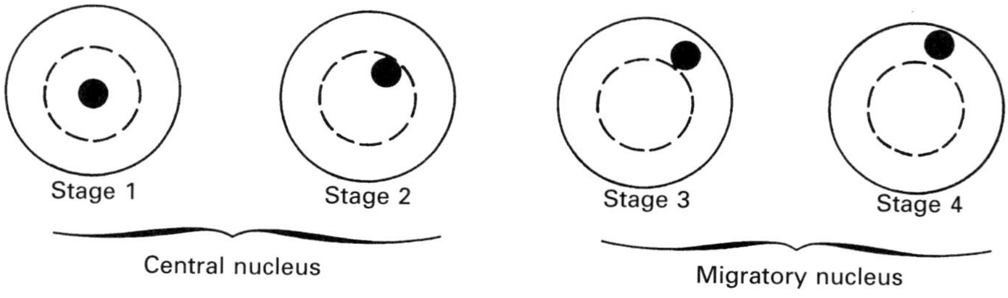

Fig. 3.4. Diagram of the stages of migration of the germinal vesicle in the carp oocyte. Migration proceeds in the direction of the animal pole, where the microphyle, the hole through which the spermatozoa passes during fertilisation, is found.

(oocytes and their surrounding envelopes) removed. The yolk is cleared by placing them for 15 min in a mixture of 95% alcohol (or 40% formalin), glycerine and acetic acid in the proportion 6 : 3 : 1 in order to determine the position of the germinal vesicle. When the nucleus has begun its migration (stage 3), the chances of ovulation are at their maximum (i.e. around 70%). From the very beginning of the spawning season it is a good idea to follow the development of the germinal vesicle after the first injection; if this has induced the passage from stage 2 to stage 3 the second injection can be given with a reasonable chance of success. If, on the other hand, the first injection has not brought about a change from stage 2 it is pointless to continue; the chance of ovulation is low and the female should be returned to the rearing unit for at least a week before repeating the operation.

Hypophysation in carp

The first hormones used to induce spawning in captive fish were forms of HCG (Human Chorionic Gonadotrophin) which is mammalian in origin and, for cyprinids, only really effective for goldfish and Chinese carps (HCG is ineffective in common carp).

The most frequently used gonadotropin hormone preparation currently in use is dry carp pituitary powder, which is available commercially. However, its biological activity is not always known (it can vary between 5 and 60 μg GTH_2/mg powder depending on the preparation) and the injection of 3 mg/kg live weight female used is generally excessive and given to ensure an effective dose. This 3 mg dose has probably been derived with reference to the weight of the hypophysis of a carp, which would weigh around 1 kg. If the fish farmer has a batch of hypophyses of unknown activity available or obtains them from processors who are filleting fish, tests should be carried out, reducing the doses to 2 or even 1 mg to assess the biological activity. It is also necessary to remember that the activity of hypophysis dried in acetone and kept at ambient temperature in a dry atmosphere decreases with time (Figure 3.5). There are more sophisticated preparations which are available in the lyophilised form or dissolved in protective substances such as glycerol, whose biological activity is known and more stable, such as "calibrated carp pituitary extract" which is produced in Israel.

Some indications of doses and response times for several species are given in Table 3.4 and a diagram of the procedures is given in Figure 3.6. Frequently, but not always, two injections are given. This can be justified for female fish at the start of the spawning season, when the receptivity of the ovaries to the gonadotropin maturation hormone is low; in mid-season a single injection is likely to be sufficient, simplifying operations considerably. During the spawning season the volume of sperm which can be collected without hormonal stimulation is low (a few ml) unless the males are in the presence (even indirectly) of ovulating females. For this reason, a hormone injection most frequently of carp pituitary extract is given in the hatchery and the volume of sperm collected is proportional to the dose injected into the carp, of the order of 5 to 8 ml/kg from samples taken after 24 h for a dose of 2 mg carp

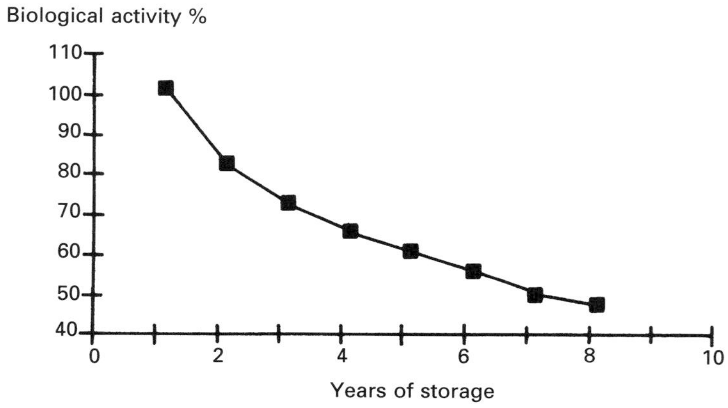

Fig. 3.5. Change in biological activity of dried carp hypophysis stored in acetone at ambient temperature in a dry atmosphere (after Jaczo, 1989, *Pol. Arch. Hydrobiol.*, **36**: 373–383).

hypophyseal powder at temperatures of 20-25°C (Figure 3.7). Analogous values have been obtained for Chinese and Indian carp but at higher doses (3–6 mg hypophysis/kg) and higher temperatures (25–30°C). In carp the first sample can be taken 12 to 24 h after injection, a second sample can be removed on the next day but this will be a lower volume (3–5 ml sperm) (Figure 3.7) but of similar quality. Injections can be given every day or every other day to prolong the duration of the response. It is even possible to obtain sperm between October and May from males kept at above 15°C and injected periodically with powdered carp hypophysis. The male receives only one dose, most frequently administered at the time the second injection is given to females although it is also possible to give it at the time of the first one; maximum spermiation is maintained for 24 h (Figure 3.7).

Use of GnRH-A and the LINPE method

Hypophysation uses exogenous gonadotropin hormones but it is also possible to induce the release of hypophyseal hormones by supplementing GnRH-A. These are small molecules (10 amino acids) which can be synthesised and prepared more easily than the gonadotropins. Such a product has been used since the 1970s in China as an analogue of mammalian GnRH (LHRH-A = D-Ala^{6}LHRH). Since then, analogues of salmon GnRH have been available (D-Arg^6 Pro^9-NET-sGNRH). It appeared that the doses needed were relatively high (>50 μg/kg) but, after the discovery made by Peter and his collaborators of the inhibitory dopaminergic mechanism described above, it has been possible to reduce the dose to 5–10 μg by combining LHRH-A or GnRH-A with an antagonistic compound, domperidone, which acts on the dopamine receptors, in a solution of propylene-glycol. Trials of applications on various cyprinids have been carried out in China by Lin; this has resulted in the LINPE method (from a combination of the names of the two authors) details of which are given in Table 3.5. A commercial preparation

Table 3.4. Doses of hormone preparation used to induce ovulation or stimulate spermiation, CH: dried carp hypophysis.

Species and sex		Preparation	Temp. (°C)	Doses (mg/kg or IU)		Interval between 1st and 2nd injection	Response after 2nd or only injection	
				1st injection	2nd injection	(h)	(h)	(degree-hours)
Common carp	F	CH	21–25	0.3	2.7–3	12	11–13	240–260
	M	CH	19–25		1–2	–	12	240
Silver carp	F	CH	22–28	6	6	6	5–7	160
	M	CH	25–30		3	–	4	
	F	CH	28–34	4	6–10	5	4–5	145
Grass carp	F	CH	22–28	6	6	5	7–9	200
	M	CH	25–30		5	–	4	
	F	CH	28–31	3–5	7–10	–	5–7	180
Bighead carp	F	CH	22–26	6	6	6	6–8	200
	M	CH	25–28	5	–	–	4	
Goldfish	F	HCG	18–25	1000 IU	–	–	20–24	
Rohu	F	CH	27–31	2–3	5–8	4–6	4–6	125
	M	CH			2–3	–	6	
Catla	F	CH	27–28	3–4	8–10	4–6	4–6	125
	M	CH			2–3	–	6	
Mrigal	F	CH	24–31	2–3	6–7	4–6	4–6	125
	M	CH			2–3	–	6	

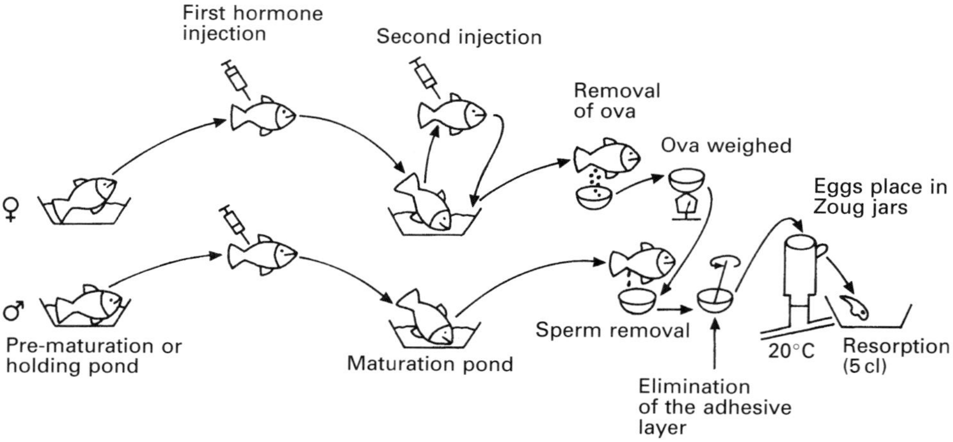

Fig. 3.6. Diagram showing operations for the induction of ovulation and stimulation of spermiation (through hormone injection), artificial fertilisation and incubation.

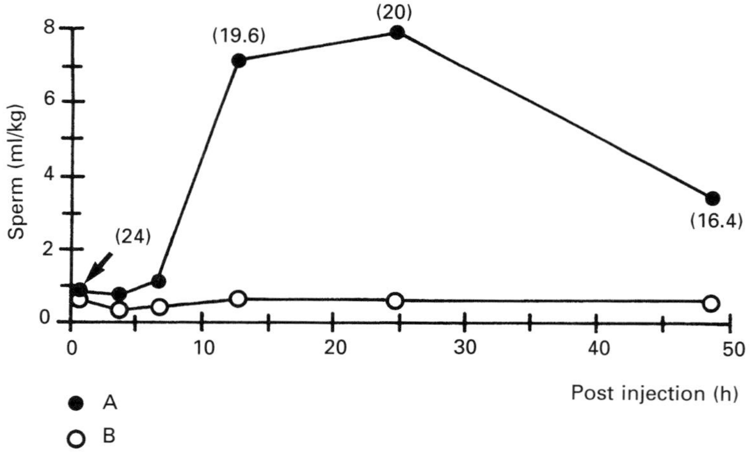

Fig. 3.7. Production of sperm by male carp which have been injected at a rate of 2 mg/kg with powdered carp hypophysis (A) or solvent only (saline solution B). Figures in brackets show the concentration of sperm in billions of spermatozoa/ml) (after Saad and Billard, 1987).

(OVAPRIM, SYNDEL, Vancouver, Canada) combines domperidone and sGnRH-A (10 mg and 20 μg/ml respectively) in a stable solution; this has been successfully tested in China and India (Table 3.6).

The LINPE method can also be applied to males. A combination of 10 mg domperidone and 20 μg LHRH-A /kg induces the production of over 5 ml sperm per day for 3–4 days. Doses of 100 μl OVAPRIM/kg induce spermiation in Chinese and Indian carps.

Table 3.5. Combined doses of domperidone and gonadotrophin releasing hormones for the induction of spawning in various species of cyprinid (LINPE method developed in China) after Lin *et al.*, in Billard and Marcel (1986).

Species	Water temperature (°C)	Domperidone (mg/kg)	Gonadotrophin releasing hormones: Type	Dose (μg/kg)	Mean interval injection–ovulation (h)
Common carp	20–25	5	LHRH-A	10	14–16
	20–25	1	sGnRH-A	10	14–16
Silver carp	20–30	5	LHRH-A	20	8–12
	20–30	5	sGnRH-A	10	8–12
Mud carp	22–28	5	LHRH-A	10	6
Grass carp	18–30	5	LHRH-A	10	8–12
Bighead carp	20–30	5	LHRH-A	50	8–12
Black carp	20–30	t o: 3	LHRH-A	10	
		t 6 h: 7	LHRH-A	15	6–8
Bream	22–30	5	LHRH-A	10	8–10

Table 3.6. Comparison of reproductive performance in female Indian carp after the administration of preparations of carp hypophysis (CH) and OVAPRIM (10 mg domperidone and 20 mg sGnRH/ml) (after Nandeesha *et al.*, 1990, Sp. Publ. Asian Fish. Soc., 41 pp).

	Catla		Rohu	
Treatment	Ovaprim	CH	Ovaprim	CH
Temperature (°C)	25–31	25–31	27–30	27–30
Dose/kg	0.4–0.6 ml	20–24 mg	0.25–0.35 ml	18 mg
Number of females	74	68	68	67
% Total ovulation	93	78	100	90
% Non-responders*	0	12	0	0
% Fertilisation	83	77	95	83
% Hatch	95	92	95	90

* The difference between the two columns corresponds to incomplete ovulations.

Various attempts have been made to implant capsules containing GnRH-A or other hormones into the body of the fish. The hormones are released progressively and, thus more closely resembles the normal physiological process. Trials have been carried out on several potential substitutes for the classic hypophysation technique which is still widely used but all of them still have the same problems (availability, stability, specificity of hypophyseal preparations) and ovulation rates failing to exceed 70–80%. It is likely that the various forms and presentations of gonado-

Fig. 3.8. Spawning substrate (strips of raffia attached to a plastic mesh) for natural spawning of carp in tanks (Carpeix Pollenca, Majorca, Spain, photo R. Billard, INRA).

tropins will, hopefully, eventually offer a cheaper widely available non-immunogenic substance with a defined, stable biological activity. The many manipulations which are part of the hypophysation technique appear to be traumatic and cause significant mortalities in broodstock after spawning which can affect 3% of the males and 10% of the females in Chinese carps. Research should focus on the development of less traumatic techniques, such as dosing orally, which limit the handling of the broodstock.

3.2.3 Induction of spawning through manipulation of environmental factors

In the wild, cyprinid spawning is determined by various physical environmental factors such as temperature, photoperiod, the presence of a suitable substrate for spawning or biotic factors including social aspects such as the presence of other spawners. For a long time fish farmers have taken advantage of this and utilised shallow spawning ponds with vegetation (Dubisch or other kinds) where the water warms up quickly, and with males and females stocked together. Reproduction occurs naturally without the need for hormonal stimulation; incubation and hatching takes place in the same pond and the eggs stick to the plants. However, this system is very dependent on climatic conditions and it is difficult to collect the larvae after they have hatched. Variations can be made to this arrangement, in better controlled conditions in hatcheries spawning takes place naturally over artificial substrates which have been placed in temperature-regulated tanks (Figure 3.8). After spawning the broodfish are removed and the eggs allowed to hatch and develop to the first larval stage in the spawning tank. An analogous method is used for spawning the Chinese carps: this consists of placing both male and female broodfish (which have usually been given the classical hormone treatment by a single injection) in circular tanks fed by a strong current of water which mimics the conditions found

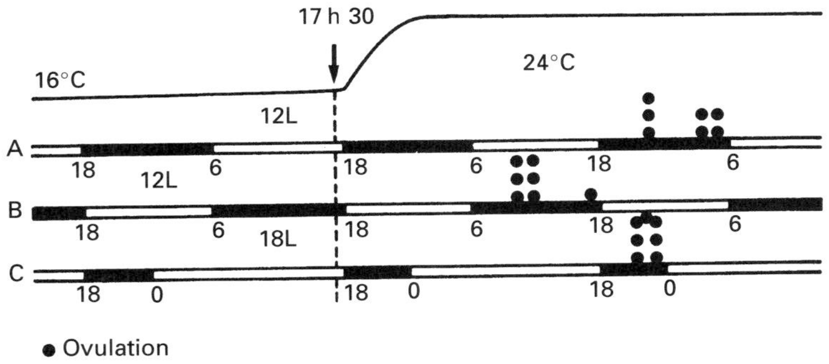

Fig. 3.9. Diagrammatic representation of the procedure for spawning induction in captive carp using temperature shock at the start (A–C) or the end of the dark period combined with a regulated photoperiodic regime (Japanese method). The dots indicate ovulation time; 12L, 18L is the duration of the light period in hours. The temperature shock is initiated at 17 h 30 (after Davies *et al.*, 1986).

in the natural environment; spawning takes place naturally and fertilised eggs are collected at the outflow of the tank.

Recent experiments on carp performed in Japan have shown that it is possible to induce ovulation simply by submitting the broodstock to a temperature shock (from 16 to 24°C in 6–8 h) beginning in the late afternoon as the light is fading (Figure 3.9); this is followed by a rise in gonadotropin in the blood circulation at the end of the following day which in turn is followed by ovulation in the middle of the following night. The timing of the rise in gonadotropin and ovulation depends on the timing of the temperature shock in relation to darkness. These results have practical applications in hatcheries, which have a source of refrigerated water or spring water available to maintain the broodstock at 16°C before the expected spawning period. The broodstock can be left to spawn spontaneously by offering them an artificial substrate for spawning as in the preceding scenario. If the farmer wishes to collect the gametes for artificial fertilisation the sexes must be separated (at least physically) as it is necessary that males and females be in the same water system so that the pheromones released by the females induce spermiation in the males.

3.3 PROCEDURE FOR THE CONTROL OF SPAWNING IN THE HATCHERY

3.3.1 Carp

Capture and grading of broodstock (Figure 3.3)

The broodfish are removed from the maturation ponds and the sexes are held separately, in tanks in the hatchery (biomass 10 to 30 kg/m^3) for several days before spawning is induced. The day before injection the females to be induced are selected; those in the most advanced state of maturity are retained. During all handling linked to injection procedures a current of water flows through the tanks to maintain the concentration of oxygen close to saturation. The broodfish should not be disturbed between injections and should always be handled with extreme care during injections. The fish are lightly anaesthetised during injection and for the removal of gametes (see Box 3.1 and Table 3.7). Handling and transfer operations are aided by the use of a folded cloth (Figure 3.10) or a landing net (made from knotless netting) in the form of a muff, open at both ends; the upper opening is attached to a rigid metal hoop (30 cm in diameter) and the other end is closed by hand during capture and opened for release once transfer has been completed.

If the pre-spawning broodfish have to be transported the duration of the transfer should be short (maximum of a few hours), the water should be well oxygenated or aerated and the density no more than 20–30 individuals/m^3. These fish are not fed for 2 to 3 days prior to injection in order that faeces do not accumulate in the tank and that there is less gamete contamination during stripping.

Preparation of hormones and injection

Hormonal preparations can sometimes be obtained in the form of a solution, which can be injected directly. If carp hypophyses are used they must be weighed and diluted prior to injection. Weighing in milligrams can pose a problem in a working hatchery. Sometimes a whole hypophysis is referred to as a unit; a dried one weighs between 3 and 6 mg and comes from a donor fish of 1 to 2 kg weight. In large hatcheries it is possible to prepare all the necessary hypophyses at the start of the season and to freeze them and keep them at −30°C until required. It is convenient to use a final concentration of 3 mg/ml as a standard dose of 1 ml per kilogram body weight for females. For example, grinding 100 mg hypophysis in 33 ml makes it possible to treat around 20 kg of females by allotting 22 ml to the females and 10 ml to the males, based on the use of a 1 : 1 ratio of the sexes at fertilisation. The solvent used is a physiological solution (6 g NaCl/l) which the fish farmer can buy from a pharmacy or make up himself using boiled water and sodium chloride.

It is preferable to time the injections so that ovulation occurs in the morning: this facilitates the work of the hatchery staff. The first injection should therefore be given at around 8.00 h and the second at around 18.00 h. It is important to be able to predict the moment of ovulation precisely in order to avoid the risks of ageing of the ova. For this, the temperature of the water must be known and remain constant through the night. The temperature sum, between the second injection and ovulation, is between 240 and 260 degree-hours; it is therefore easy to predict and achieve ovulation at around 06.00 h for a second injection at 18.00 h at a temperature of 20°C.

This injection schedule mimics the natural ovulatory peak, which begins in the evening followed by ovulation early the following day. There is wide variation between individuals in the timing of the response and the best way of detecting the moment of ovulation is by placing a male in a tank of females: the male begins to follow the females and demonstrate spawning behaviour from the moment ovulation occurs. This technique is particularly useful for Chinese and Indian carps for which

Box 3.1
Anaesthesia of fish

Anaesthesia is currently practised to facilitate the handling of fish including that of breeders. The anaesthetics most used are MS222 (to 100 mg/l of water), quinaldine (0.25 ml/l of water) and phenoxy ethanol (0.2 ml/l of water) (cf. Table 3.7). At these doses and at a temperature of 22–24°C anaesthesia takes effect in minutes. The fish must be withdrawn from the anaesthetizing bath as soon as there is a loss of balance and before the respiratory movements of the gills cease (stage II 2). If the latter should occur during handling, the mouth of the animal must be placed intermittently under the running water of a tap in such a way as to restore artificially, by a see-saw motion, the movement of the gills, which re-establish spontaneously within minutes. In order to inspect the fish individually it is preferable to anaesthetize a limited number at a time. The anaesthetizing bath must be oxygenated and renewed periodically on days when a large amount of fish are anaesthetized. The volume of the bath will be between 5 and 10 times the volume of fish anaesthetized at one time.

The stages of anaesthesia and recovery are as follows:

Stages of anaesthesia	State of fish	Reactions and behaviour
0	Normal	Reaction to external stimuli, balance and tonicity
I 1	Light sedation	Slight loss of reaction to external stimuli, visual and tactile
I 1	Deep sedation	Total loss of reaction to external stimuli apart from strong manual pressure Slight reduction of gill movements
II 1	Partial loss of balance	Partial loss of muscular tonicity Reaction only to strong tactile or vibratory stimuli Preservation of rheotaxis but ability to swim greatly affected Increase in the frequency of gill movements
II 2	Loss of balance	Total loss of muscular tonicity Reaction only to strong manual pressure Decrease in the frequency of gill movements to below normal
III	Loss of reflexes	Total loss of response Weak gill movements Weak heart rhythm
IV	Medullary collapse	Cessation of respiratory gill movements Heart stoppage some minutes later

Stages of recuperation	State of fish	Reactions and behaviour
II 1	Recovery of muscular tone	At least partial recovery of tone of swimming muscular and of appetite
II 2	Recovery of balance	Total recovery of balance and partial recovery of cutaneous sensitivity but no evasive behaviour
I 1	Recovery of swimming	Slow swimming without contact with side walls, but still as evasive behaviour
I 2	Restoration of evasive	Responses still slow
0	Total recovery	Occasionally vigorous swimming action just before total recovery

Table 3.7. Characteristics of the principal fish anaesthetics (see Box 3.1).

Common name of anaesthetic	Form	Index of toxicity (1)	Dose per litre of water (2)	Duration (in min) Removal (stages II-III)	Duration (in min) Return	Observations
Tertiary amyl alcohol	Liquid		0.2–1 ml	2–10	20	Usable for transport 0.2–0.4 ml/l, strong odour
2-Amino-4-phenylthiazole	Crystals		20–40 mg	1–3	5–10	
Amobarbitol			100 to 150 mg			Reduces intake of oxygen, usable for transport
Chloroethane	Crystals	1.6	200 mg	2–3	3–8	Used in salmon culture in Norway
MS 222 (methane sulphonate of Tricaine)	Crystals	20–100 mg	1.3	3–5		
Methylpenthyl	Liquid	>3	1–8 ml	2.4	2–4	Usable for transport
Phenoxyethanol	Liquid		0.1–0.5 ml	1.5	3–6	
Propoxate® (Jansen)	Crystals	>3	0.2–1 mg	5	10	Does not reduce oxygen intake
Quinoline (3)	Liquid	>3	5–10 µl	1–6	1–10	Does not pass stage II 2
Quinoline sulphate			20–30 mg	0.5–3	1–13	

(1) Index of toxicity I_t = LD 50 (lethal dose 50%)/ED 50 (efficient dose 50%); this index measures the toxicity of the product: the margin of accuracy is too weak when $I_t < 1$.

(2) The intensity and speed of the anaesthesia are increased when larger doses are given.

(3) Strong odour, insoluble in water; dissolve first in acetone 1/8.

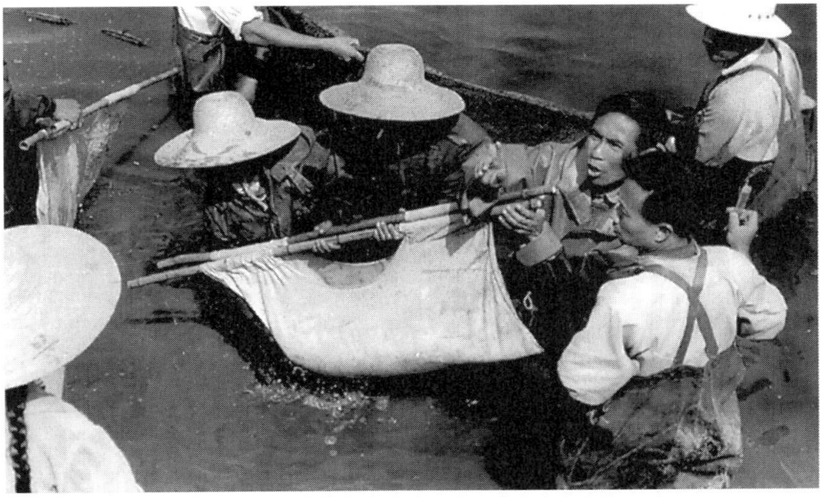

Fig. 3.10. A method of handling broodfish in a sort of hammock made of tulle which retains sufficient water that ensures the fish is never totally out of the water (*Pisciculture de Wuxi*, China, photo R. Billard, INRA).

the detection of the exact time of ovulation is crucial because of the low survival rate of the ova.

In order to remove the risk of spontaneous ovulation in the tanks it is common practice to suture the genital orifice of females immediately after the second injection.

Artificial fertilisation (Figure 3.11)

Broodfish are anaesthetised again for gamete removal; they are held either in the operator's hand or on a table covered with a damp, padded surface (polyurethane foam for example). It is important that the gametes (sperm or ova) are not contaminated by water; the body of the fish should be patted dry but not wiped, as this would remove mucus. One drop of water in 10 ml of sperm or ova will not have much of a negative effect on the gametes. It is particularly important to avoid any faecal contamination by applying a plug of flexible paper or damp tissue into the rectum to prevent the expulsion of faeces and avoid the release of urine during the extraction of sperm. If significant quantities of urine contaminate the semen it can be decanted, rejecting the supernatant which is likely to have a low concentration of spermatozoa.

In order to facilitate the work of the hatchery, all the males needed for fertilisation can be treated in the morning. The removal of sperm is aided by the use of test tubes or small glasses but storage should be in large diameter vessels (becher or erlenmeyer type) which give a large area of contact between the air and the milt which should be no more than a few centimetres deep. Because they are very fragile, the tails of carp

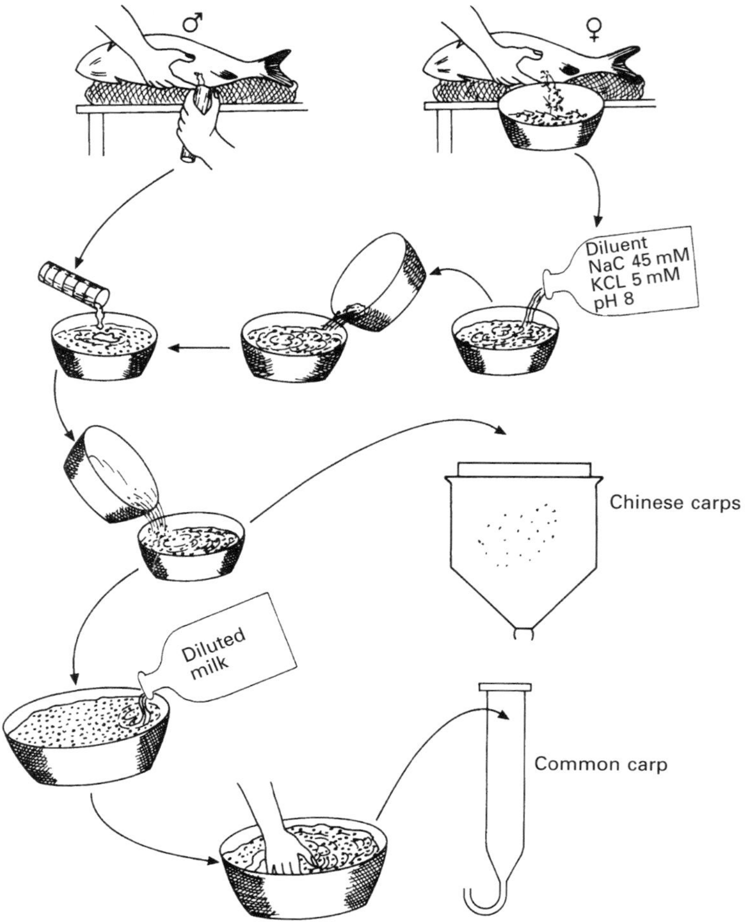

Fig. 3.11. Diagram showing the fertilisation process for Chinese and Indian carps and common carp; (1) Removal of gametes (2) fertilisation (3) removal of the adhesive layer by mixing in a dilute solution of milk (1 : 5), only for the common carp which has sticky eggs (an alternative method is to put the eggs into a solution of urea, 20 g and NaCl, 4 g/l, (4) placing in an incubator. The eggs of Chinese and Indian carp are not sticky and can be transferred directly to an incubator.

sperm become detached from the head easily. Hence it is not good practice to remove sperm and fertilise using a syringe.

Unless there are contra-indications linked to programmes of genetic improvement, sperm from all of the males can be mixed on condition that the same volume is removed from each individual and the contribution is identical. There should be equal numbers of males and females. In order to avoid inbreeding there should be at least 20 males and 20 females and if possible 50 to 100 (p. 105).

The pooled sperm is used up as the females are stripped. It is best to fertilise the eggs from each female individually and to incubate them separately so that the performance of each female can be followed (this should be recorded in the stock record book). Incubation of individual batches also allows the rapid withdrawal of a batch if the eggs appear to be of poor quality. For the stripping of ova and insemination, a series of 3–4 litre bowls should be used: these should be either weighed or graduated by volume in order to quantify the ova available to be fertilised.

Before fertilisation the ova are mixed with the diluent. The traditional Woynarovich diluent made up of NaCl and urea can be beneficially replaced by a buffered diluent made up as follows:

NaCl	2.63 g
KCl	0.37 g
Tris	0.3 g
Glycine	1.4 g

After mixing in one litre of water (spring water or water taken from the hatchery supply), and being allowed to settle – the pH is automatically adjusted to 8 because of the tris/glycine combination. If the fish farmer has a pH meter it is only necessary to add Tris and adjust the pH to 8 with a few drops of HCl.

The operations associated with fertilisation should be performed as quickly as possible: one volume of diluent is added to one volume of ova and the mixture is obtained by simply transferring into another container. Immediately after this, the sperm, measured in a graduated test tube, is added to the ova/diluent mix at a rate of 1 ml sperm for each litre or kilogram of ova. These are mixed rapidly by two successive transfers of the complete sperm/ova/diluent mixture using another bowl. The mixture is then left for 2–3 min during which fertilisation occurs. The next operation is to eliminate, while it is still forming, the adhesive layer that develops around the egg while hydration is taking place. Various techniques are used. One consists of agitating the eggs for 10 min by hand continuously in a solution of 20 g urea and 4 g NaCl per litre of water and then, occasionally over 40–45 min, renewing the solution three or four times; this operation is performed in a 10/15 litre bucket, remembering that a litre of fertilised eggs will, after an hour's hydration, occupy six to nine litres. Another technique commonly applied in Europe uses milk diluted 1 : 5 in water into which eggs are placed after fertilisation and stirred periodically over 30 min. This can be simplified by placing the mixture of eggs and milk in Zoug jars through which air is bubbled for 30–60 min; after this water is added and progressively eliminates the milk (Heymann technique).

Storing sperm

After the stimulation of spermiation the sperm can remain in the genital tract for 2–3 days and retain their viability. During this time they can be removed and stored at ambient temperature or in the refrigerator. Spermatozoa from cyprinids, like most fish species only retains its ability to fertilise for a period of time between a few hours and a few days. This probably occurs because of the problems of anoxia, which results from the high density of spermatozoa in the milt; the sperm rapidly exhaust the available oxygen from the seminal fluid. Storage is improved if there is a large surface area of the fluid exposed to air, for example by having a shallow layer of fluid in the bottom of a wide-mouthed erlenmayer flask covered with damp paper to avoid desiccation. Storage for several days is possible in non-porous plastic bags, which have been inflated with oxygen. It is also possible to transport sperm under these conditions. The technique of using frozen sperm is not yet used in commercial cyprinid culture.

Embryogenesis and yolk resorption

Incubation

Carp eggs are traditionally incubated in Zoug jars (or bottles). Those used most frequently have a capacity of 7 litres, are supplied by an ascending water current and an outflow towards the top which is in the form of a spout emptying into a gutter (Figure 3.12). The water-supply-valve system is often placed above a rack of bottles and linked to their bases by a flexible-pipe system; this avoids sucking eggs into the basal-inflow system if the inflow valve is not functioning. The bottle has a conical-shaped base, 30 cm deep which creates a suitable pattern of movement in the water for incubation; 1.5 to 2 litres of hydrated eggs (corresponding to 250–300 g of ova)

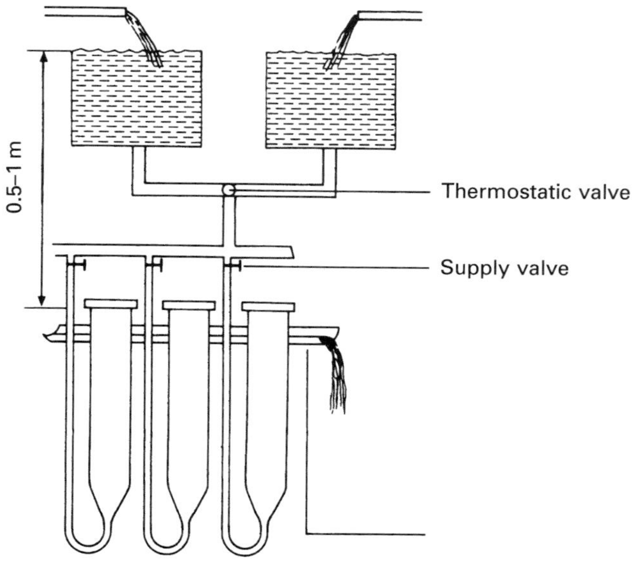

Fig. 3.12. Diagram of a water-supply system to Zoug bottles in an open circuit. Temperature control is ensured by a thermostatic valve wich mixes water from two header tanks, one with heated water from outside (e.g. from a thermal power station). If water is heated electrically only one header tank is required (information from L. Horvath).

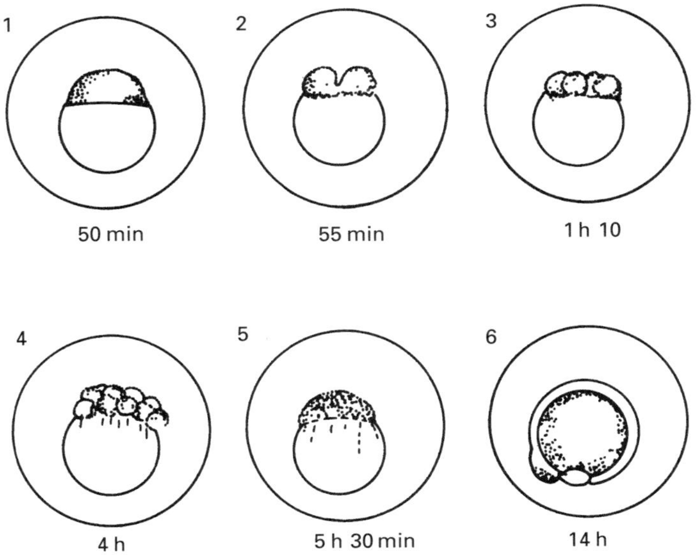

Fig. 3.13. Timing of the early embryonic developement of carp at 25°C. (1) Fertilised egg; (2) first-cell division; (3) four-cell stage (blastomeres); (4) morula; (5) blastula; (6) start of myotome segmentation.

are put into a bottle which has been half-filled with water, then the valve is opened so as to provide a flow of 0.6 to 0.8 l/min of water which is saturated with oxygen. The flow is increased to 1–1.2 l/min when the eggs reach the eyed stage, which occurs after 24 h of incubation at 20°C (see Figure 3.13 for details on the developmental stages). The total duration of incubation is 60–70 degree-days; the optimum temperature is between 22 and 24°C (Table 3.1). 200 litre capacity incubators (see below) can also be utilised.

During incubation treatment with malachite green can be given every 6–12 h, as a 5-min bath, to prevent saprolegniosis. The concentration used is 5 mg/l (Table 3.1). This can be performed by closing the inflow valve and then adding and mixing in the bottle at a rate of 10 ml of a stock solution of 500 mg/l which is kept made up in the hatchery (Figure 3.14). The use of malachite green is likely to be regulated. Formalin can also be used for treatment (0.1 to 0.2 ml/l) to which a little methylene blue can be added (Table 3.1).

Poor-quality eggs or those that are not fertilised become whitish and should be removed as they provide a substrate for the development of saprolegnia. When the normal eggs have reached the eyed stage in the bottle, on the second day of incubation, the water inflow is shut off and the eggs are allowed to settle for a few minutes. The infertile, non-developing eggs are lighter and accumulate at the surface where they can be easily siphoned off. If necessary, this operation can be repeated 10 h later, before hatching. To facilitate this sorting and for reasons associated with genetics and rearing, (p. 103, 142) it is preferable that each batch should be incubated separately.

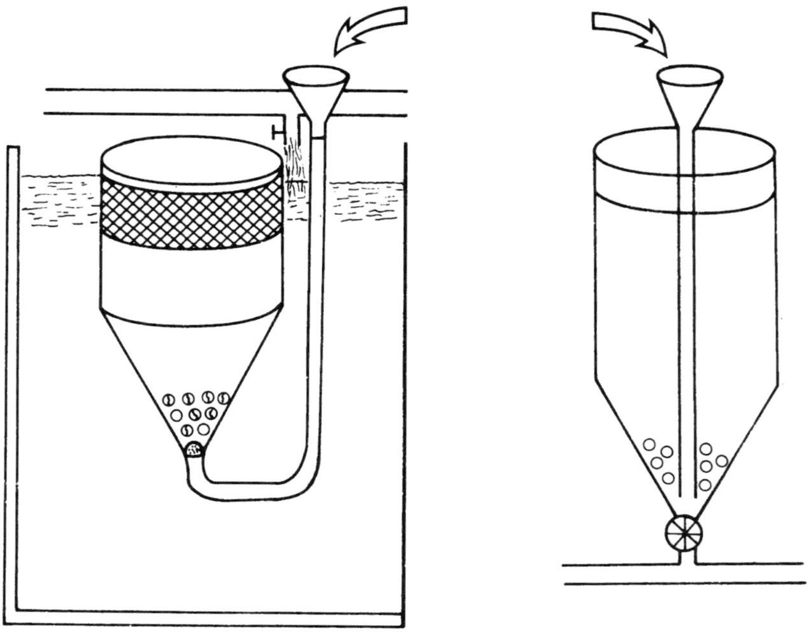

Fig. 3.14. Anti-fungal treatment (malachite green* or other) which delivers the treatment compound to the base of the bottle and gives a mixture of even concentration through the incubator (after Woynarovich and Horvath, 1981). On the right, in the water inlet pipe at the top, a funnel is placed through which the solution is introduced. On the left, a pipe with a funnel above is introduced directly at the base of the incubator.

* Malachite green is the most effective treatment for saprolegniosis but it leaves residues in the eggs and fish and the outside environment; it should be used in moderation until an acceptable substitute becomes available. The use of malachite green is now restricted in some countries.

Hatching and the resorption of yolk

When the first signs of hatching have been observed in the bottles, all the embryos are siphoned, carefully, into 3–4 litre bowls in which they sediment to the bottom where conditions quickly become anoxic. This induces rapid movement and leads to quasi-simultaneous hatching. This operation should not last more than 10 min at 20°C. It is also possible to achieve mass hatching in Zoug bottles by stopping the inflow of water for 5 to 10 min. The newly hatched larvae are then transferred to 200 litre capacity cylindro-conical incubators (Figure 3.15) with an inflow of water at a rate of 12–15 l/min, operating on the same principles as the Zoug bottle, where yolk absorption takes place (taking around 60–70 degree-days). A mantle of filtering material is placed at the top of the incubator (0.5–0.6 mm mesh); water can overflow through this without removing larvae. The filter should be cleaned continually, particularly at the start, when the eggshells accumulate after hatching. The stocking density of the incubators is around 500 000 larvae for 200 l of water. This corresponds to the contents of seven Zoug bottles filled to their normal capacity.

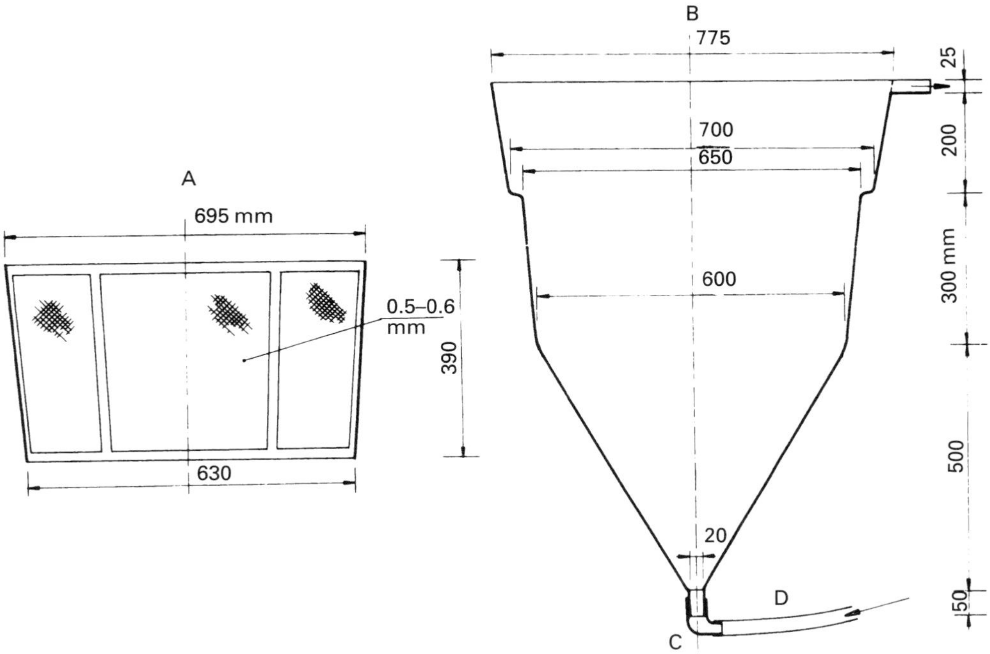

Fig. 3.15. Diagram of a 200 l tank which can be stocked with carp larvae which are resorbing the yolk. Part A which is mesh (0.5–0.6 mm) fits in the upper part of cone B (model by Woynarovich, Agrober, Hungary).

The water should be permanently saturated with oxygen. During transfer the temperature difference should not exceed 0.5 to 1°C. For the first 36 h the larvae attach themselves to the walls of the incubator. They then detach themselves and for the next 36 h swim freely, coming to the surface where they take in air to inflate the swim bladder. At this stage they are able to ingest food and can be put into rearing systems, either in outdoor ponds or in indoor tanks in controlled condition nurseries. It is common practice to first feed in the incubation vessels, where yolk resorption has been completed for about 36 h before transfer to outdoor ponds. The traditional diet for first feeding is finely ground hard-boiled egg yolk that provides an introductory meal and fills the digestive tract. Nowadays well-balanced artificial diets are available and are preferred to egg yolk (p. 139).

Several systems are available for the hatching and collection of larvae. Cylindro-conical incubators can be combined with a system for harvesting the larvae (Figure 3.16). The Indian double happa system makes it possible to separate the larvae from the egg shells after hatching (Figure 3.17) but the cuboid form is not recommended as the larvae accumulate in the corners where the water becomes anoxic.

Post-larvae are fragile: they should be handled with great care and never taken out of the water (Figure 3.18). When they are to be removed from the alevin vessels they should be siphoned through a 6–8 cm tube into a tank of about 1 m^3 volume.

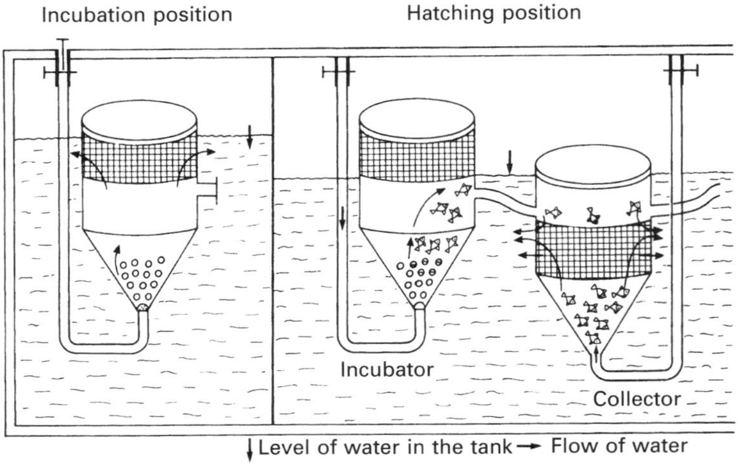

Fig. 3.16. Cylindro-conical incubation system which can be, after lowering the level of the water, coupled with a collector to concentrate the newly hatched larvae (after Woynarovich and Horvath, 1981).

This can be emptied through an inclined pipe placed on the trolleys that are used in hatcheries. Inside this is placed a sort of Indian happa in the form of a fine mesh (0.5 mm) net container attached to a light metal frame in which the post-larvae are placed. These can then be transported easily to the first fry ponds nearby or to special tanks or plastic bags for long distance transport (see p. 252).

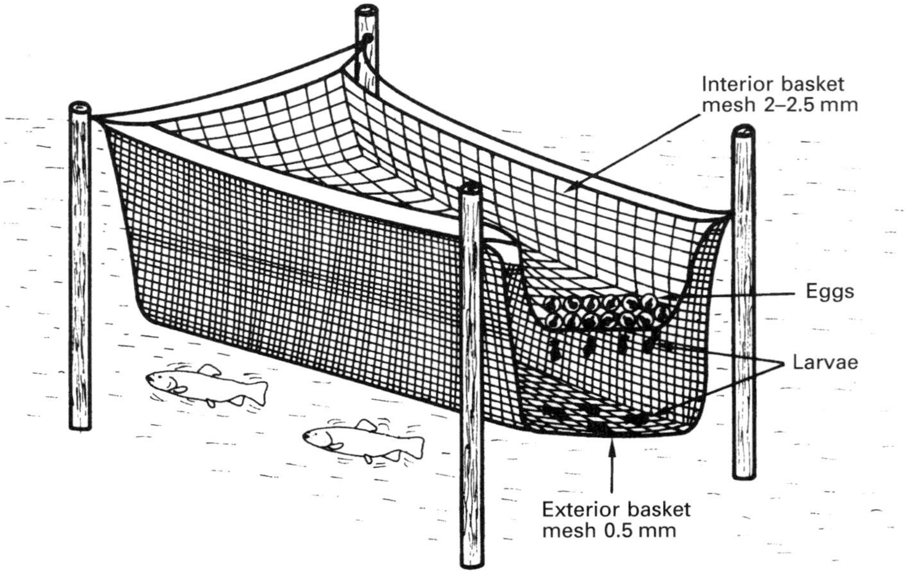

Fig. 3.17. Indian happa with a double layer of mesh of different sizes which allows the separation of egg shells from larvae during hatching.

Fig. 3.18. Chinese technique for harvesting fingerlings and fry in hatching and first-fry ponds, the juveniles are concentrated in the net which is held under water and removed with a bowl (photo R. Billard, INRA).

3.3.2 Chinese and Indian carps

Methods for the artificial spawning of other farmed cyprinids, particularly the Chinese and Indian carps, are very similar to those used for the common carp. Some of the characteristics are listed in Table 3.1 and in the appendix on the other species (p. 305). The procedure is identical for tench although a single injection of carp hypophysis is sufficient, at the same dose for females and a higher dose for males (9–10 mg per male). The doses for Chinese and Indian carps are slightly higher than those used for common carp (Tables 3.4 and 3.6). For these species the critical point for artificial spawning is the detection of ovulation: fertilisation must be immediate as the ova can only survive for a very short time (see above). Frequently, broodstock which have been hypophysised are placed with both sexes together in circular tanks (Chinese carps) or in a happa (Indian carps) (Figure 3.17) where spawning occurs naturally. Circular tanks, from 10–20 m^2 are also used for both groups of species. The eggs are not sticky and it is not necessary to place spawning substrates in the tanks. The number of broodfish placed in the spawning tank depends on its size. For each female, two males whose combined weight equals that of the female are usually placed in the tank: three females and six males are thus placed in an 8 m^3 tank. Reproduction generally occurs 6 h after the final injection and is accompanied by active spawning behaviour. If there is no evidence of spawning behaviour it is necessary to check whether the females have ovulated completely and, if this is the case, to strip the fish by hand (artificial fertilisation).

Artificial fertilisation uses the same procedure and the same diluents as for carp but there is no need to remove the sticky outer layer. The eggs can be transferred to the incubator immediately after fertilisation, a few minutes after the gametes have

Fig. 3.19. Net cages installed in a pond for the incubation, resorption and first feeding of carp (Ovidiu hatchery, Romania) (photo R. Billard, INRA).

been mixed and after a quick wash to eliminate excess sperm. It is important to remember that eggs increase greatly in size after they have been transferred to the incubator (6000 hydrated eggs per litre of water in the incubator). For this reason, Zoug bottles are no longer used except in small hatcheries. In bigger hatcheries, larger incubators of up to 200 litre capacity and even bigger are used. These are the cylindro-conical type as described for the yolk resorption stage of carp larvae (3–15) which can hold 1 200 000 eggs. During incubation formalin treatments at a ratio of 1 : 5000 to 1 : 10 000 are given every 4 h (Table 3.1) and dead eggs and debris are siphoned off after hatching which occurs after 30–35 h at 22°C. An alkaline protease is sometimes added to the incubator to dissolve the eggshells and other organic waste. Yolk resorption takes place in the same incubators.

Incubation, resorption and first feeding can also take place in fine mesh cages in ponds (Figure 3.19) or tanks where the climatic conditions are suitable (temperature >20°C), but oxygen levels are not always favourable. Some aspects of the spawning of other pond species are given in the appendix at the end of the book.

3.3.3 Different types of hatcheries and their management

There are several strategies used for controlled spawning and for the establishment of a hatchery. In practice, spawning and first rearing of juveniles are most frequently managed by the on-growers themselves, either through natural spawning in a spawning pond or a Dubisch pond, or using a hatchery–nursery system that is sometimes shared by several producers. There are a few specialised hatcheries, which supply on-growers; these are engaged in genetic improvements, keeping well-defined lines and offering a selected product, which is well adapted to the conditions of on-growing. There are likely to be more of this type of hatchery in the future because of the requirements for genetic improvements. There is a wide range of types of these

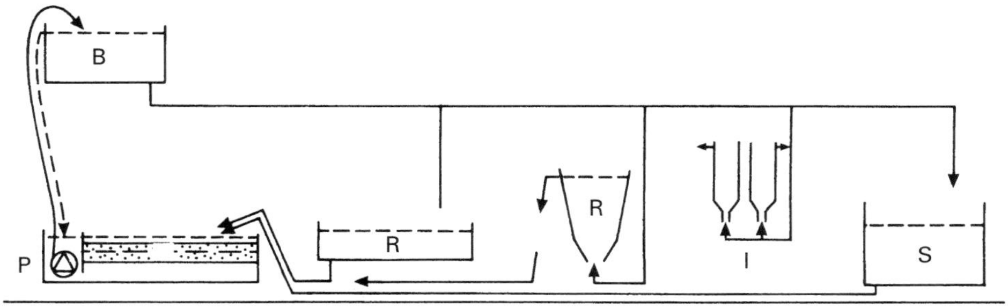

Fig. 3.20. Diagram of equipment and water circulation in a land-based hatchery (CEMAGREF model) (cf. Table 3.8).

Table 3.8. Material required for holding carp broodfish, eggs and larvae in a simple closed circuit hatchery. Holding capacities are given for a series of spawnings: four series are possible in a season. Channels can be added for reproduction of carp and catfish (information supplied by O. Schlumberger).

Material	Holding capacity (per batch)	Densities	Containers required and volumes
Maturation and holding tanks	20–25 kg females + 5–10 kg males	10–30 kg/m^3	2 × 2 tanks 1000 l
Incubation (in Zoug bottles)	Around 15 l eggs	Vol. eggs 50% of the vol. of the bottles	10–12 bottles 5–7 litres
Cylindro-conical tanks for resorption	1.5–2 millions larvae	5 000 larvae per litre	8–10 tanks 60 l

hatcheries from small rustic ones constructed by the farmers themselves (p. 94) to modular "turnkey" ones which are sold by companies such as Agrober in Hungary and Dwivedi in India. The hatchery usually operates on a through-flow water system, i.e. the water is eliminated after a single use, but such types are used mainly in summer, during the normal spawning season, without heating or temperature control of the water. This system has the advantage that neither dissolved organic matter nor the suspended solids including faeces or those resulting from fish handling, incubation and hatching do not accumulate in the system. When incubation and larval rearing take place outside the normal season it is necessary to regulate the temperature of the water, which requires the use of recycling systems. Such systems are not generally economically viable for on-growing but may be worthwhile in hatcheries where biomass is lower and the value of the product is higher. Characteristics of closed-circuit systems are given for a simple hatchery (Figure 3.20) but these can be extended and scaled up for a much larger unit (Figure 3.21).

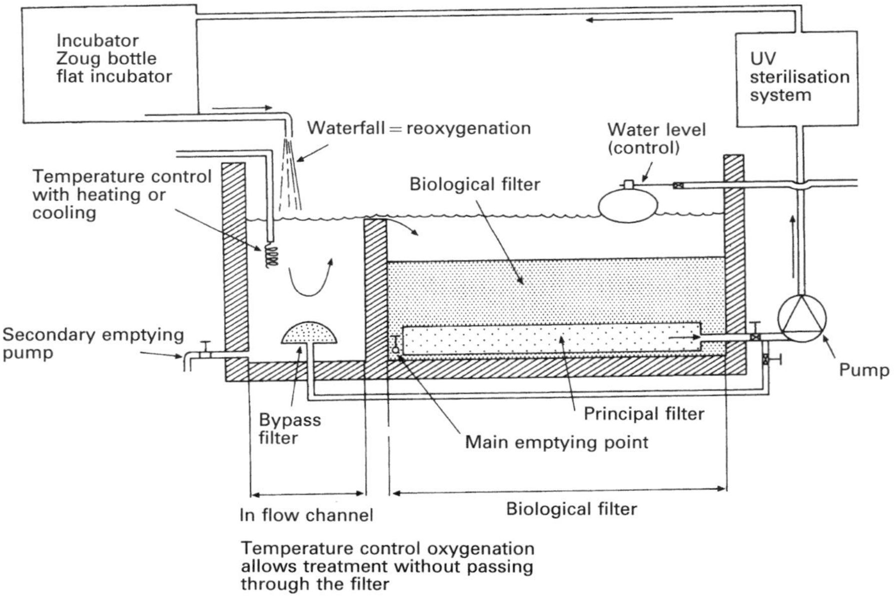

Fig. 3.21. Diagram of a recirculating hatchery system with temperature regulation and water purification. In such a system it is essential to respect a well-defined relationship between the area of the filter and the flow which is theoretically $3\,m^3/h/m^2$. In practice it is better to keep this flow down to $2–2.5\,m^3/h$ for a filter with a surface area of $1\,m^2$. For more details of this recycling system see Petit (in Barnabé, 1989).

Water for the hatchery can come from a variety of sources; the desired characteristics have been described on p. 51. Water from reservoirs should be filtered through gravel and through sand to eliminate suspended matter, eggs and large zooplankton such as cyclops which can attack the fish eggs and larvae (see p. 146). Water coming from springs or other groundwater sources will usually need to be heated and oxygenated. Drinking water from the mains supply will sometimes also need to be heated. Hot and cold water can be fed into separate tanks and mixed through a thermostatic valve (Figure 3.12). There are more elaborate systems of thermoregulation associated with purification systems (Figure 3.21). Water from deep wells should be reoxygenated and sometimes treated to eliminate elements such as iron. Storing water in a reservoir large enough to hold a 24 h supply before use provides a means of sedimenting the suspended solids (SS) and unwanted colloids.

3.3.4 An example of a simple recirculating hatchery[(2)]

It can sometimes be beneficial to operate a hatchery with recycled water to respond to the needs of a fish farmer or group of producers who are able to look after its operation and maintenance themselves. The successful functioning of the hatchery

(2) From O. Schlumberger 1993.

depends on the relationship between the size of the filter and the volume and biomass in the tanks and the incubators.

The example described below corresponds to a unit with a capacity of 5 million carp larvae at the yolk-sac resorption stage, each season, split between four spawnings. It was installed in 1981 in the south of France at Arles (Pisciculture Bacoupharis) and has been in continuous operation. Because of its versatility, this installation can be used over a large part of the year for pike, zander, carp, tench, catfish and Chinese carp. The hatchery is shown diagrammatically in Figure 3.20 and the components and their operation are as follows:

Header tank (B). Volume around 200 l with:

—Outflow to distribution system
—Inflow/outflow of warm water from a heating unit
—Outflow to filter

Filter (F). Submerged biological filter, one-quarter of the total water volume, area $1\,m^2$ for $2\,m^3$ (polyethylene foam support; $2 \times 10\,cm$ layers) placed in two connected tanks measuring $2\,m \times 1\,m \times 0.60\,m$. These have the filtering/purifying capacity to maintain a batch of broodstock (carp, tench) through a maturation period lasting several weeks with manufactured diet (around 1% live weight per day). For the filter dimensions see Figure 3.21.

Circulation pump (P). Lifts the water 2 m into the header tank; the flow at this level is equivalent to the volume of the installation, around $10\,m^3/h$.

Other equipment (*S*). Holding tanks for broodfish, (*I*) incubation system, (*R*) yolk resorption system.

Heater. Power = temperature drop × hourly flow. In this example: $2°C \times 10 = 20\,kcal \times 1.16 = 23.2\,kcal$. The heat exchanger is linked to the header tank. The installation is placed in a temperature-controlled environment with a concrete floor and an outflow channel.

Starting up the system. The system is started up progressively. The purifying capacity of the filter increases as bacteria colonise the support medium. The "increase in power" of the system continues over a period of around 3 weeks (at 20–22°C) during which time different groups of bacteria (*Nitrosomonas* and *Nitrobacter* mainly) develop progressively and equilibrate in proportion and density for optimum degradation of organic matter. The "maturation" of the system requires a few weeks longer and leads to a "conditioned" system. At the same time as nitrification, there is production of H^+ ions, which can lead to acidification of the water.

In practice. After filling and starting up the system the filters are seeded with the output of septic tanks and some fish are introduced into the holding tanks where they are fed pellets at regular intervals. This results in the production of organic wastes and, simultaneously, progressive colonisation of the filter by bacteria. Daily measurements of water quality show:

—a progressive increase in the concentration of NH_4^+
—then an increase in the concentrations of NO_2^- at the same time as the concentration of NH_2^- decreases.
—finally, after around 20 days, the levels of nitrites bottom out and stabilise while those of ammonia are maintained at low levels, below toxic concentrations. Simultaneously, the concentration of nitrates, which have a very low toxicity, increases progressively. After this stage, around 3 weeks after the start up at 20–22°C, it becomes possible to increase the biomass of fish in the system, introducing broodstock and beginning spawning operations.

Comments

It is prudent to check water quality regularly (NH_4^+, NO_2^- and pH). The presence of bacteria actively working on the filter makes it impossible to use antibiotics to maintain the health of the fish. Formalin and malachite green can be used without adverse effect in the system while it is operating (stock solution of 3 to 10 g malachite green/litre of formalin applied at the rate of 1 litre per 50 m^3).

Only disease-free fish should be transferred to such a system; there should be a quarantine period before transfer to the recirculation system.

A hatchery guaranteeing the production of juveniles can only be justified where there are fry ponds nearby into which they can be stocked. For a system similar to the one described here (capacity: 5 million carp larvae/season) there should be around 10 000 m^2 of fry ponds. Their proximity to the hatchery reduces transport time and facilitates co-ordination between hatchery operations and the preparation of the ponds: the maximum density of plankton should occur at the time when a batch of larvae are completing the resorption of the yolk (p. 79 for a description of artificial spawning and incubation and p. 125 for the first fry stage).

3.3.5 Natural spawning on artificial substrates

Some farmers prefer to allow fish to spawn naturally by placing broodstock in ponds with artificial substrate over which the carp can spawn. After spawning the broodstock are removed and the first stage of larval rearing completed in the same ponds. A good example of this practice is at the Carpeix Pollenca farm in the Balearic Islands.

This farm is fed from an underground supply of water (from a depth of 30 m) with a constant temperature of 17–18°C. This water is partially recycled in summer and totally recycled in winter. The broodfish (around 25 females and 15 males weighing between 5 and 10 kg) are reared throughout the year in a 50 m^3 pond and all removed at the end of March or beginning of April and placed into rectangular concrete tanks measuring 5 × 2 m with a water depth of 80 cm. Each tank is divided in two by a grid with two males placed at the upstream side (i.e. the water inflow end) and three females downstream. If the broodfish are smaller (<3 kg), five females and four males are placed in each of the compartments. The broodstock remain here throughout April and are fed daily at around noon with carp granules (25–30% protein) until satiated. The flow rate in each tank is 4 l/min and aeration ensured by a venturi system. The temperature of the water increases slightly throughout April going from 15–16°C (minimum 14°C, maximum 17°C) to 18–20°C. During the first few days of May the females begin to show some signs of activity and move close to the grid which separates them from the males. At this point artificial substrates are placed on the bed of the tank; these are made from strips of raffia around 20 cm long which are attached to plastic mesh stretched over a PVC frame (Figure 3.8) and the central separation grid is removed. The males and females are reunited and rapidly start to display spawning behaviour. Reproduction takes place within 24 h: in 80% of cases in the morning. About 80% of the eggs are fertilised; these are stuck to the raffia and the walls. The raffia is not absolutely essential; spawning will take place over the plastic mesh alone. After spawning the broodfish appear sluggish and can be removed from the tanks easily (in the late morning); they are guided to the downstream end of the tank using a grid and removed with a net for treatment in a salt bath (50 g NaCl/litre). The anaesthetised broodfish are placed for around 5 min (two fish at a time) in a 100 l saline bath—this is strongly aerated up to the point where the mucus on the surface of the fish appears slightly whitish. At this point the female's abdomen can be massaged to remove any eggs which have not been spawned. Where there are significant wounds salt is applied directly. The broodfish are then replaced into their 50 m^3 rearing ponds which, between times, have been dried and disinfected. All of the females spawn in this way over a period of 1 to 2 weeks. Those that do not spawn are placed in another tank where spawning has not yet taken place and finally almost all females spawn spontaneously without the need for any hormonal stimulation.

After removing the broodfish the flow is increased to 8 l/min to eliminate mucus, dead floating eggs and other debris. The eggs are left to incubate in the spawning tank and hatching takes place 3 days later at 20°C. The spawning grids are then removed and cleaned. Five days after spawning (the day before the completion of resorption) rotifers are introduced into the tanks. The rotifers, besides being the first source of food for the larvae, attack the saprolegnia fungus and prevent its development.

The rotifers which are introduced have been cultured in a 50 m^3 tank (7 m × 7 m) from the end of April, a week before the predicted start of spawning, by filling the tank with water with a high concentration of organic matter drawn from another rearing tank stocked with growing carp at a biomass of 20 kg/m^3. A uniform layer of cut grass is added to the surface of the water. This results in a rapid development of a rotifer bloom and the water becomes black by the time it is ready for introduction into the fry ponds. The rotifer tanks are supplied through the base with water from the intensive carp rearing pond that was used to fill the tank initially. From the outflow of the rotifer tank a flow of about 5 l/min goes to each fry pond (rotifer density at least 3000/l). This flow rate is altered according to the density of the rotifers and the density of the carp, which can vary between 50 000 and 100 000 larvae per tank. The rotifers are distributed to the tanks in this way for a period of 8 days (7 days after the completion of yolk resorption).

During the final two days of rotifer distribution, daphnia are added; they are given to the fry for 10 to 15 days. These daphnia come from a lagoon-type pond on the farm (see p. 211) and are captured at night by concentrating them with the aid of a light source (100 W light) which is switched on for 10 min. The daphnia are captured with the aid of a fine net and at the start three ladles of 100 ml strained daphnia are distributed to the fry tanks every day. The numbers are increased in order to maintain a large excess of daphnia in the fry ponds, thus preventing cannibalism. After 10 days (15 days post-spawning) artificial diet (crumb, 42% protein) is progressively substituted over a period of around 10 days; the food is given *ad lib* and reduced if wastes appear. From the 30th day after yolk resorption the fry eat Number 1 granules (0–5 mm diameter) and the size can be progressively increased. After 2 months a granule with a lower protein level (18–20%) is introduced. This is used until the fish reach market size. At the 50th day of rearing the juveniles are 2–3 cm long and can easily be netted out and stocked into 50 m^3 tanks (around 4000 individuals/tank) in which they are reared until they reach a size of 250–400 g after around 1 year (this is the commercial size in the Balearic Islands). Such a system could be coupled to normal production methods achieving a filletable fish of 600–1200 g in the second year.

BIBLIOGRAPHY

Billard R., Marcel. J., 1980. Quelques techniques de production de poissons d'étangs. *La Pisciculture Française*, **59**: 9–49.

Billard R., Alagarswami K., Peter R.E., Breton B., 1983. Potentialisation par le pimozide des effets du LHRHa sur la sécrétion gonadotrope hypophysaire, l'ovulation et la spermiation chez la carpe commune *Cyprinus carpio*. *C. R. Acad. Sci. Paris*, **296**: 181–184.

Billard R., Cosson J., Perchec G., Linhart O., 1995. Biology of sperm and artificial spawning in carp. *Aquaculture*, **129**: 95–112.

Chang J.P., Peter R.E., 1983a. Effects of dopamine on gonadotropin release in female goldfish, *Carassius auratus*. *Neuroendocrinology*, **35**: 351–357.

Davies P.R., Hanyu I., Furukawa K., Nomura M., 1986. Effect of temperature and photoperiod on sexual maturation and spawning of the common carp III. Induction of spawning by manipulating photoperiod and temperature. *Aquaculture*, **52**: 137–144.

Linhart O., Kudo S., Billard R., Slechta V., Mikodina E.V., 1995. Morphology, composition and fertilization of carp eggs: a review. *Aquaculture*, **129**: 75–93.

Renard Ph., Billard R., Christen R., 1987. Formation of the chorion in carp oocytes. An analysis of the kinetics of its elevation as a function of oocyte ageing, fertilization and the composition of the dilution medium. *Aquaculture*, **62**: 153–160.

Rothbard S., 1981. Induced spawning in cultivated cyprinids, the common carp and the group of chinese carps. I. The technique of induction spawning and hatching. *Bamidgeh*, **22**: 42–47.

Saad A., Billard R., 1987. Composition et emploi d'un dilueur d'insémination chez la carpe *Cyprinus carpio*. *Aquaculture*, **66**: 329–345.

Sokolowska-Mikolajczyk M., Mikolajczyk T., 1991. Control of spawning in cyprinids. *Riv. Ital. Acquacol.*, **26**: 209–215.

Weil C., Fostier A., Billard R., 1986. Induced spawning (ovulation and spermiation) in carp and related species 119–137. In: R. Billard, J. Marcel (eds), *Aquaculture of Cyprinids*. INRA, Paris.

Woynarovich E., Woynarovich A., 1980. Modified technology for elimination of stickiness of common carp (*Cyprinus carpio*) eggs. *Aquacultura Hungarica (Szarvas)*, **11**: 19–21.

Woynarovich E., 1982. Hatching of carp-eggs in "Zugger" glasses and breeding of carp larvae until an age of 10 days. *Bamidgeh*, **14**: 38–46.

Woynarovich E., Horvath L., 1981. La spawning artificielle des poissons en eau chaud: manuel de vulgarisation FAO. *Doc. Tech. Pêche*, **201**: 191 pp.

4

Genetic improvements[1]

4.1 DOMESTICATION

The common carp is considered to be a single species, *Cyprinus carpio* L, with different populations adapted to local environments forming a range of geographical races.

The origins of the domestication of the common carp are back in the dawn of aquaculture, 4000 years ago in China. In comparison to birds and mammals, fish can hardly be said to be truly domesticated. In practice, there is still little difference between wild populations and stocks used in rearing. Carp show many different morphological characteristics in their body shape, scales, body composition and other features but fertile crosses between these races are always possible.

Ever since fish for aquaculture were first taken from the wild, the characteristics of domesticated races have changed progressively because of both broodstock selection and rearing methods (association with other species, type of fertilisation, preferred harvest size, etc.).

In Europe, carp are harvested by completely emptying the ponds where common carp is the dominant species. When the pond is being emptied, the broodfish are selected by the operator who will make a choice on the basis of preferred characteristics. The offspring of these broodfish will then be used for restocking the system. From the 16th century there have been many exchanges of carp stocks between different regions and countries in Europe. During the last few decades, the pace of progress in rearing techniques has accelerated, with the construction of specialised ponds for keeping broodstock (Dubisch ponds) or for on-growing using juveniles produced in hatcheries where spawning is induced hormonally, diluents for sperm and controlled incubation techniques are used. The selection pressures operating through these techniques, parallel to those performed intuitively by the farmers, have led to a rapid differentiation of cultured stocks.

---

[1] M.G. Hollebecq and P. Haffray.

In contrast, carp in Asia have usually been cultured in association with other species and even with other types of agriculture such as rice growing. Animals for human consumption are harvested from the pond with a net. Recruitment of juveniles takes place because some of the broodfish are able to evade capture: this method of selection contrasts markedly with European practices.

For around 50 years there has been progressive improvement in the performance of carp farms through the use of artificial spawning techniques which have entailed the selection and holding of broodstock with desired phenotypic characteristics; this has led to organised methods of selection. In practice the qualitative and quantitative improvement of fish production has utilised a diverse range of genetic techniques. However, as in all fish-farming systems it is often difficult to evaluate the likely future performance of broodfish because of variations in the environment. These reduce the precision of the estimation of the hereditary potential of the fish, particularly when all the different stages of the rearing cycle are not fully managed. It becomes extremely difficult and expensive to rear the few tens of related family groups needed for an independent estimation of hereditary potential.

Because carp have been domesticated for so many years and because of the constraints listed above, the setting up of a programme for genetic improvement of a carp line is based on the following different stages:

—the rational choice of strains;
—crossing and hybridisation;
—optimisation of the management and selection of broodstock;
—control of sex and sexual maturation.

4.2 THE CHOICE OF STRAINS AND THEIR MANAGEMENT

Studies on the genetic identification of populations or stocks are an essential prerequisite for defining a rational strategy for management or genetic improvement.

Identification of stocks

Several techniques for the identification of stocks of cultured fish can be used with varying degrees of effectiveness for cyprinids:

—The study of *meristic characters* (number of vertebrae, scales along the lateral line, branchiostegal rays or morphometrics (length, body weight, body depth, etc.) shows that there is a gradual variation in these characters from China to the west of Europe. However, if these populations from extreme ends of the geographical range are reared under the same environmental conditions, some of the meristic differences, such as the number of vertebrae, disappear (Komen, 1990). The expression of these characteristics is very frequently influenced by environment. Their use as part of a study of local genetic variation should therefore be approached with caution.

—*Cytogenetic techniques* such as "banding" (differential coloration of chromosome fragments) allow the reassortment of karyotypes between populations to be identified: this appears to be of little use in cyprinids.

—Observation of *protein polymorphism* often shows low levels of variability; however, this may be enough to characterise carp stocks.

—Recent advances in *molecular genetics* allow the study of genomic, ribosomal or mitochondrial DNA and the more detailed analysis of genetic variability.

Evaluation of performance

The knowledge of the characteristics of stocks is extremely useful technical information for the development of programmes for genetic improvement because in many studies, particularly in Israel, Asia and Europe, it has been shown that there are significant differences in growth performance between races of the common carp.

In order to test the genetic value of different races of carp, Wohlfarth and Moav (1985) developed a procedure for testing growth performance in a common environment or "communal testing" (Figure 4.1) adapted to the constraints of the pond environment. The relative effectiveness of the method can be improved by the use of specific electrophoretic markers. The technique makes it possible to compare

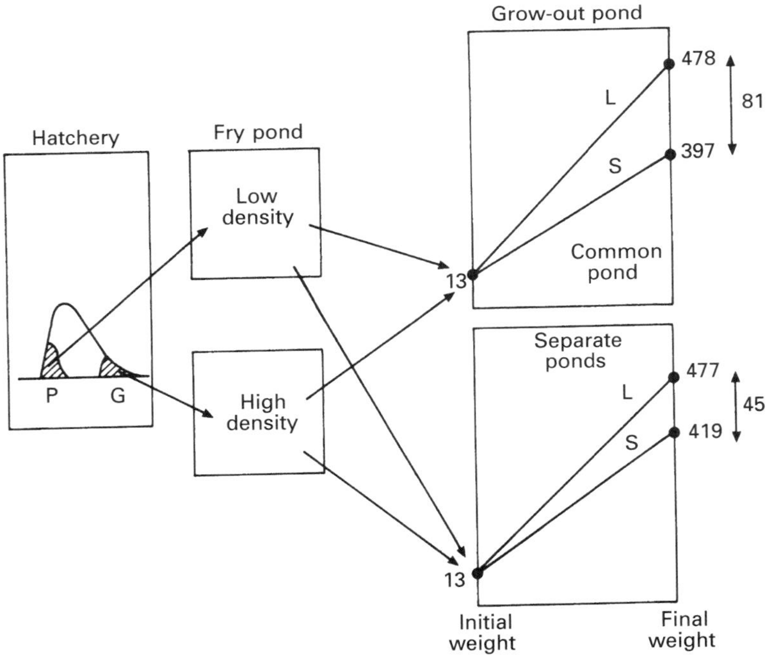

Fig. 4.1. Technique for mass fry rearing used to overcome differences in initial weights separates the largest (L) and smallest (S) individuals from a single carp spawning. This induces some variabiliy of environmental origin in the initial weight (mg) of fish introduced to the grow-out pond. The communal testing technique is used to allow the expression of growth potential in the different fractions where there is competition in the same environment (after Moav and Wohlfarth, 1968; Wohlfarth and Moav, 1971).

different races of carp in different latitudes and rearing conditions. Wohlfarth and Moav (1991) have shown that the races which perform best in floating cages are not always the ones which do best in ponds.

According to Kirpichnikov (1987), there are significant differences in growth potential at different stages of development. The Russian race, Ropsha, grows more rapidly in the first year than in the second. According to Hulata *et al.* (1985), the Chinese races have a high growth capacity for the first 60–80 days which reduces thereafter. This is the reverse of the growth pattern for the European races in Israeli rearing conditions.

Performances appear to be highly influenced by the level of intensification of the production method used. At low densities, European races have significantly higher growth rates than the Chinese "Big Belly". When rearing techniques intensify, not always using artificial diets, differences diminish (Figure 4.2).

Breeding characteristics also vary between races: the Asiatic Big Belly race shows precocious sexual maturity as well as a higher gonado-somatic index than the range of European races. At the same time, the method of rearing has produced a marked

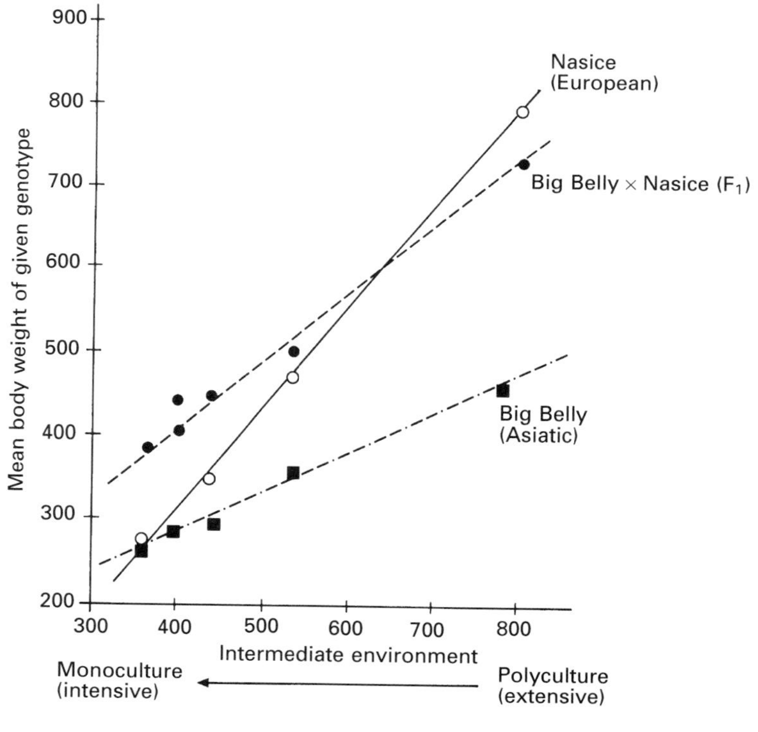

Fig. 4.2. Growth of European and Asiatic strains of common carp and of their F_1 cross. Increasing environmental values correspond to decreasing densities in the pond (after Moav *et al.*, 1975).

capacity to escape capture by fishing net in races which are harvested by this method (Wohlfarth *et al.* 1975).

In addition to its use as a food, the carp is highly appreciated in sport fishing. In many countries, races displaying "wild" phenotypes, i.e. oblong shape, completely covered with scales, are crossed with "domesticated" races to supply the market. Suzuki *et al.* (1978) have shown that wild stocks are characterised by a lower rate of capture and an ability to exert a greater tension on the line used for capture.

The high fecundity of cyprinids may be a factor in the rapid development of inbreeding in a population as a very low number of broodfish are required to produce the next generation; this may sometimes be less than 10. During experiments, a decrease of 15% in growth performance has been demonstrated after a single generation of brother–sister crosses. It is thus important to rationalise the management of broodstock and recognise the risk of genetic drift.

The progression of inbreeding is managed through planned matings and the use of sufficient broodfish for each generation. This is designed to transmit from one generation to the next the maximum percentage of genetic variability necessary for the genetic management of stocks and to avoid the negative effects of inbreeding.

For example, crossing 50 males and 50 females in each generation allows the conservation of 99% of the genetic variability of the initial stock. Crossing 20 males with 20 females conserves 95% of this variability.

As a precaution, there is a range of measures which can be implemented; these include choosing from a non-inbred stock, using equal numbers of males and females and separate incubation of the output of each female (Chevassus, 1989).

4.2.1 Crossing and hybridisation

After stocks have been characterised, crossing allows genetic improvements to the species to be made. This method depends on the exploitation of the effects of heterosis such that the mean performance of the descendants is superior to that of the parents either by crosses between species (hybridisation) or between strains (crossing). The effects of heterosis are linked to the relationships of dominance between homologous genes. These are expressed by the increase in the rate of heterozygosis in the offspring in relation to that of the parents and are even more effective when the parental lines are genetically distinct.

Crosses between stocks

The traditional carp-rearing lines appear to demonstrate that over the years there has been successful selection for growth characteristics. In practice, crossing between different races of common carp rather than selection techniques has been the basis for the genetic improvement of this species.

Today, the practice of crossing stocks has developed and become one of the few methods used in on-growing. The technique requires the independent maintenance of several pure lines. Various programmes for crosses have shown significant heterosis for growth rate (Figure 4.3) and resistance to disease and low temperatures.

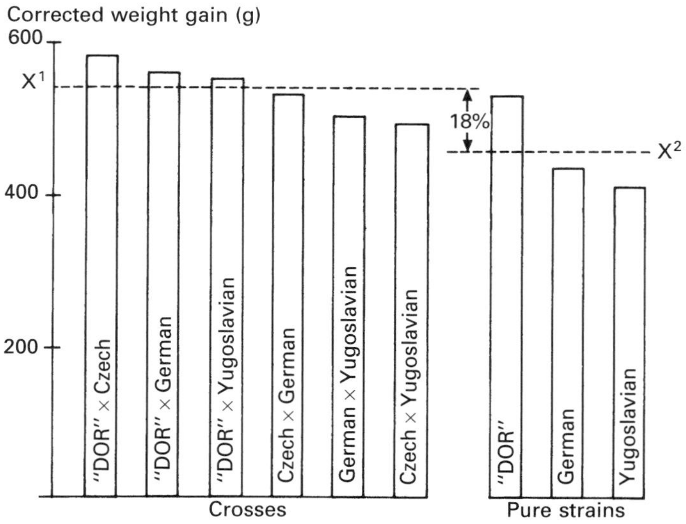

Fig. 4.3. Comparison between the growth rates of three pure strains of common carp reared in the same pond and their crosses (after Wohlfarth *et al.*, 1984).

In Israel the most frequently used crosses are between the "DOR70" race and European races originating in Yugoslavia or the former Czechoslovakia. Crosses between different European races and the Chinese race known as "Big Belly" have exhibited growth performances 10 to 20% greater than the mean for the pure races (Wohlfarth *et al.* 1984). In Hungary, commercial crosses have been developed: the Sz215 and Sz31 (Figure 4.4). These show improvements in growth rates of 15 to 20% and are widely used in on-growing. Testing of commercial stocks in the former Czechoslovakia confirms the value attributed by Pokorny (1990) to these crosses. In Russia many crosses show promising characteristics (Kirpichnikov, 1987). The "basic" stock is often the "Ukraine" race which is crossed with the Ropsha, Ljuben and Politva geographical races and others. In Estonia, Gross (1991) showed the superiority of crossing the Ropsha race with Estonian or German strains for the characteristics of survival and resistance to cold.

It is often good practice to cross "improved" lines (either imported or local) with local races. The maternal line is representative of the local race, which is maintained in the traditional manner. The paternal line (the pure line) is maintained at a test centre: this reduces overall costs.

As a general rule, in all of these studies, while the performance of the progeny of crossing is sometimes better than that of the mean of the parents, it is often equal to or worse than the mean of the better parent. Wohlfarth (1993) showed that the degree of progress may be limited by the interstrain aptitude to the cross, the interactions between genotype and environment and depends on the age and weight at which the measurements are taken.

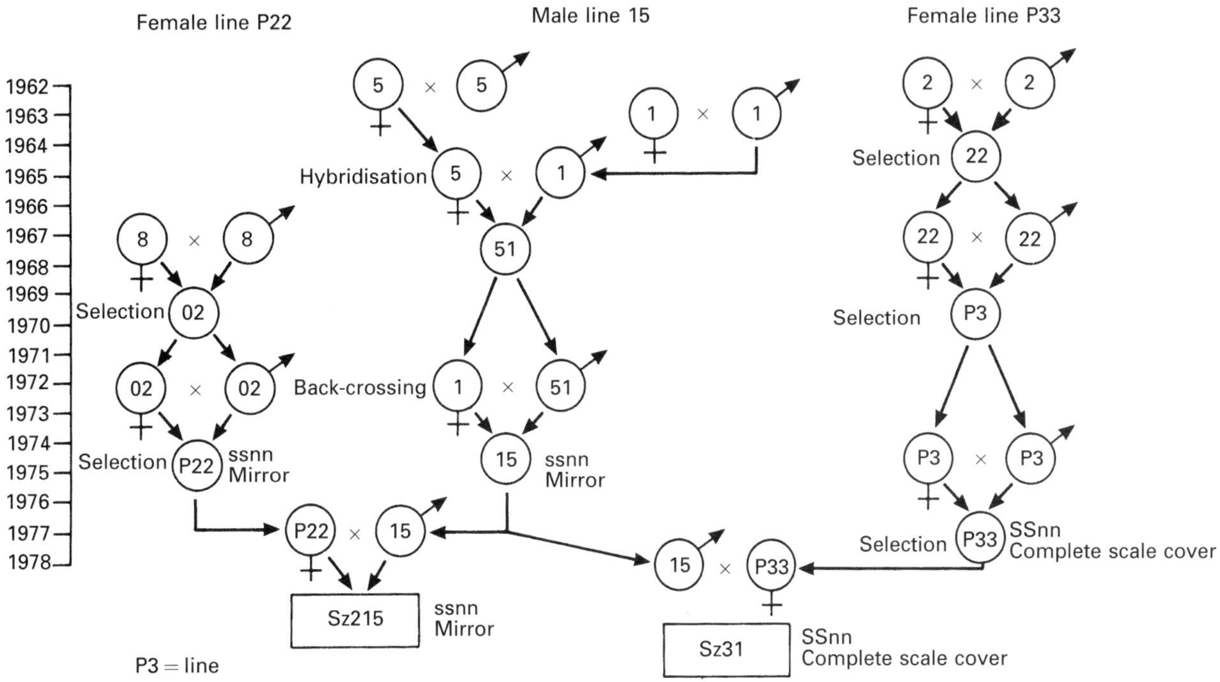

Fig. 4.4. Diagrams showing crosses between strains for common carp in Hungary for the establishment of parental lines (reference to scale pattern, Figure 4.5) (after Bakes, 1990).

Interspecific crosses (hybridisation)

Crosses between species are one of the oldest methods of achieving genetic improvements. The practice is based on the observation that spontaneous hybridisations occur in nature. Research is centred on the effects of complementarity between characteristics or the sterility of the crosses obtained. The latter characteristic is particularly interesting when the fish are to be used for stocking of sport fisheries.

In their natural environment the common carp and goldfish can produce hybrids which are almost all sterile. Under controlled conditions hybrids are obtained from the Asiatic cyprinid species such as the silver carp, the grass carp and the bighead carp (Bakos *et al.*, 1978). Trials with tench and Indian carps have also been performed. The hybrids with the silver carp are currently the only ones which seem to provide benefits for improving the quantitative production of carp.

Hatching and survival rates of hybrids are generally poor. Some of the offspring appear to be allotriploids and/or gynogenetic diploids. Cytogenetic techniques (see below) applied to hybrid fertilisation provide a range of triploid and/or tetraploid genotypes which remain at an experimental stage in carp, but, in salmonids, often allow a significant improvement in the viability of hybrids by permitting the combination of the advantages of the species, such as resistance to diseases.

4.2.2 Selection

Methods of genetic improvement based on additive genetic variability utilise different methods of selection. These methods are based on the choice of the individuals which are most suited to transmit the quantitative and qualitative characteristics, defined by their selector, to their offspring.

For common carp certain characteristics such as scale pattern or skin colour have a simple Mendelian-type inheritance based on the expression of one or two genes which have been the subject of mutations. Other characteristics such as viability or growth rate are determined by a greater number of genes. Their improvement requires more complex methods of estimation of hereditary potential.

Character with simple determinism

The scale pattern of carp is determined by two autosomal genes (i.e. not sex-linked) present on two chromosomes and of incomplete dominance: S (Scaly) and N (Nude). The range of possible combinations is shown in Figure 4.5. The N gene is lethal in the homozygous state and any larvae with this genetic composition die at hatching. The heterozygous state of the N gene gives rise to an incomplete covering of scales, sometimes reduced to a single row along the lateral line, as well as negative secondary effects (pleiotropy) on a number of functions (viability, increased susceptibility to diseases, perturbation of metabolism, slow growth, etc.). These disadvantages have led to the elimination of individuals carrying this mutation in many production stocks in Israel, the former USSR, Poland and Hungary.

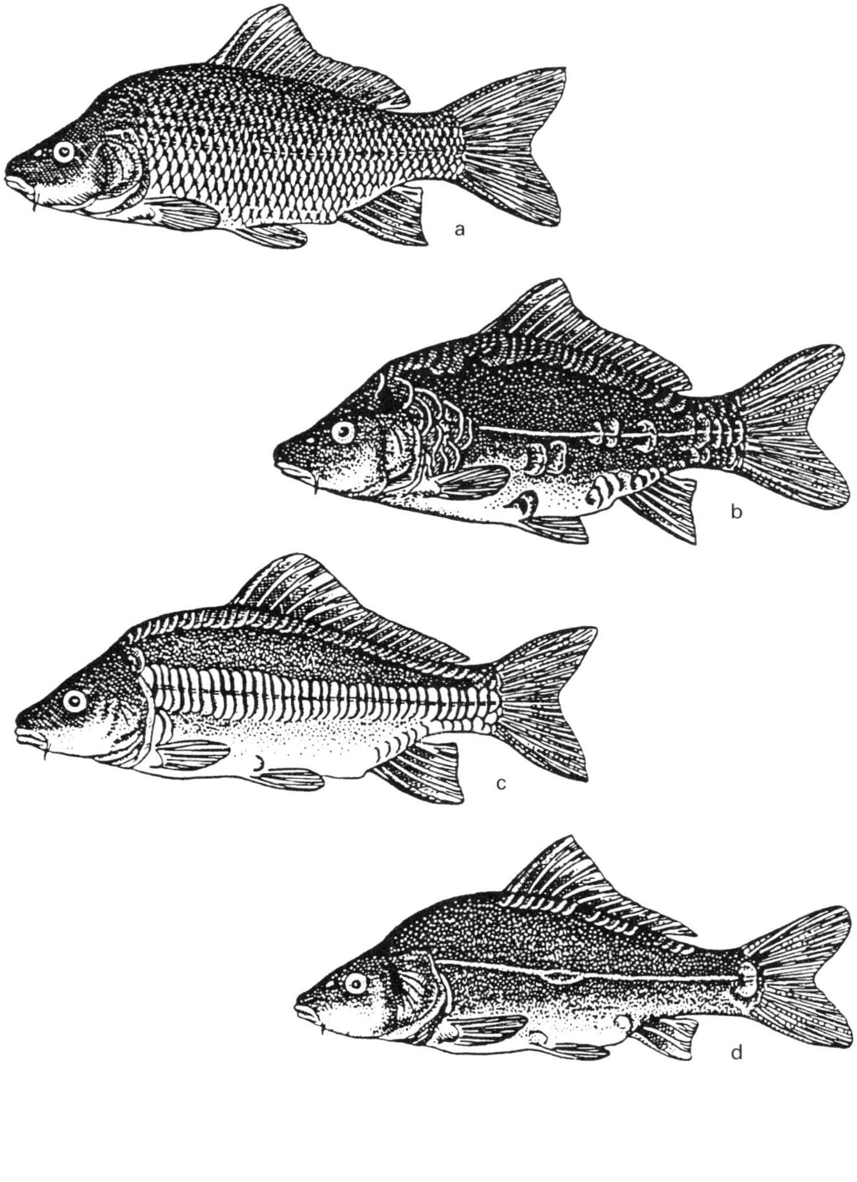

Fig. 4.5. Types of scale pattern for common carp *Cyprinus carpio* (after Kirpichnikov, 1981).
(a) SSnn and Ssnn: scaled carp
(b) ssnn: mirror carp
(c) SSNn and SsNn: line of scales
(d) ssNn: leather or naked carp.

The *colouration of the skin* can be determined by both environmental and genetic factors. Blue, gold or grey colours are determined by Mendelian-type mutations which influence the development of the cells responsible for pigmentation.

According to the review by Komen (1990):

—*red colouration* is due to the dominant gene R in the homozygous or the heterozygous state. The recessive state (rr) corresponds to the steel grey or silver-grey colour because of the absence of xanthophores;
—*blue colouration* giving a transparent or matt appearance is found frequently in Israeli (bi), Polish (bp) or German (bd) stocks and is due to the recessive mutation "tp" which determines the absence of guanophores;
—the *blond phenotype* (orange to yellow, sometimes with black spots on the lateral line) is due to a defect in the melanophores determined by a double recessive mutation of the B (black) gene, designated as b1b1/b2b2;
—the *blue–grey* colour comes from the association of rr and tptp;
—the *white colour* corresponds to the rr b1b1/b2b2 genotype (used as a marker in Israel);
—the *golden colour* (luminous orange) is due to a recessive mutation of the G (Gold) gene.

Some genotypes for colours are associated with reductions or improvements in performance, probably attributable to the fact that they are close to genes for quantitative characteristics.

Thus the L gene (Light) which adds luminosity to the coloration induces a lower viability to the heterozygous state Ll and lethality in the homozygous state LL.

The genetic determination of coloration in ornamental varieties of carp (koi, igoi, etc.) is very complex and the subject of commercial secrecy, notably in the countries of origin where selection programmes are implemented (p. 301).

Characteristics with complex determinism

Methods

The selection of complex characteristics rests on the exploitation of the additive genetic effects such that the mean performance of the progeny is equal to that of the parents. The variance of the phenotypic performance of a family $V(P)$ is the sum of the variance of the genetic effects of the hereditary potential $V(G)$ and the variance of the effects linked to the farming environment $V(M)$:

$$V(P) = V(G) + V(M) + \mathrm{cov}(GM)$$

where the covariance between the genotype and the environment $\mathrm{cov}(GM)$ is most frequently taken to be nil.

For a given characteristic, the gain expected, ΔG is directly proportional to i, the selection intensity ($i = S/\sigma P$ where S = the difference between the value of selected individuals and that of the individuals from which the selection is made), to h^2, its heritability which corresponds to the proportion of the performance of the selected animals which is genetically transmitted to their descendants and to σP, the amplitude of the variation of the character whose improvement is sought:

$$\Delta G = ih^2\sigma P$$

or:

$$\Delta G = Sh^2$$

The speed of development of this improvement will be inversely proportional to the interval between generations (generation time). However, the environment frequently masks the part of the performance which is attributable to the hereditary potential and the probability of keeping an individual based on its true genetic value is low.

To improve the genetic performance of a farmed stock different methods of selection are used to choose individuals. Among these are:

—*individual selection* or mass selection which consists of choosing an individual based on individual performance without using any information on related animals or the effects of environment. This method can be applied to strongly heritable characteristics ($h^2 > 0.4$) such as skin colour, age at first maturing and the timing of sexual maturation.

—*combined family selection* which consists of choosing for one or more characteristics the best individuals from the best families. This allows for the improvement of characteristics with a low heritability (e.g. growth rate) but requires a selection procedure using between 150 to 200 families for each generation.

Applications

For all types of animal production systems the main interest is in improving growth rate. However, for carp, this characteristic has not been improved, in spite of many attempts. An experiment in Israel using divergent individual selection over five successive generations has resulted in descendants showing a continuing decrease in performance and an upper line which is equal or slightly inferior to the control (Figure 4.6). Heritability of growth has been estimated as nil by several authors. Different explanations have been advanced to account for this lack of response to selection:

—farmed races already have a significant level of inbreeding, due in particular to a low number of genetic variations in the initial stocks and/or historical practices which have led to highly inbred races;

—the methods used do not allow the correct estimation of the hereditary potential of the individuals for this characteristic. Thus, in culture, there appears to be a significant heterogeneity in growth rate which manifests itself by the appearance of a group of individuals with a weight significantly above the mean, the "shoot" carp whose selection value is zero but which affect the attainment of the hereditary potential of other individuals (Figure 4.7).

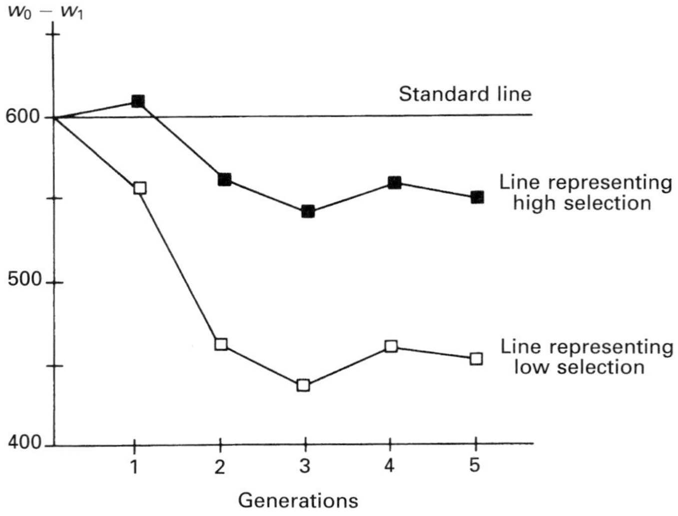

Fig. 4.6. Effects of bidirectional selection of the individual level on the growth of common carp in a pond expressed as adjusted weight gained after one summer ($w_0 - w_1$) in relation to the number of generations over which selection has taken place.

However, in other experiments where the first of these hypotheses is not fulfilled, some authors have been able to achieve positive estimates of heritability for growth rate. Several trials carried out in Russia have produced positive estimates of the transmission of growth traits in carp under different environmental (temperature) conditions by analysing lipid metabolism. The capacity for storing and releasing lipids according to season is thus a reliable indirect criterion for measuring this potential.

In contrast to growth rate, mass selection for the improvement of body conformation has been remarkably effective. For a long time, empirical mass

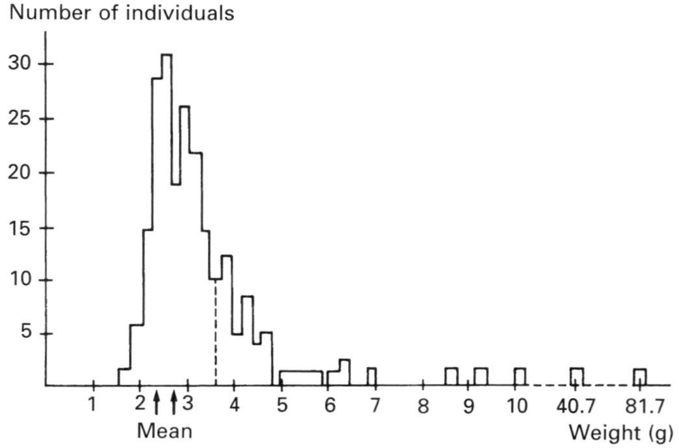

Fig. 4.7. Distribution of the weight of 4-month-old common carp (after Moav and Wohlfarth, 1973).

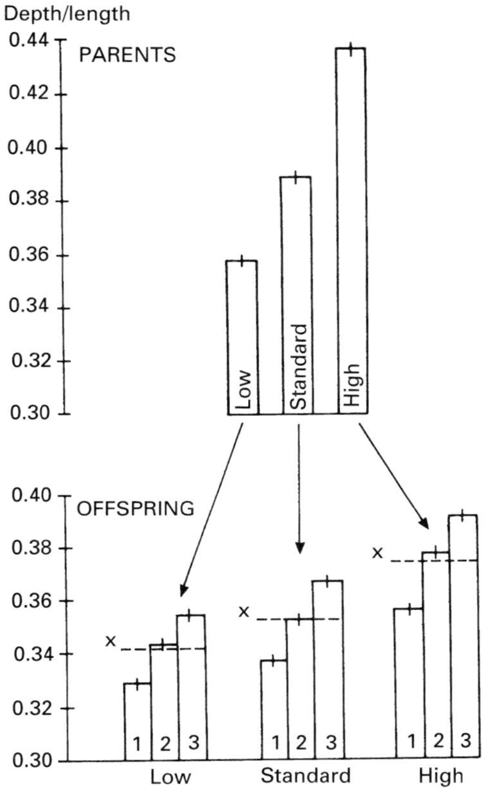

Fig. 4.8. Individual selection for body shape (depth/length) in common carp. The short vertical bars represent the variance of the mean. The dotted horizontal lines represent the means of replicates in three different fish farms 1, 2 and 3 (after Ankorion *et al.*, 1992).

selection has produced carp with very different condition factors. Thus, the European races, Aisghgrund and Galician, have been selected, based on their growth in depth in comparison to standard phenotypes which are more elongated. It is possible to change the height/length relationship (Figure 4.8, h^2 obtained $= 0.3$) but, paradoxically, the deepest bodied carp do not have the highest growth potential.

At the same time attempts have been made to improve growth performance, there has been notable success in selection for disease resistance in trials carried out by Wohlfarth (1986) and Kirpichnikov *et al.* (1993).

4.2.3 Management of sex and induction of polyploidy

Genetic improvements in carp can also be based on the use of methods where chromosome number is manipulated. The common carp was one of the first species of fish subjected to chromosomal manipulations but more progress has been made with the salmonids. Techniques of gynogenesis and sex-reversal have been used in salmonid culture since the early 1980s and are now considered to be standard practice, allowing the production of monosex stocks of triploid females which remain immature, in contrast to triploid males which, even though they are sterile, show a degree of gonad development.

Sexual dimorphism and growth rate

There is a significant difference in growth rate between male and female carp. This is to a large degree due to the fact that sexual maturation occurs 1 year later in females than in males.

In male carp in the tropics, first maturation is at 1 year old (500–600 g) and in temperate zones at 2 or 3 years old, depending on temperatures. Sexual maturation is accompanied by a decrease in growth performance (Figure 4.9). The development of the testes in relation to the weight of the body (gonadosomatic index, GSI) reaches 8 to 10% in the spawning season and seriously reduces the flesh yield at harvest.

In female carp, the first sexual maturation may be at a body weight of 1.5 to 2 kg which is at 18 to 24 months in the tropics and 3 years in temperate regions. The GSI can represent 20–30% of the body weight throughout most of the year as for this species gametogenesis recommences at the end of the spawning season (May–June). By the autumn the stock of ova has been replenished and makes up 15–20% of the live weight. In short farming cycles, the use of monosex female populations will be beneficial as it overcomes the effects of precocious maturation in males. For on-growing fish over a longer cycle, the complete sterility of all of the stock overcomes the problems linked with sexual maturity of the females.

Induction of gynogenesis and triploidy

In fish, sperm irradiated with ultra-violet (UV) light are capable of stimulating the embryonic development of the ova but contribute no genetic material to the embryo. Such insemination results in gynogenetic, haploid larvae ($= n$ chromosomes) which die at or a little after hatching. The production of viable larvae requires the doubling of the number (diploid $= 2n$ chromosomes) of maternal chromosomes (gynogenesis) through the retention of the second polar body (inhibition of the second meiotic division) by treating the ova which have been fertilised in this way by thermal or pressure shock.

When such shocks are applied after fertilisation with intact spermatozoa, triploid individuals ($3n$ chromosomes) are produced with two sets of maternal and one set of paternal chromosomes. When the shock is applied after the expulsion of the second polar body the doubling of the maternal chromosome set can be obtained by the suppression of the first cell division of the egg (mitosis). These two types of shock lead to the production of:

—gynogenetic diploid individuals which are totally homozygous through endomitosis when insemination has been with irradiated sperm; or

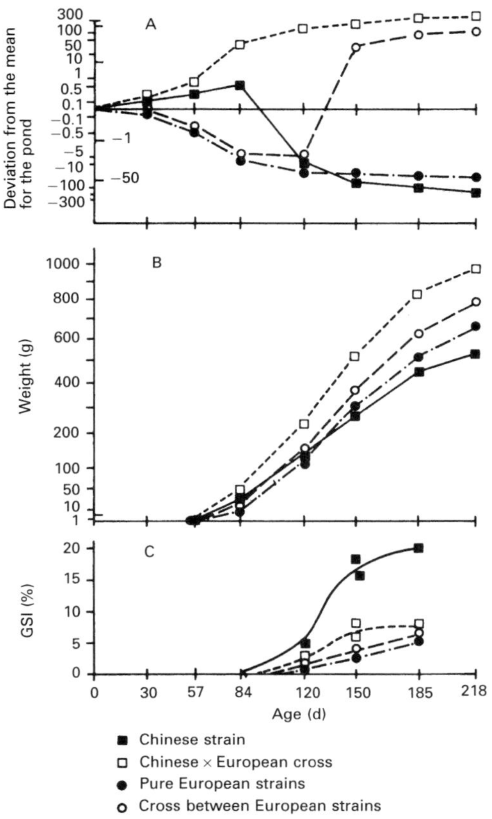

Fig. 4.9. Relationship between sexual maturity and growth rate in different strains of common carp (after Hulata *et al.*, 1985).

(a) growth curves expressed as deviation in relation to the mean for groups in a pond.

(b) the same growth curves expressed in the conventional form of weight gain: $W^{0.67}$.

(c) development of the male gonad shown as the gonadosomatic index (GSI).

—tetraploid individuals when the shock is applied to normally fertilised eggs.

Several techniques for the retention of the second polar body have been tested experimentally.

A cold temperature shock (0–4°C) lasting for 1 h applied 5 to 10 min after insemination at 20°C with irradiated sperm gives a yield of 20 to 50% gynogenetic embryos. The application of different shocks after fertilisation with non-irradiated sperm gives a triploid yield of around 50% with a cold shock. Linhart *et al.* (1991) obtained survival rates of 93.5% and triploid yields of 80% to the fry stage by applying a colder shock (0–2°C) 5 min after fertilisation (24°C) lasting for 40 min.

A 1 to 2-min long, warm thermal shock (39–40°C) applied 5 min after insemination with irradiated sperm at 20°C, leads to rates of gynogenesis of between 20 and 60% and triploid yields of 80 to 100%, if the shock is given after fertilisation with non-irradiated sperm (Hollebecq *et al.* 1986, 1988).

A high-pressure shock of 7110 to 8225 psi beginning 2 to 5 min after fertilisation and lasting for 5 min, induces a triploid yield of 80% (Linhart *et al.* 1991). Similar temperature and pressure shocks were applied by Komen *et al.* (1988, 1991) to induce the inhibition of the first mitotic division in carp.

In common carp, gynogenetic diploid individuals are all female (Nagy *et al.* 1978) which leads to the assumption that the female is homogametic (XX).

Monosex female lines

Spin-offs from the induction of gynogenesis are the establishment of inbred lines (see below) and above all, the production, after fertilisation, of monosex female lines or populations.

Commercial production of all-female stocks is only possible after the masculinisation of genetically female future broodstock at the alevin stage through the feeding of masculinising steroids (17α-methyltestosterone) in the diet. Nagy *et al.* (1981) achieved this inversion at an application rate of 100 mg hormone/kg of food between day 40 and day 80; the exact timing of application should be between day 8 and day 98 after hatching. More recently, Komen *et al.* (1989) induced a production rate of 92.7% males using a concentration of 50 mg/kg between the 6th and 11th week after hatching. According to the same authors, treatments of 50 mg/kg and 100 mg/kg applied later or earlier induced a high proportion of sterile individuals. The procedure induces the development of individuals which are genetically female but show all of the external characteristics of males. The "neo-males" produced this way are used as the male parents in the commercial production of monosex stocks of females.

Performance of monosex diploid and triploid lines

The potential benefits of monosex female populations and the characterisation of stocks has been demonstrated indirectly in experiments in Israel where 7-month-old female carp are 40% bigger than their male counterparts for the Chinese "Big Belly" race and 16, 13 and 10% for the European races "Dor 70", "Nasice" and "Gold" respectively.

To our knowledge there are very few comparative studies of the performance of triploid carp in relation to diploid individuals, but the techniques are beginning to be applied on a production scale in countries including Russia. Gervais *et al.* (1980) monitored individuals up to the age of 20 months. The GSI showed values of 0.7% and 10 to 20% respectively for triploid and diploid carp. The growth of the two groups was similar for the first three weeks of rearing. According to Taniguchi *et al.* (1986), 20-month-old triploid carp have greater perivisceral adipose deposits than diploid carp. The same authors showed that the growth rate of diploids is greater than that of triploids at 9 and 16 months. The GSI of non-monosex triploids was 0.2% at 20 months while that for diploids was 3.1% for females and 1.35% for males.

However, in contrast to observations made on salmonids, sexual maturation of triploid female common carp is not eliminated but appears to be delayed while triploid males appear to be sterile as their gonad development remains incomplete (Wu *et al.* 1993).

Crosses between inbred lines

The outcome of a cross is theoretically more effective if the parental lines are genetically distinct; thus there are attempts to increase the "individualisation" of stocks. In fish, as in plants, the increase in inbreeding within different parental lines may provide a potential technique for the rapid differentiation of two populations.

In the years between 1975 and 1985, crosses between related individuals, autofertilisation of artificially created hermaphrodites and gynogenesis through the retention of the second polar body enabled the rapid development of the degree of inbreeding. However, in practice, the crossing of these has been slower than predicted in theory; the mechanisms of meiosis prevent any stabilisation of the genotype (Nagy *et al.* 1984).

Subsequently, completely homologous inbred lines (all alleles homologous) have been obtained by inhibition of the first division of the egg (mitosis). Crossing of unrelated homozygous inbred lines is carried out in the hope of obtaining a completely heterozygous commercial product (hybrids between the first generation coming from the best heterotic cross) and completely homogeneous (totally identical siblings). But inbreeding brings with it a deterioration in the suitability for rearing because of the increased appearance of rare recessive alleles (see above). In addition, the build-up and maintenance of these rare inbred parental lines is not without its problems for the improvers. In addition to this many operations are required to evaluate reliably the proportion of the effect due to heterosis after crossing between inbred populations.

In summary, cytogenetic techniques can allow appreciable improvements to be made to the productivity in rearing systems:

- by simplifying the management of broodfish;
- by reducing the heterogeneity of growth of groups placed in on-growing systems;
- by improving flesh yield after gutting;
- by standardising the organoleptic quality of the flesh of the animals which do not become sexually mature.

4.2.4 Aspects of molecular genetics and transgenics

Recent developments in biotechnology have opened up several potential medium- and long-term applications:

—*early establishment of genetic potential*. An initial application might consist of determining the genetic potential of an individual at an early stage by a survey of the genes specific for the expression of certain characteristics using genomic DNA analysis. However, the scale of the scientific programmes needed to detect these genes suggests that progress towards commercial application will be slow.

—*production of transgenic individuals*. A second application might consist of improving the performance of an individual by inserting a gene for a given character into the genome (resistance to disease, resistance to cold, production of a particular molecule, growth rate, etc.).

The introduction of foreign genes can be performed manually in the egg using a micromanipulator. In carp and other cyprinids foreign genes can be incorporated between 10 and 30 min after fertilisation. As a preliminary the chorion of the egg is digested by trypsin. The dechorionation makes it easier to see the position of the second polar body area where the male and female pronucleus are located.

The mean integration rate of these genes is around 10% at the present time if the number of transgenic descendants after crossing the first transgenic generation produced by injection with standard animals is taken into account. Another method of introducing genes has been recently developed in loach: electroporation results in a 10% yield of transgenic animals. The replication of the introduced genes begins immediately after the micro-injection of the egg.

The recognition, isolation, multiplication and insertion of genes has now been successfully achieved for several species of fish including carp. The expression of these genes is more difficult to induce but recent results from common carp have shown a significantly higher production of growth hormone. Zhu and Cui (1992) showed that 10–25% of the population had an improved growth rate, up to around four times that of the controls.

The introduced gene was passed on to the second generation which showed an improved growth rate. The growth of F2 transgenics is not inhibited following hypophysectomy, demonstrating that the exogenous gene for growth hormone is expressed in tissues other than the hypophysis.

Some transgenic carp show a modification in morphology attributable to differential growth rates of diverse tissues (skeleton, modified protein synthesis) which causes an increase in the depth/length ratio.

One of the current problems of transgenesis remains the lack of control of the site of integration in the chromosomes and their expression. In practice, in carp some transgenic individuals (growth hormone) have shown a reduced growth rate in comparison to the controls. This unwanted phenomenon may be associated with the site of integration of the foreign DNA into the host chromosome (insertion into a "house-keeping" gene or gene with vital function, the over-expression of which causes inhibitions).

In order to target the integration of foreign genes into functional regions of the host-fish chromosome, trials have been carried out adding homologous sequences (regions flanking known functional genes and repeat DNA sequences).

Once the techniques for producing transgenic fish have been established, the production of transgenic fish of commercial benefit will require great care; notably the study of the possible impact of these animals on natural populations.

4.3 CONCLUSION

There have been many studies published on the genetics of the common carp. While the biochemical genetics of carp populations have only rarely been the subject of exhaustive studies comparable with those on salmonids, they do, however, allow the definition of the level of genetic variability in the domestic or wild populations and the characterisation of stocks. The conclusions of these studies on genetic identification allow useful advice to be given to farmers on the genetic management of their stock.

When comparisons are made between different stocks of carp a large interstock genetic diversity is shown for most of the characteristics which are of biological or commercial interest: growth rate, body shape, spawning characteristics, hardiness. This variability should be exploited, in the first instance, to choose the best stocks with the most desirable production characteristics. It may also be utilised through the rearing of the products of crosses which benefit from the advantages of each of the parental stocks and from hybrid vigour, as has been demonstrated in the achievement of improvements in growth rate. However, it should always be remembered that the product of a cross is rarely better than the best performance of the parental stocks.

The expression of a large number of characteristics is controlled by several genes. The colour of the skin and the pattern of the scales are examples of criteria which allow the differentiation of a commercial product or significant improvement in the competitiveness of rearing units. In this way, the eradication of the N gene, lethal in the homozygous state and responsible for the "leather" phenotype, should be strongly encouraged in order that rearing units only contain individuals whose performance is not diminished by the presence of this mutant gene that has numerous negative pleiotropic effects (secondary effects of a gene).

In practice, selection is only effective in adapting the conformation of animals and improving the genetic resistance to some pathogens.

As in salmonids, on-growing of monosex stocks of fertile female diploids or sterile triploids appears to show interesting possibilities. Standardisation of flesh quality to give an improved yield after gutting and of growth rate through the rearing of immature individuals should encourage farmers to set up monosex female populations.

Carp production systems utilise a range of techniques to improve the genetic characteristics of the stocks while guaranteeing a consistent quality of product to the consumer.

BIBLIOGRAPHY (includes additional references not listed in the text)

Ankorion Y., Moav R., Wohlfarth G. W., 1992. Bidirectional mass selection for body shape in common carp. *Genet. Sél. Evol.*, **24**: 43–52.

Bakos J., 1987. Selective breeding and intraspecific hybridization of warm water fishes. In: K. Tiews (ed.), *Selection, Hybridization and Genetic Engineering in Aquaculture*. Berlin, Vol I: pp. 303–311.

Bakos J., 1990. *Productivity improvement of common carp by hybridization of different landraces* (Abstract). *Symposium on Carp Genetics* SZARVAS.

Bakos J., Krasnai Z., Marian T., 1978. Crossbreeding experiments with carp, tench and Asian phytophagous cyprinids. *Aquacultura Hungarica (Szarvas)*, **1**: 51–57.

Babouchkine Y. P., 1987. La sélection d'une carpe résistant á l'hiver. In: K. Tiews (ed.), *Selection, Hybridization and Genetic Engineering in Aquaculture*. Berlin, Vol I: pp. 447–454.

Cataudella S., Sola L., Corti M., Arcangeli R., La Rosa G., Mattoccia M., Cobolli-Sbordoni M., Sbordoni V., 1987. Cytogenetic, genic and morphometric characterization of groups of common carp, *Cyprinus carpio*. In: K. Tiews (ed.), *Selection, Hybridization and Genetic Engineering in Aquaculture*. Berlin, Vol I: pp. 113–130.

Cavari B., Funkenstein B., Moav B., Hong Y, Schartl M., 1993. All fish gene constructs for growth hormone gene transfer in fish. *Fish Physiol. Biochem.*, **11**: 345–352.

Chevassus B., 1983. Hybridization in fish. *Aquaculture*, **33**: 245–262.

Chevassus B., 1987. Caractéristiques et performances des lignées uniparentales et des polyploïdes chez les poissons d'eau froide. In: K. Tiews (ed.), *Selection, Hybridization and Genetic Engineering in Aquaculture*. Berlin, Vol II: pp. 145–161.

Chevassus B., 1989. Aspects génétiques de la constitution de populations d'élevage destinées au repeuplement. *Bull. Fr. Pisc.*, **314**: 146–168.

Chourrout D., 1982a. Gynogenesis caused by ultraviolet irradiation of salmonid sperm. *J. Exp. Zool.*, **223**: 175–181.

Chourrout D., 1982b. Tetraploidy induced by heat shocks in the rainbow trout (*Salmo gairdneri*). *Reprod. Nutr. Dévelop.*, **22**: 569–574.

Falconner D. S., 1974. *Introduction á la génétique quantitative*. Ed. Masson et Cie., Paris.

Gervai J., Peter S., Nagy A., Horvath L., Csanyi V., 1980. Induced triploidy in carp, *Cyprinus carpio* L. *J. Fish Biol.*, **17**: 667–671.

Gjedrem T., 1983. Genetic variation in quantitative traits and selective breeding in fish and shellfish. *Aquaculture*, **33**: 51–72.

Gross R., 1991. A comparison of genetic, morphometric, productional and carcass traits of the common carp strains and strains crossed in Estonia (Abstract). *Symposium of Fish Genetics*, Wuhan, China.

Hines R., Wohlfarth G., Moav R., Hulata G., 1974. Genetic differences in susceptibility to two diseases among strains of the common carp. *Aquaculture*, **3**: 187–197.

Hollebecq M.G., Chambeyron F., Chourrout D., 1988. Triploid common carp produced by heat shock. In: B. Breton, Y. Zohar (eds), *Reproduction in Fish—Basic and Applied Aspects in Endrocrinology and Genetics*. Ed. INRA, Paris, **44**: 208–212.

Hollebecq M.G., Chourrout D., Wohlfarth W.G., Billard R., 1986. Diploid gynogenesis induced by heat shocks after activation with UV-irradiated sperm in common carp. *Aquaculture*, **54**: 69–76.

Hulata G., Wohlfarth G., Moav R., 1985. Genetic differences between the Chinese and European races of the common carp IV: Effects of sexual maturation on growth patterns. *J. Fish Biol.*, **26**: 95–103.

Kirpichnikov V.S., 1981. *Genetic Bases of Fish Selection*. Springer Verlag, Berlin, 410 pp.

Kirpichnikov V.S., 1987. Selection and new breeds of pond fishes in the USSR. In: K. Tiews (ed.), *Selection, Hybridization and Genetic Engineering in Aquaculture*. Berlin, Vol II: 449–460.

Kirpichnikov V.S., Ilysasov Ju.I., Schart L.A., Vikhman A.A., Ganchenko M.V., Ostashevsky A.L., Simonov V.M., Tikhonov G.F., Tjurin V.V., 1993. Selection of Krasnodar common carp (*Cyprinus carpio* L.) for resistance to dropsy: principal results and prospects. *Aquaculture*, **111**: 7–20.

Komen J., 1990. Clones of common carp, *Cyprinus carpio*. New perspectives on fish research. PhD thesis (in English, with Dutch summary), Agricultural University, Wageningen, The Netherlands.

Komen J., Duynhouwer J., Richter C.J.J., Huisman E.A., 1988. Gynogenesis in common carp (*Cyprinus carpio* L.). I. Effects of genetic manipulation of sexual products and incubation condition of eggs. *Aquaculture*, **69**: 227–239.

Komen J., Lodder P.A.J., Huskens F., Richter C.J.J., Huisman E.A., 1989. The effects of oral administration of 17α-methylestosterone and 17β-estradiol on gonadal development in common carp, *Cyprinus carpio* L. *Aquaculture*, **78**: 349–363.

Komen J., Bongers G., Richter C.J.J., Muiswinkel Van, Huisman E.A., 1991. Gynogenesis in common carp (*Cyprinus carpio* L.). II. The production of homozygous gynogenesic clones and F1 hybrids. *Aquaculture*, **92**: 127–142.

Komen J., Wiegertjes G.F., Van Ginneken V.J.T., Eding E.H., Richter C.J.J., 1992. Gynogenesis in common carp (*Cyprinus carpio* L.). III. The effects of inbreeding on gonadal development of heterozygous and homozygous gynogenatic offspring. *Aquaculture*, **104**: 51–66.

Komen J., Eding E.H., Bongers A.B.J., Richter C.J.J., 1993. Gynogenesis in common carp (*Cyprinus carpio* L.). IV. Growth, phenotypic variation and gonad differentiation in normal and methyltestosterone-treated homozygous clones and F1 hybrids. *Aquaculture*, **111**: 271–280.

Linhart O., Flajshans M., Kvasnicka P., 1991. Induced triploidy in the common carp (*Cyprinus carpio* L.): a comparison of two methods. *Aquat. Living Resour.*, **4**: 139–145.

Moav R., Wohlfarth G.W., 1968. Genetic improvement of yield in carp. *FAO Fish. Rep.* (44) Rome, **4**: 12–29.

Moav R., Wohlfarth G.W., 1973. Carp breeding in Israel. In: J. Moav (ed.) *Agricultural Genetics Selected Topics*, Wiley, New York.

Moav R., Wohlfarth G.W., 1976. Two-way selection for growth rate in the common carp (*Cyprinus carpio* L.). *Genetics*, **82**: 83–101.

Moav R., Hulata G., Wohlfarth G., 1975. Genetic differences between the Chinese and European races of the common carp. I. Analysis of genotype-environment interactions for growth rate. *Heredity*, **34**: 323–340.

Moav R., Brody T., Wohlfarth G., Hulata G., 1976. Applications of electrophoretic genetic markers to fish breeding. I. Advantages and methods. *Aquaculture*, **9**: 217–228.

Nagy A., Bercsenyi M., Csanyi V., 1981. Sex reversal in carp (*Cyprinus carpio*) by oral administration of methyltestosterone. *Can. J. Fish. Aquat. Sci.*, **38**: 725–728.

Nagy A., Rajki K., Horvath., Csanyi V., 1978. Investigation on carp, *Cyprinus carpio* L., gynogenesis. *J. Fish Biol.*, **13**: 215–224.

Nagy A., Csanyi V., Bakos J., Bercsenyi M., 1984. Utilization of gynogenesis and sex reversal in commercial carp breeding: growth of the first gynogenetic hybrids. *Aquacultura Hungarica (Szarvas)*, **14**: 7–16.

Poko J., 1990. Results of rearing and performance testing in imported strains of carp (*Cyprinus carpio* L.). *Prace Vurh Vodnany*, **19**: 34–46.

Recoubratsky A.V., Gomel B.I., Emelyanova O.V., Pankratyeva E.V., 1992. Triploid common carp produced by heat shock with industrial fishfarm technology. *Aquaculture*, **108**: 13–19.

Saud A., Billard R., 1987. Spermatozoa production and spermiation yield in the carp *Cyprinus carpio*. *Aquaculture*, **65**: 67–77.

Suzuki R., Yamaguchi M., Ito T., Toi J., 1976. Differences in growth and survival in various races of the common carp. *Bull. Freshwater Fish. Res. Lab.*, **26**, 2: 59–69 (English summary).

Suzuki R., Yamaguchi M., Ito T., Toi J., 1978. Catchability and pulling strength of various races of the common carp caught by angling. *Bull. Jap. Soc. Sci. Fish.*, **44**: 715–718.

Taniguchi N., Kijima A., Tamura T., Takegami K., Yamasaki I., 1986. Color, growth and maturation in ploidy manipulated fancy carp. *Aquaculture*, **57**: 321–328.

Thien T.M., Thang N.C., 1991. Selection of common carp (*Cyprinus carpio* L.) in Vietnam (Abstract). *Symposium of Fish Genetics*, Wuhan, China.

Wohlfarth G.W., 1986. Selective breeding of the common carp. In: R. Billard, J. Marcel (eds), *Aquaculture of Cyprinids*. INRA, Paris, pp. 195–208.

Wohlfarth G.W., 1993. Heterosis for growth rate in common carp. *Aquaculture*, **113**: 31–46.

Wohlfarth G., Moav R., 1971. *Genetic investigations and breeding methods of carp in Israel.* Rep. F.A.O./U.N.D.P. (TA) (2926): 160–185.

Wohlfarth G., Moav R., 1985. Communal testing a method of testing the growth of different genetic groups of common carp in earthen ponds. *Aquaculture*, **48**: 143–157.

Wohlfarth G., Moav R., 1991. Genetic testing of common carp in cages. 1. Communal versus separate testing. *Aquaculture*, **95**: 215–223.

Wohlfarth G., Moav R., Hulata G., 1986. Genetic differences between the Chinese and European races of the common carp. 5. Differential adaptation to manure and artificial feeds. *Theor. Appl. Genet.*, **72**: 88–97.

Wohlfarth G.W., Nagy A., Mcandrew B.J., 1991. A mistaken method for correcting potential bias in genetic testing of common carp, *Cyprinus carpio* L., and tilapias, *Oreochromis* spp. *Aquaculture and Fisheries Management*, **22**: 309–316.

Wohlfarth G., Moav R., Hulata G., Beiles A., 1975. Genetic variation in net escapability of the common carp. *Aquaculture*, **5**: 375–387.

Wohlfarth G.W., Feneis B., Von Lukowicz M., Hulata G., 1984. Application of selective breeding of the common carp to European aquaculture. *Eur. Maricult. Soc. Spec. Publ.*, **8**: 177–193.

Wu C., Ye Y., Chen R., Liu X., 1993. An artificial multiple triploid carp and its biological characteristics. *Aquaculture*, **111**: 255–262.

Zhang P., Hayat M., Joycee C., Gonzalez-Villasenor L.I., Lin C.M., Dunhan R.A., Chen T.T., Powers D.A., 1990. Gene transfer, expression and inheritance of pRSV-rainbow trout-GH cDNA in the common carp, *Cyprinus carpio* (Linnaeus). *Molecular Reproduction and Development*, **25**: 3–13.

Zhu Z., Cui Z., 1992. Hormonal replacement therapy in fish-human growth hormone gene expression and functioning in hypophysectiomized fish. *2nd International Symposium of Fish Endocrinology*, St Malo, France, 1–4 June 1992.

5

Juvenile rearing

5.1 REARING JUVENILES IN A HATCHERY–NURSERY SYSTEM

The rearing of fry is a key stage in fish farming. During this stage, fragile, newly hatched larvae (cutaneous respiration, because of the lack of gills, no exogenous feeding because the mouth has not yet opened) are taken through to the early on-growing phase. Newly hatched larvae are very vulnerable to "bioagressors" (p. 143). From the point of view of terminology a larva is morphologically different from an adult; this stage is completed when the swim bladder fills with air. At the fry stage which follows this, the animal has the morphology and swimming behaviour of an adult fish and takes in food. At the fry stage the fish is very sensitive and it is essential to provide a good diet which leads to high survival and a satisfactory growth rate. Larval rearing follows directly after spawning and usually takes place in the same buildings. Fry rearing is always practised as a monoculture.

The hatchery–nursery is a homogenous entity. In order to make best use of the huge number of gametes produced by cyprinids and convert them into fry suitable for on-growing, it is essential to protect the very vulnerable early stages and keep them in controlled rearing conditions. The objective is to protect them against predators and parasites (p. 146) while feeding them with a diet that is qualitatively and quantitatively adapted to their needs. It is only after this larval stage that the fry can be released into the natural environment or into on-growing ponds, at a size when they are big enough to overcome the problems of climate, food and predators. Management of the nursery appears to be highly dependent on that of the hatchery, as it is essential that the system should be ready for the introduction of the larvae in the days following hatching (immediately after the swim bladder has been filled) into an environment which is rich in prey and free from predators. In order for this to happen the preparation of fry ponds must be synchronised with the induction of spawning which is why these ponds are attached to the hatchery. Co-ordination is difficult when the larvae are distributed to more distant sites. It is also logical to combine the hatchery and nursery, as both structures require a high degree of technical competence on the part of the staff. Several species are often reared successively in the units, thus prolonging their period of operation.

There are a number of systems for rearing juveniles, each producing a different-sized individual at the end of the stage, varying from a few grams to a few tens of grams. One approach is to use small outdoor ponds in which the food web is managed, to produce the necessary food for the first few days (complementary feeding with artificial diets follows later). Another approach is to retain the larvae in indoor tanks and feed them from the start with an exogenously derived diet, either plankton harvested from outdoor ponds or artemia or artificial diets which have a limited availability on the market. Several authors have described techniques for juvenile rearing (Horvarth *et al.* 1984, 1986, 1992; Jhingran and Pullin, 1988) (see General Bibliography).

5.2 FRY REARING IN OUTDOOR PONDS[(1)]

This often takes place in two different types of ponds:

—the first, used for early rearing, are small (<1 ha). Fry are reared in these up to a weight of a few mg in 3–5 weeks;

—others, called second-fry ponds, which are much larger (several hectares); these are used for the rest of the season and produce, by the autumn, individuals of a predetermined size, depending on their density and the rate at which they have been fed in order to meet the requirements of the production system (p. 182 and 186 and Table 5–1 for production over 3 years). Water quality is described on p. 39.

5.2.1 The first-fry stage (Figure 5.1)

Ponds and their preparation

Techniques for the management of the first- fry-stage ponds have been particularly well established for carp using the knowledge gained in Central Europe, Israel and China.

These ponds are small in area, no bigger than 1 ha with a depth of 1 to 1.5 m. This facilitates their management, allowing complete and rapid emptying, the possibility of covering them with nets to exclude avian predators (p. 209), facilities for distributing feed and for capturing juveniles for sample weighing or examining for health status. It is good practice to raise the banks (1–2 m) to provide shelter from the wind. Inflowing water must be filtered to eliminate predators (insects, piscivorous fish) (Figure 5.2); frogs should be excluded by installing a mesh barrier 45–60 cm high. The most usual practice is for the pond to be filled a few days before the larvae are stocked. The sequence of operations is summarised in Figure 5.1. Before or at the same time as filling with water, various cultural operations are performed (including

(1) R. Billard.

Table 5.1. Standard information on stocking rate and performance of carp-production ponds based on a 3-year cycle in the climatic conditions of central Europe (Hungary) (after L. Horvath, pers. comm.).

	Length of the rearing stage	Density of fish (ind/ha)	Fish weight (g)		Survival %	Biomass (kg/ha)	Final production (kg/ha)	
			Initial	Final			Gross	Net
1st year								
First-fry stage	3–4 weeks	1–6 millions	0.004	0.2–0.3	35–60		100–600*	100–600*
Second-fry stage	2.5–3 months	10 000–100 000	0.2–0.3	20–100	60–80	30–60	1 400–1 800	1 300–1 700
2nd year	1 season	5–10 000	20–30	200–300	70–80	100–150	1 400–2 400	1 300–2 200
3rd year	1 season	1 000–3 000	200–300	1 000–1 500	80–90	300–500	1 600–2 500	1 300–2 000

* (kg/ha/month).

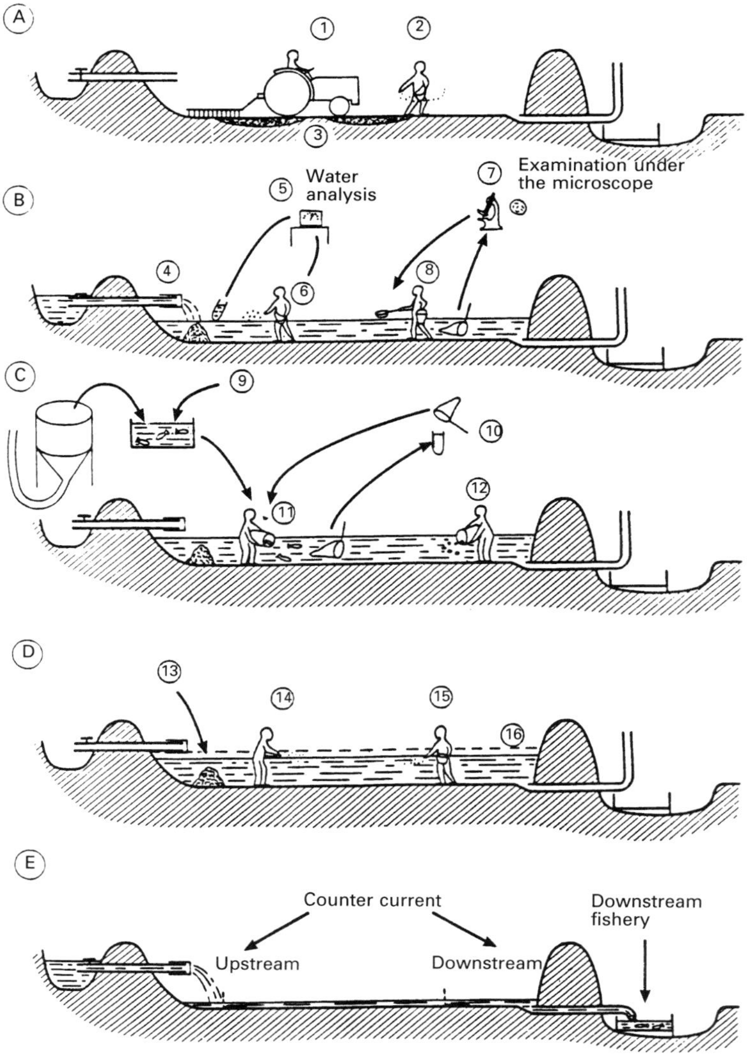

harrowing and removing plants) and the profile of the pond is adjusted in order to ensure that it can be emptied completely. Fertilisers are applied as necessary to the bottom of the pond before filling up begins:

- soil improvement, 200–300 kg/ha quicklime (CaO) if the layer of silt is thin, double if it is thick.
- organic manure (3–7 t/ha) cattle or pig manure or poultry wastes or 12–15 t of diluted pig manure (5% dry matter).

Fig. 5.1. Diagram of the preparation and use of carp first-fry ponds:
A: 8 to 10 days before stocking fry at the end of yolk-sac resorption, the bottom of the pond is flattened and quick lime is applied (1, 2 and 3).
B: partial filling (4) with filtered water; 6 days before the larvae are stocked (the same day as spawning), measurement of levels of minerals in the water (5) and spreading of organic and mineral fertilisers (6). Examination of different forms of zooplankton under the microscope (7) and, if necessary, destruction of the larger forms through application of insecticides such as 'Dipterox'® (3 kg/ha for a depth of 50 cm of water) (8).
C: The fingerlings are moved from the hatchery to the pond (9) (temperature of the water during transport and in the pond should be close to that during resorption (±1 °C); quantification of the biomass of rotifers present in the pond (filtration of 100 l of water should produce a minimum of 3 to 4 ml rotifers wet volume (or 3000 rotifers/l) (10). Stocking of larvae and reintroduction of zooplankton (large varieties, preferably daphnids (11, 12).
D: For a few days of rearing more fertiliser is added and, eventually, artificial diet (13, 14); the pond is filled completely (15, 16).
E: Fry are harvested after 4 to 6 weeks of culture. The pond is emptied very slowly to allow all the fish to drop into the external fishery. Fresh water can be added to the pond when only a few cm of water remain and there is a counter-current fishery established upstream or downstream of the pond. It is also possible to fish in open water (p. 91).

Water can be introduced at any moment after this preparation but the exact time is dependent on the stage of the larvae and, above all, on climatic conditions. Mineral fertiliser is applied at a rate of between 100 to 150 kg ammonium sulphate at 43% nitrogen per hectare, spread while the pond is filling up. This ensures that the fertiliser dissolves evenly throughout the pond (see Figure 7.9). Superphosphate is spread on the pond in solid form after it has been filled at a rate of 100 kg per hectare (18% P_2O_5) or, better, in liquid form (p. 163).

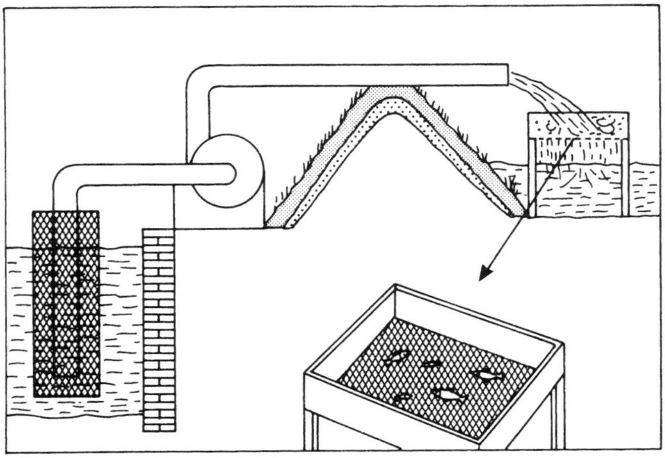

Fig. 5.2. Example of a system which prevents the introduction of unwanted fish into a pond when it is being filled by pumping in water: a mesh tube is placed around the inlet pipe and a sieve under the water inflow (after Horvath, in R. Billard 1983).

One of the best strategies is to establish the food chain (p. 11) with the bacterial, phytoplankton, protozoan, rotifer and crustacean succession. The larvae feed at first on rotifers and even protozoa (ciliates) (Table 5.2) and should be stocked into the ponds when the prey has reached maximum density. Because of this it is necessary to synchronise the production of larvae in the hatchery and the development of the appropriate plankton in the fry pond. In order for this to happen, the pond should be filled with water on the same day as hypophysation is carried out (p. 79), so long as the water temperature exceeds 16°C. If the temperature increases regularly to around 20°C, ciliates and rotifers will be present in sufficient quantity when the larvae reach the stage of swim-bladder inflation, which occurs after around 10 days. Because of climatic vagaries it is necessary to follow the temperature of the water and, in the event of a drop and consequent slowing down in the development of the community, it is possible to delay hatching and yolk resorption to a limited extent by reducing the temperature in the hatchery; a decrease of a few degrees is possible without damaging embryos or larvae.

If the pond is filled with groundwater there is little likelihood of introducing parasites and larvae such as copepods or aquatic insects. However, if water is taken from a river or pond, such parasites can be introduced. This can occur even with filtration at the inflow (Figure 5.2) and when sampling and identification is carried out immediately after filling the pond with water and before the release of the larvae. If there appear to be too many predators (see Figures 2.5, 2.6); larvae should not be released as they would rapidly be attacked. It is possible to adopt a method which is frequently used in Central Europe based on a treatment with an organic ester of phosphoric acid, Neguvon® or Dipterex®, insecticides which selectively destroy insects, copepods and cladocerans without affecting rotifers and thus favour their development. Doses are applied in order to achieve a final concentration of 0.25 to 0.3 mg/l of water. Such insecticides break down fast (within a few days) and do not impede the ultimate development of the food chain although it is likely to be beneficial to reseed with cladocerans, *Moina* for silver carp and tench and *Daphnia magna* and *Daphnia pulex* for common carp. This inoculation is done 2–3 days after stocking of the fish for *Moina* and later for *Daphnia*. The plankton is obtained by fishing in untreated ponds using nets of the appropriate mesh size. Stocking rates are of the order of 10 litres of plankton per hectare. Reseeding with cladocerans can be carried out by introducing water from a neighbouring untreated pond. For this the level of the water should be lowered before treatment; this reduces the quantity of insecticide required. However, some people may be hesitant about the use of insecticides and prefer to keep the larvae in controlled conditions in indoor tanks for a few weeks (p. 138) until they have reached a more advanced stage of development and are better able to resist predatory plankton.

When larvae are to be introduced at the end of the yolk-sac stage into ponds which cannot be drained, the same technique is used: control of predators, treatment with organophosphates, or delaying the release of larvae until they have developed to a more advanced stage. In carp-rearing systems it is sometimes even advised to treat the pond systematically with insecticides from the time it is filled; this has the

Table 5.2. Characteristics of live prey for common and herbivorous carp fry.

Species	Fry age (days)	Chief prey types	Temperature (°C)		Development of prey (d)
			Before the fry stage	During the fry stage	
Common carp	1–11	Protozoa, rotifers, small crustaceans	15–20	18–25	5–7
	11–18	Crustaceans (small and medium), insect larvae	15–25	18–25	10–15
	19–33	Large crustaceans, insect larvae, artificial diet	15–25	18–25	15
Herbivorous carp	2–15	Protozoa, rotifers	18–25	18–25	2–3
	15–30	crustaceans, artificial diet	18–25	18–25	10–15

advantage of eliminating all crustaceans and promoting the development of the rotifer population because the rotifers are no longer preyed upon by crustaceans.

Stocking the larvae

Larvae are introduced to the ponds in sunny, warm weather (water temperature above 18°C) with no wind (which might cause waves) and near to a protected bank in shallow water, which is saturated with oxygen.

The stocking density varies according to temperature, degree of intensification and the desired size of the larvae after 4 weeks. Figures of one million larvae per hectare are current in Central Europe but this can rise to 2–5 million. Density can be higher in small ponds with a surface area of a few hundred m^2, which are more easily protected than larger ones. Numbers released will also be related to the quantity of phytoplankton available. During the stocking of the larvae into the pond it is essential to estimate the biomass of rotifers present in the water by filtering 100 litres with a plankton net (mesh 60–80 μm). A quantity of plankton of around 1–3 ml or cm^3 (wet volume) for 100 litres filtered allows densities of 400 to 600 larvae per cubic metre of which 35 to 50% may be captured four weeks later as fry. If the volume of zooplankton is below 1 ml/100 l the number of larvae should be less than $100/m^2$.

It is important that the larvae are able to catch prey as soon as they are released into the pond. It is good practice to give them, just before release, several meals 3 hours apart made up of finely milled artificial diets for juveniles (p. 139). In Central Europe the current practice is to distribute cooked yolk from hens' eggs placed on gauze suspended in the water. The larvae are thus surrounded by a high density of food particles. Food availability should be controlled continually as there will be a rapid decrease in prey numbers as the biomass of fry increases because their mortality rate will probably be very low. From the first week, supplementary feeding will be necessary.

Many fish farmers consider that the intensification of production with additional feeding during the first weeks is very profitable. The acceleration of growth rates which results and the possibility of stocking the second-fry ponds earlier, with larger fry, contribute to reducing the length of the production cycle. Where complementary feeding continues, commercial feeds or various mixtures such as yeast, dried blood, fermented soya and fish meal are used: all of these must have a particle size of 50–100 μm for 10 days and 400–500 μ after this. A simple diet can be made from wheat flour (25%), fishmeal (25%), bloodmeal (25%) and soya flour (25%). More sophisticated diets are becoming increasingly available and are preferable (p. 139). The feeding rate, expressed as kg per day for 100 000 fry is around:

• 1 to 1.5 kg	throughout the first week,
• 2 kg	throughout the second,
• 3 kg	during the third,
• 4 kg	during the fourth.

These quantities will vary according to the richness of the environment in plankton: changes in the plankton should be monitored. When there is a drop in numbers, more feed should be given and/or plankton production stimulated by the distribution of organic and mineral fertilisers, the quantity of which can be established roughly after measuring N and P. Additions of 60 kg ammonium sulphate (43% nitrogen) and 30 kg superphosphate can be made each week after the larvae have been introduced and will complete the first application. Organic manure should only be spread in moderation as an excess of organic matter can damage the gills. Dried poultry manure (100 kg/ha) can be applied 2 weeks after the fry have been introduced. During the first-fry stage the volume of water in the pond is increased progressively: for a 500 m^2 pond the volume of water goes from 200 m^3 at the time of stocking to 300–350 m^3 after 2–3 weeks. Because of the high biomass, the rearing time should not exceed 4–5 weeks. After this the fish are harvested. However, examination of the development of the community and the condition of the fry may advance the harvest if the plankton numbers decrease and the fry are thin and not growing.

Harvesting juveniles

When the fry are being harvested the general practice is to drop the level of the water by half and then capture the fry in mid-water using a seine net (10–15 m × 1.2 m) with a small mesh size, handled by two people. After being drawn through the water the net is raised by the base and stretched out horizontally, the fry are progressively collected and harvested in the same way as the larvae, with the aid of a cup so that they are never removed from the water (Figure 3.18). After the net has been passed through several times the pond is emptied and the remaining fry are concentrated in the fishery where they are captured as above or directed into a hoop net placed in the downstream end of the outflow channel (p. 245). It is important to allow clean water to flow during the harvest, especially during the final phase when the fry that have escaped the nets are concentrated in the fishing device after the pond has been emptied. If fresh water is not available, the water from the fishery can be recycled using a pump; the objective is to maintain a high level of oxygen (close to saturation) in order that the fry never experience conditions of anoxia.

The hoop net at the downstream end should be kept permanently in place from the start of the pond-emptying procedure in order to retain any escaping fish. These should be continuously removed so that they are not exposed for too long to the strong currents passing through the net. It is also possible to use counter-current fisheries (Figures 5.1E, 7–22).

After harvesting the fry should be counted in order that the correct number can be stocked into the second-stage fry ponds. An acceptable method of assessing numbers is to count the fry retained in a small sieve, which serves as a measure, and multiply by the number of sieves counted. At this time, depending on the health of the fry, disease-prevention measures may be required. A standard treatment uses a NaCl bath (30 g/l) for 30–40 s to eliminate ectoparasites such as *Trichodina*. Matters relating to diseases are covered on p. 142 for juveniles and p. 195 for fish during the on-growing stage.

Table 5.3. Characteristics and performance of first- and second-fry stages in outdoor ponds.

	First-fry stage	Second-fry stage
Ponds		
Size (ha)	≤ 1	≥ 1
Depth (m)	0.4–1.5	1.0–1.5
Treatments applied to ponds		
—Initially		
CaOH kg/ha	200–300	
Superphosphosphate (18% P_2O_5) kg/ha	100	100
Ammonium sulphate (43% N) kg/ha	100	150
Manure t/ha	5–7	
—During rearing		
Superphosphate (18% P_2O_5) kg/ha	30 (3 times)	20–30 every 2 weeks
Ammonium sulphate (43% N) kg/ha	60 (3 times)	20 every 2 weeks
Organic manure kg or m^3/ha	100 kg (poultry manure)	10 m^3 every 2 weeks*
The fish		
Length of the stage	3–5 weeks	3–4 months
Density (ind./ha × 1 000)	1000–6000	10–100/200
Initial size (cm)	0.6–0.7	2.5–5
Initial weight (g)	0.004	0.2–2
Final size (cm)	2.5–5	—
Final weight (g)	0.2–0.3	—
Size in relation to density/ha—100 000	—	15–20 g
— 20 000	—	50–70 g
— 10 000	—	100–150 g
Survival (%)	35–60	60–80
Yield (kg/ha/cycle)	100–600	1000–2000
Feed		
Natural	Rotifers, cladocerans, copepods	Large zooplankton, chironomids
Artificial (balanced diet)		
particle size mm	0.1–0.2	up to 3
Feeding rate (kg/h/100 000 fry)	1 to 4	5–10

* Manure 5% dry matter.

At the end of the first-fry rearing phase the size of the carp fry depends on their age (3 to 5 weeks generally), water temperature, stocking density and the food available from endogenous or exogenous sources. In temperate regions (Central Europe) the usual size is 2–3 cm and 200–300 mg after 3 weeks and 750–1500 mg at 4 weeks. In Israel the fish generally reach a weight of 2 g in 1 month; the percentage survival is between 35 and 60% for carp. A summary of operations and performance in the first-fry stage is given in Table 5.3.

Techniques for the rearing of grass carp and tench are similar to those described above for carp. The stocking density of larvae can be significantly higher for phytophagous species (up to 5 million/ha) and survival may reach 90%.

5.2.2 The second-fry rearing stage in ponds

The objective of the second-fry rearing stage is to prepare fry for on-growing at clearly defined, controlled densities in order to 'programme' the final size by supplying the appropriate quantity of feed in relation to the biomass of fish present. Juveniles come either from the first-fry ponds (see above) or indoor tanks in controlled conditions (p. 138). The ponds for the second-fry stage are generally larger than those for the first stage, between 1 and 15 ha and 1 to 1.5 m deep. Plants or woven branches should be kept on the banks, particularly those exposed to the prevailing winds, to prevent erosion and the resultant water turbidity which would limit the development of algae. The ponds are left dry over winter and prepared in spring: dykes repaired, dead vegetation removed from the bed and the banks and lime (150 kg quick lime/ha) spread on the bottom.

The goal of pond management depends on the degree of intensification and the chosen method of production. When production exceeds 4–5 t/ha, the contribution of plankton and benthos to the diet will be negligible and production controlled by the exogenous food supply. The management of the trophic web will be limited to maintaining a population of unicellular algae to provide at least part of the oxygen and for removal of ammonia. When production is around 1 to 3 t/ha the contribution of the trophic web to the feeding of the fish is such that management is required: the system should be forced by the application of fertilisers and by giving complementary feeds when the quantity of plankton is insufficient. This distinction can also be made for the on-growing phase in the following years (p. 157). Only the details relating to the second-fry stage are considered here, and only the example of semi-intensive production with management of the trophic web to produce food and oxygen. Intensive production based on the quasi-exclusive supply of artificial diet is covered on p. 179.

The timing of filling of the ponds depends on the method by which the 4-week-old fry have been produced. If they have been produced in outdoor ponds, the second-fry stage cannot begin until the end of May or the beginning of June in temperate regions. If spawning and the first-fry stage have been under controlled temperature conditions and feeding is entirely artificial in the hatchery, fry may be ready for the second-rearing stage at an earlier date; as soon as the water temperature reaches 15–18°C. Under these conditions the ponds can be filled and fertilisers applied earlier (when water temperatures reach 15–16°C) in order to kick-start the trophic web.

Chemical fertilisers (150 kg ammonium sulphate and 100 kg of superphosphate/ha) are spread immediately after the pond has been filled. Supplementary doses of 20–30 kg/ha of each are applied every 2 weeks. Organic manure is distributed at the same frequency and may, for example, be pig manure with 5% dry matter (10 m^3/ha).

Introducing the fish depends on many parameters which have been reviewed previously (temperature, nutrient status of the pond) but the most important factor is the size the farmer hopes that the fish will reach at the end of the first summer. Under optimum conditions it may be possible to obtain 100 000 20 g individuals, 20 000 50 to 70 g individuals or 10 000 100–150 g individuals (Table 5.3). The essential complementary feeding will be adjusted to the number of fry. At the density of 50 000 fry ha plankton contributes at least half of the fishes' feed during the first weeks of feeding. The complementary feeding may be made up of cereals (10% of the biomass of fish per day) with added amino acids or granules with a low level (20–25%) of proteins (3–4% of biomass of fish per day during warm weather) (p. 167). At concentrations of 100 000 individuals/ha the contribution of plankton and benthos to the diet rapidly becomes negligible and feed consists of 3 mm diameter dry granules with a 20–25% protein content, given at a daily rate of up to 5% biomass. It is possible to replace a part of the granular diet with cereals. The feed given to carp fry during their first summer is similar to that used during the on-growing phase (p. 167).

If there is much food remaining 5–6 h after it has been distributed, the next day's ration should be reduced. Because of the high labour costs associated with this method of feeding there is a tendency to use automatic feeders (p. 240). If the intensification of production remains relatively modest (1–2 t/ha), continuous aeration will not be needed as long as the turbidity is not too high and that unicellular algae are still present (at a concentration of 50 000 cells/ml).

Samples of fry should be taken at regular intervals from the region of the automatic feeder; the mean weight can be calculated from a sample of 100 fry. This enables the farmer to calculate the food conversion efficiency. If this is greater than 1.5–2 for granule based diets with a protein content of 20–25% or above 3–5 for cereals, a problem should be suspected. This may be caused by poor management of the farm and feeding or by a disease problem, particularly due to the presence of parasites. When samples are being taken the health of the fish should be assessed, looking for signs of anorexia (thin fish), anaemia (assessed from the gills), the presence of parasites and the contents of the digestive tract. Pressure on the abdomen forces out faeces that can be examined for indications of the dietary regime: there should be an alternation of dark and light grey zones, corresponding respectively to an intake of plankton or benthos and artificial feeds. If this succession and the relative size of the zones do not correspond to the availability of the different diets (respective amount of plankton and distributed feed in the pond) there is an anomaly in the feeding behaviour or availability of the prey. A decrease in dark zones may suggest that the plankton is insufficient and that more artificial feed should be given. Levels of oxygen should be monitored throughout the rearing phase particularly during warm, cloudy weather (p. 40) and if necessary aerators should be used (p. 234).

In temperate regions it is usual to capture carp fry in the autumn at the end of the first summer of rearing using a seine and hoop net, after lowering the level of the water and concentrating the fish in the capture channel or fishery. Clean 'fresh' water should flow constantly through the channel or aerators should be placed so as to maintain sufficient oxygen in the water (>3–4 mg/l). Grading out of small fish takes place either during harvesting or at the end of the process. During harvesting and grading the fish should be out of the water for as short a time as possible. Because of the reduced biomass in comparison to fish of commercial size it is possible to use equipment such as Archimedes screws or fish pumps which transfer fish in water (p. 248). After fishing and grading, a prophylactic treatment should be applied routinely. This consists of exposing the fish for 5 m (with aeration) to a combination of:

- NaCl (cooking salt) 2 kg/m^3
- Dipterex® or Neguvon® insecticides: 200 g/m^3
- Malachite green at 100 mg/m^3 (0.2 l of hatchery stock solution which has been made up at 500 mg/l (p. 87).

It should be noted that piscivorous fish have a poor tolerance to insecticides and that the dose rates given above are only applicable to cyprinids. Chloramine T is a useful disinfectant and antiparasitic treatment with a low toxicity but it must not be used in normal metal tanks; it should only be used in plastic or stainless steel tanks. Doses can be up to 70 g/m^3. This treatment can be given during the last minutes of transport or just before stocking into overwintering or holding ponds (p. 202). The emptying of treatment baths into the pond or natural environment should be avoided.

5.2.3 Overwintering

During winter it is preferable to keep the fish in small ponds where they can be monitored effectively and protected. Ponds with an area of around 500–1000 m^2 and 2 m deep are treated with quicklime (200 kg CaO per hectare) before filling with water and with malachite green (5–10 g/m^3) afterwards. Once the malachite green has dispersed, fish are introduced at a density of 5 to 10 kg/m^3 and a permanent flow of water ensured (5–10 l/min for 100 kg of fish or 1–2 m^3/min for a 2000 m^3 pond). During winter, a weekly treatment with malachite green (0.1 mg/l) prevents the development of saprolegniosis.

Samples are taken each week in order to check on the health of the fish. If the temperature is below 7°C, no feeding is necessary but above this food should be offered. Protection against avian predators should be installed (nets, bird scarers, p. 203). When there are heavy frosts the formation of ice at the inflow should be prevented by placing a plate to disperse the water under the inflow pipe and removing any ice that forms at the monk.

In less-extreme climates where temperatures generally remain above 7°C carp fingerlings can overwinter in larger ponds, which can be given a light dressing of organic fertiliser (p. 165) and food can be distributed. It should be remembered that sampling and also protection from avian predators is more difficult in large ponds; harvesting operations are also more complex.

5.2.4 Use of plastic sheeting or glasshouses above ponds to limit temperature losses[(2)]

Simple use of plastic sheeting directly on the water surface

This is a new technique that consists of stretching a sheet of bubble wrap close to the surface of the water in order not only to limit the losses of heat but also, more importantly, to transfer solar energy to the water efficiently. Experiments carried out in the former Czechoslovakia in ponds of 80 m^2 surface area for 59 days (18/4–16/6) have shown that a strong thermal stratification forms below the bubble wrap (5 to 8°C increase in the first 5 cm). Marked stratification has also been observed for oxygen. Because there is no mixing, the thermal gain for the whole body of water remains very limited (+1°C). The development of algae observed at the surface is short term (10 days) after which the plastic becomes opaque because of algal fouling and the transparency rapidly reaches 100 cm compared with 30–40 cm in the control pond. After 48 days the performance was significantly lower in the plastic-covered pond, the weight of individual fry 3.9 times lower and survival 2.6 times lower than in the control, uncovered pond. This approach, which seems attractive at first, is not therefore appropriate. An improvement might be possible by only covering half of the pond with plastic: positive results have been achieved in Germany using this technique.

Use of plastic greenhouses above unheated ponds

Fry ponds are covered with plastic sheeting 2 m above the surface of the water in cold temperate zones in order to limit the heat loss from the water. Differences of +3.1°C and 6.8°C were obtained in the former USSR in the growing season and cold season respectively, in comparison with neighbouring uncovered ponds. Other experiments, performed in Germany in the new Landers, have shown that it is possible to keep the temperature at a constant 23°C in June–July in small fry ponds (1000 m^2) covered by a plastic greenhouse while in neighbouring uncovered ponds the temperature fluctuates much more and does not exceed 20°C. This technique is also used in Israel to allow the production of carp fry early in the season. Apart from the costs, the technique has the disadvantage of the risk of anoxia; in practice the increase in temperature and the absence of wind limits the oxygenation of the water mass. Oxygen levels should be monitored and, if necessary, an aeration system installed.

Installation of a greenhouse with heated water

There are many examples of the use of plastic or glass greenhouses above small ponds, whose water is warmed by various sources of heat (oil boilers, geothermal

(2) Information from R. Berka.

water). A classic example is that used on eel farms in Japan. Eels were traditionally cultured in small outdoor ponds but farming has switched progressively to the use of heated water and the covering of the ponds with glasshouses. It is possible to build up to a high degree of intensification by using devices for aeration and for water purification. Such intensification, with the resulting increase in production costs, cannot be justified for cyprinids under the present market conditions except for ornamentals (koi carp, varieties of goldfish such as "bubble eyes", sarasa (p. 301)) and for the production of juveniles. Techniques of recycling–purification are most frequently described in works on salmonid rearing.

5.3 FEEDING CARP LARVAE ON ARTIFICIAL DIETS(3)

It is possible to have eggs and larvae of species such as the common carp in the hatchery throughout the year by the use of temperature control and hormonal treatment of broodstock (p. 70). However, the potential for rearing these in normal on-growing facilities (managed ponds) is limited by outdoor climatic conditions, particularly temperature. To extend the time of juvenile production, larvae should be reared indoors in temperature-regulated conditions. Feeding of these larvae is the main problem. The harvest of zooplankton from the wild is often irregular in both quantity and quality and its distribution carries the risk of introducing parasites to the rearing unit. *Artemia* nauplii are generally an expensive source of food. Another solution is to use artificial diet, which allows the rearing of carp larvae without the use of any live prey (Charlon *et al.* 1986).

The use of dry diets in closed or semi-closed systems imposes particular constraints; the installation of automatic feeders, specially adapted rearing systems and cleaning systems. However, this type of feeding allows the regular production of juveniles, independent of season, in a similar way to the production of fry in salmonid culture.

5.3.1 Feed

Manufacture

The formulation of starter diets is empirical in nature, as the nutritional requirements of larvae are still not fully known. The following formulation and protocol gives good results; it should be noted that the proportions can be modified and that other basic materials can be added:

- Mix 1 kg of chopped beef liver (fresh or frozen), 500 g of industrial yeast, 70 g vitamin mix and 50 g mineral mix (mixtures as used for trout) to obtain a homogenous paste.

(3) P. Bergot and P. Kestemont.

- Pass the paste through a mincer with the 2 to 3 mm diameter blade in place.
- Dry the moist granules in a ventilated oven (40°C) for 24 to 48 h.
- Sieve to separate into 100–200 μm, 200–400 μm and 400–600 μm fractions.
- Keep cool (4°C) in a closed container until use which should preferably be within 3 months.

Use

The size of the food particles should be matched to the size of the larvae. If particles are too big they will not be eaten and will fall to the bottom of the tanks and contribute to a decline in water quality. The same is true of particles which are the right size initially but which stick together to form clumps. Small particles also contribute to water pollution and it is important to increase the size of food particles up to the next grade as soon as possible, bearing in mind that the larvae are growing constantly. The change in size should be based on the average size of the batch and not on the size of the largest individuals. If the change is made too early the small larvae will be at a disadvantage and the heterogeneity of size will increase (Table 5.4).

From day 21 (and sometimes even day 15) it is possible to give commercial trout diet, sieved to a suitable size, in place of starter feed.

At 19–20°C carp larvae are capable of ingesting feed on the 3rd day after hatching. To be certain that no larvae starve initially it is best to distribute some feed at the end of day 2. Larvae capable of feeding can be recognised by the presence of an inflated swimbladder and by their exploratory swimming behaviour. Before this stage they stay in contact with the walls and move erratically.

It is almost essential that the feed be distributed from an automatic feeder as it is offered several times an hour for over 12 h each day. A quasi-continuous distribution pattern maintains a good concentration of particles floating on the surface around the larvae. Micro-granules float for a moment and then drop to the bottom of the tank. Their trajectory in the water and distribution in the tank is dependent on the circulation of the water within the tank, as is the behaviour of the larvae. The inflow

Table 5.4. Changes in the size and the quantity of feed distributed during the first stages of carp rearing (24°C).

	Length of the juveniles (mm)	Mean weight of juveniles (mg)	Particles size (μm)	Quantity for 10 000 larvae (g/week)
1st week	6.5–9	1.2–8	100–200	150–200
2nd week	9–15	8–40	200–400	500–600
3rd week	15–19	40–200	400–600	850–1 200

of water is regulated to produce the homogenous behaviour of larvae and to avoid turbulence.

At 24°C, an extended period (16 h out of every 24 h) of feed distribution combined with a period of illumination (e.g. 6 h to 22 h) gives good results. The optimum artificial daylength is not yet known. The increased feeding time allows an increased food consumption and, consequently, growth rate. It has been found that a night time rest is beneficial. The total quantity of feed given in the first 3 weeks can be roughly calculated based on a conversion index of between 0.7 and 1 (conversion index = weight of dry diet given/wet weight gain of the larvae). Adjustments in feed distribution during rearing are made by checking every day that all the larvae have food-filled digestive tracts (Table 5.4). The feeders described by Charlon (1990) allow the regulation of the amount of feed given and also the length of time over which it is given. Belt feeders are suitable for large batches (p. 242).

5.3.2 Rearing equipment

Rearing tanks designed for the use of live prey which can remain in the water for a long time with little resultant pollution are not suitable when dry diets are used. These particles should be distributed in excess, which results in a high concentration of organic matter, both dissolved and particulate, which in turn favours the proliferation of microorganisms. This proliferation is more rapid at higher temperatures. The wastes cannot be evacuated in the same way as with trout alevins or older carp fry using fast currents. Such currents would wash away these small larvae which have a limited swimming capacity. Cleaning by siphoning is prolonged and it is difficult to avoid sucking up the larvae: it is always fairly ineffective.

An effective system (INRA) consists of mobile tanks installed in fixed tanks as described by Charlon (1990). During rearing a significant turnover of water dilutes dissolved substances and removes fine suspended particles. The larger particles are removed daily after transferring the larvae to a clean tank. When the larvae have been removed the dirty tanks can be cleaned rapidly, for example with a high-pressure warm-water jet. Larvae are transferred to clean tanks manually for small tanks (a few litres) or using mechanical assistance (hydraulic jack) for bigger rearing volumes (a few hundred litres). Transfer and cleaning can be automated.

The complete system for larval rearing comprises a reserve of clean temperature-regulated water that is sterilised by UV tubes before being fed into the rearing tanks. At the outflow of the tanks, the water is filtered and recycled by a pump (p. 94, Figure 3.21).

Several small, independent circuits appear to be better than one single circuit: the effects of breakdown are reduced; a batch which is suspect from the point of view of disease can be eliminated and its circuit disinfected without disturbing the rest of the rearing system. The larvae remain in these tanks for a short time (2–3 weeks). When they reach a weight of a few milligrams they can be transferred to structures such as those typically used in salmonid rearing.

In the hatchery, all equipment must be capable of operating without any risk to personnel, particularly the electrical systems. The protection of material and of fish should be incorporated in the designs from the beginning: gravity-fed water, protection of electrics from drying out and level alarms which give a warning of the breakdown of pumps. It should be possible to dismantle the pipework for thorough cleaning at the start and finish of the rearing of each batch. The cleanliness of the hatchery must be ensured to guarantee the consistency of the results. It is not a question of ensuring complete sterilisation but of keeping bacterial populations at a level below that which interferes with rearing.

5.3.3 Husbandry

Larvae are introduced to the rearing tanks on the second day after hatching and the temperature is increased progressively from 19–20°C (incubation temperature) to 24–25°C (rearing temperature). Larvae can tolerate a higher temperature and grow more rapidly but the risks also increase. At high temperatures, it is necessary to take rapid action when there is a breakdown, for example to counteract the risks of anoxia after pump failure.

The growth in weight of the larvae is rapid, with a doubling of the biomass possible every 2 to 3 days. This implies either a limitation of the initial density (e.g. 50 larvae/litre) when the batch is going to remain in the same tank or halving the density every 2–3 days if the initial density is high (above 100 larvae per litre). The criteria set for a successful operation are a high survival rate (80–95%), growth conforming to predictions, normal morphology of juveniles and a homogenous batch at the end of rearing. The appearance of abnormalities in the morphology of the alevins (deformed vertebrae, atrophy of the operculum) may reflect poor nutritional quality of the diet which should be changed. Studies on homogenous batches at the end of rearing have led to the recommendation that batches should be made up at the start of rearing from the larvae from one pair of broodfish, which have been incubated together and hatched at the same time.

5.3.4 Conclusion

The production of carp juveniles with the use of dry diets is similar to that of salmonid fry but with differences such as the small initial size of the carp larvae and the higher rearing temperature. The most delicate stage is at the start; difficulties decrease as the larvae grow. The goal is the production of a batch of juveniles at a certain weight on a chosen date. It seems to be more important to research improvements in spawning reliability, minimising its risks than to aim for records of growth rate or rearing density.

5.4 DISEASES: PREVENTION AND TREATMENT[(4)]

The artificial spawning of fish in hatcheries is generally very effective, but leads to conditions of high densities of offspring which favour the appearance of one or more

(4) M. Morand.

bioaggressors: parasites, bacteria or viruses which are encouraged by the sensitivity of the larvae, then fry to the quality of their environment and their feed. Already at this stage the disease status is the result of the disequilibrium between the fish, their environment and the pathogens.

There are many causes of the breakdown of equilibrium and the most obvious are not necessarily the initial cause; thus the appearance of large numbers of external parasites may be a sign of massive contamination which has the effect of reducing the defences of the larvae towards the parasite itself. This should always be remembered, however obvious the bioaggressor, and particularly if no bioaggressor can be found during pathological investigations. Whatever the cause, before developing a strategy for prevention and counteraction, two aspects of the problem must be considered: the "pathology of the environment" and the infectious pathology in its broadest sense.

Avian predation is a significant cause of mortality in farmed juveniles in outdoor ponds and is discussed in the section on on-growing (p. 203).

5.4.1 Diseases associated with the environment

In a typical land-based rearing system, the risks are relatively well known but can be a consequence of modifications of the physical, chemical or biological features of the environment.

Physical factors

Among the basic physical factors, as well as water flow and turbidity, temperature and its stability are the key elements to success. All fluctuations in the temperature of the rearing water should be avoided and infinite precautions taken during transfer. Jumps in temperature of 5 to 8°C are lethal and juveniles (alevins and fry) die showing symptoms of cardiac and respiratory failure. These symptoms are more marked and death more rapid when the fish have been fed. Digestive enzymes are affected and also food becomes stuck in the digestive tract. The food ferments and the abdominal wall becomes dilated, the fish are lethargic and die. When transfer is essential the juveniles should be acclimated stepwise, by 1°C/h, with a maximum daily change of 6 to 8°C. Precautions such as these should be taken when the fry leave the hatchery. The change in temperature can be effected progressively by adding water from the future rearing tanks, whose physical and chemical characteristics may be different from those in the hatchery. This progressive transition avoids osmotic and ionic shocks and makes it possible to observe the behaviour of the larvae in their new rearing environment (p. 132). If handling, transfer, cleaning and control of flows are badly managed mortalities are likely to occur, either directly or indirectly or there may be malformations of alevins or fry which have been subjected to shock.

Chemical factors (p. 39)

The chemical quality of the water in a hatchery is based on a series of equilibria between different substances and dissolved or undissolved gases in the water.

Minerals affect water quality through their buffering capacity but excesses or deficits can be limiting for certain species (e.g. chlorides for halophilic fish species).

Problems of deoxygenation are directly linked to temperature; up to 12–15°C the risks are limited but at temperatures which are regarded as being physiologically optimum for cyprinids, oxygen demand is raised while the organic wastes increase and encourage the spread of bacteria (flavobacteriaceae or others), protozoa and fungi.

Elsewhere, water quality can be unfavourable for alevins if there are wide fluctuations in CO_2, particularly in pond water:

—an excess of CO_2 in the water is prejudicial to the survival of fry which cannot eliminate CO_2 from the blood, the CO_2–O_2 exchanges in the blood and gills are limited, the fish experience acidosis, become lethargic, breathing rate speeds up, appetite is lost and eventually death ensues.

—a deficit of free CO_2 can also cause problems and may occur when photosynthesis is active; in these conditions the fry which are incapable of regulating their acid-base equilibrium through respiration lose CO_2 transcutaneously which leads to a fatal state of acidosis. In such a situation the fry come to the surface of the water and die showing symptoms of asphyxia even when oxygen is available.

When the larvae begin to take in food they produce metabolic wastes, among which are nitrogen compounds, and several problems have been noted which lead to abnormally high-mortality levels and mediocre-growth performance where there is excess NH_3 (0.14 to 0.40 mg/l NH_3 can be dangerous for carp fry).

Biological factors

In intensive larval rearing, the biological quality is generally stable as the techniques which have been developed most frequently "neutralise" the biological factors in the natural environment through warming, filtering or the actual source of the water (e.g. from a well). However, the biological filters in recirculating systems can develop an abundant flora and fauna which, under some circumstances, colonise the rearing tanks containing the larvae and fry causing cutaneous lesions and septicaemias leading to the onset of infectious diseases. In addition, in intensive systems, larvae and fry are directly dependent on the quality of the feed that is distributed (deficits or excesses of macro-elements, oligo-elements and the presence of pollutants).

In natural or semi-natural outdoor-pond-rearing systems it is essential to manage the food web effectively (p. 126). Painstaking precautions should be taken to avoid contamination by predators such as dytiscid beetles and their larvae and some species of cyclops.

5.4.2 Infectious diseases

Along with the mortalities induced by various environmental factors, there are mortalities associated with infective agents (bioagressors), which may be parasites, viruses and/or bacteria. It is important to remember that, apart from intensive rearing units, many known infective agents are an integral part of the environment (with the general exception of viruses): the consequence of this is that if these agents are detected in the unit the factors which have caused their "explosion" should be investigated and effective counter-measures put in place. Two hypotheses have been advanced:

—either a new infective agent has been introduced in sufficient quantity to produce morbid effects;
—or the infective agent is already present in the environment and some factor has made the fish susceptible to it.

In this review, for convenience, diseases have been divided into those associated with viruses, bacteria and parasites including fungi.

Viruses

At present viruses are seldom a problem in the larval stage, although rhabdoviruses can cause massive mortalities with haemorrhagic septicaemia lesions. Catfish fry have been shown experimentally to be sensitive to the rhabdoviruses causing spring viraemia of carp and, on farms, to iridovirus. Mortalities occurring during outbreaks of viral diseases are generally massive and occur rapidly. Lesions are initially discrete and haemorrhagic and are followed by signs of damage to the central nervous system. However, such observations are still unusual because of the limited intensive rearing of carp and the fact that disease outbreaks involving viruses may not have been properly studied.

Bacteria

These pathogens can be divided into two groups comprising specific pathogens such as *Aeromonas salmonicida novi* and opportunist pathogens such as *Aeromonas hydrophila* and *Pseudomonas* sp.

The first of these is the causative agent for cyprinid erythrodermatitis and lesions appear progressively after stress or during contamination from the exterior. The second type are ubiquitous bacteria which cause septicaemias and mortalities which occur with varying degrees of rapidity and scale in relation to the underlying cause. The lesions are those associated with septicaemias and to a greater or lesser degree petechial haemorrhages (haemorrhagic spots, cutaneous ulcers). These may (or may not) join up, developing into ulcers which are at first restricted to the skin but then spread to deeper tissues such as the underlying musculature. Gills are often discoloured, the spleen and kidneys swollen and haemorrhages visible on the internal organs. In addition, exophthalmia (swollen and protruding eyes) and ascites sometimes occur.

Apart from septicaemial bacteria, certain conditions allow bacteria from the environment to colonise the skin, the external epithelium and sometimes the internal organs of the fish. For many years the bacteria which contribute to the breakdown of organic matter have been called myxobacteria. This term has now been dropped in favour of the term "flavobacteriaceae" which is better adapted to the taxonomy of the bacteria and comprises also *Flexibacter*. These bacteria appear to be highly opportunistic and colonise fish that for various reasons have developed a thickened mucus coat. Such reasons include:

—a heavy burden of external parasites;
—a nutritional deficit;
—an environment with a heavy load of suspended matter which damages the skin and the fins (fin-rot) and, especially, the branchial (gill) epithelium (gill-rot).

These bacteria are found in young fish and are associated with "gill disease" and also cutaneous lesions; they digest the mucus and then attack living tissue causing serious lesions in the gills with oedema, haemorrhage, asphyxia or on the skin with cutaneous ulcers. For several years the flavobacterium have been found associated with septicaemias in cyprinids but their etiology is still not fully understood.

Parasites

Parasites belong to a range of taxonomic groups: those that affect larvae and fry are distinguished by their location either on or in their host.

External parasites

Intensive rearing units fed by spring or ground water are rarely affected by infesting organisms that damage the young fish. However, it is not unusual for contamination to allow the development of the protozoan *Ichtyobodo necator*, the agent of costiosis which develops on the skin and gill epithelium, leading to pruritus, signs of asphyxia and death (Figure 5.3).

Once in ponds or tanks larvae and fry are likely to encounter a variety of protozoans such as *Trichodina*, *Ambiphrya*, (ex-*Scyphidia*), *Apiosoma* (ex-*Glossatella*), *Epistylis*, *Phagobranchium* (ex-*Trichophyra*). *Ichthyophthirius multifiliis* is particularly dangerous on its own at high temperatures (>20°C). While the first of these parasites only develop significantly on weakened individuals, the last one (*Ichthyophthirius*), because of its localisation under the epidermis, particularly on the gills, can lead to pruritus and lethal asphyxia, with or without the presence of white spots (Figure 5.4).

It is also possible to find a flagellate, *Cryptobia branchialis*, on the gills. This is more of a passenger than a serious parasite but as with all other parasites it can proliferate if the fishes' defences are impaired.

In older fish, the risks from monogeneans seem to increase with *Gyrodactylus* sp. (Figure 5.3) and *Dactylogyrus* sp. and, to a lesser extent, *Diplozoon* sp. Concentrated on the skin, fins and gills, *Gyrodactylus* irritates the affected areas leading to erosion of the fins and gills and hyperproduction of mucus, leaving affected individuals in a weakened state. *Dactylogyrus* and *Diplozoon* are specific to the gills where they cause irritation, and the production of large amounts of mucus with subsequent asphyxia.

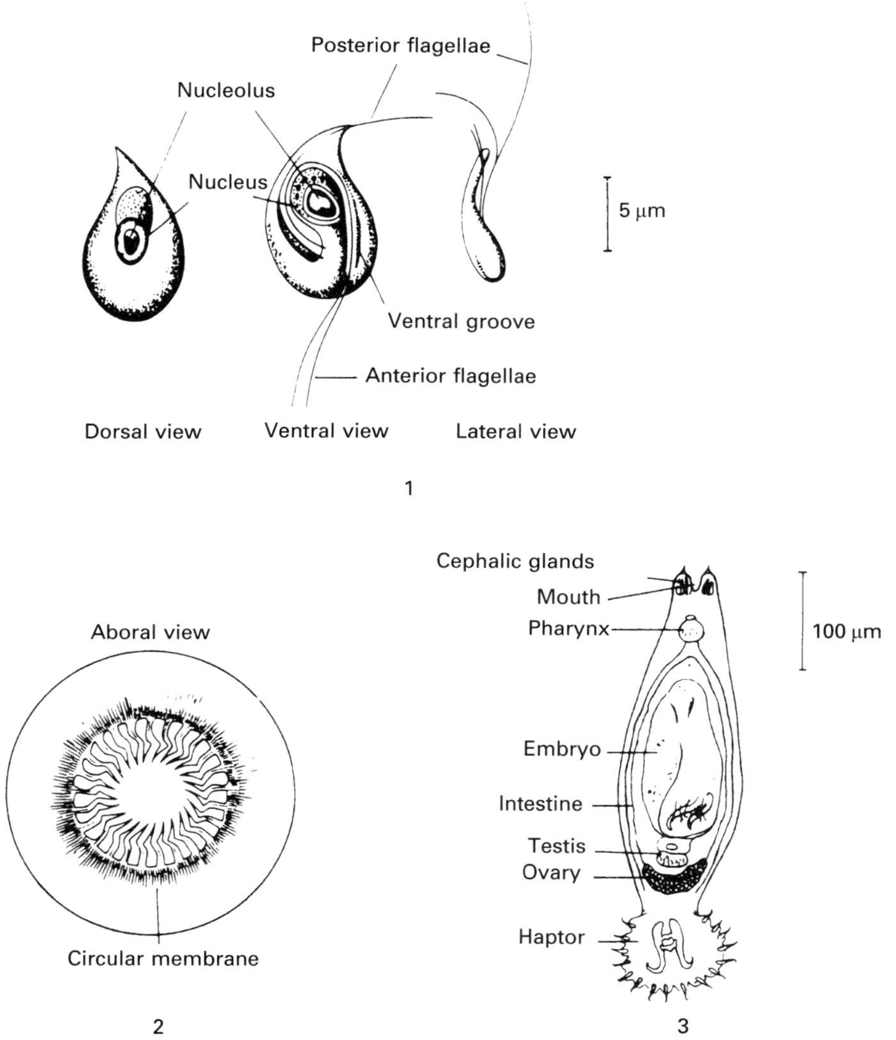

Fig. 5.3. Diagrams of cyprinid ectoparasites: (reprinted from De Kinkelin *et al.*, 1985)
1. *Ichtyobodo necator* = *Costia necatrix*; sized according to Lom *et al.* (1989): 4–6 × 9–12 μm.
2. *Trichodina* sp.; diameter 20–40 μm after Paperna, (1982).
3. *Gyrodactuylus* sp. (0.2 × 0.9 mm, after Bykovskii, 1957).

Finally, various large parasites can infest fry. These include the leeches *Geometra piscicola* and others and crustaceans such as *Argulus* sp., *Ergasilus* sp., etc. These relatively large parasites pierce the host to withdraw blood; this may provide a route for the entry of viruses, bacteria or parasites, in addition to the initial damage caused.

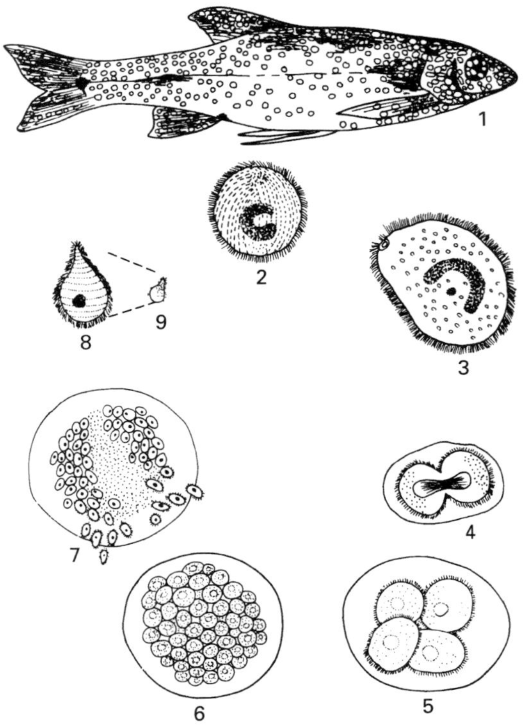

Fig. 5.4. The life cycle of *Ichthyophthirius* sp.
1. Infected fish, 2. Troiphozoite, 3. Adult troiphozoite (0.6–0.8 mm), 4 to 6. Divisions, 7. Free tomites (30–40 µm), 8–9. Infective tomites.

Alongside the parasites from the animal kingdom are those from the plant kingdom. These are the fungi or mycoses, which frequently infect lesions already present on the skin or the gills.

Internal parasites (Figure 5.5)

In young fish, internal parasites, wherever they are located, are at the start of their infestation and their effects are thus difficult to recognise. Such parasites are more dangerous in older fish because of their growth and proliferation. In addition it is not always certain whether the internal parasites which have been found and described in fish always exhibit genuine pathogenicity.

The internal parasites which infest juvenile fish have planktonic larval forms, e.g. *Bothriocephalus acheilognathi* and *Ligula intestinalis*, or have larvae which penetrate the organisms actively either through the gills (*Sanguinicola inermis*) or through the skin or eyes (*Diplostomum* sp.). A process which may be similar results in infestation by *Sphaerospora renicola*, a parasite of the blood and internal organs, which has been identified by some workers as the causative agent of inflammation of the swimbladder, a disease which can cause mortalities in June and July. Such mortalities have not yet been reported in France although the parasite is present.

Among parasites that have been found in juvenile stages of fish, mention should be made of the myxosporidians such as *Myxobolus* sp. and *Thelohanellus nikolskii* and the microsporidians. The latter can invade a range of organs, causing the formation of cysts of varying size but rarely affect survival, except for the various forms of *Myxobolus* that develop in the skeleton.

Prevention and treatment

The best methods of preventing disease depend on providing the larvae and fry with optimum conditions to avoid being subjected to stress.

Prevention

1. Use materials adapted to the size of the larvae and juveniles and disinfect regularly (see below).
2. Never introduce any broodfish which have not been treated to remove parasites and certified as bacteria and virus-free.
3. Use good-quality water and monitor the principal parameters and, possibly, add-mineral salts to increase the buffering capacity.
4. Do not mix different species and age classes.
5. Use self-cleaning tanks for indoor rearing and only use good-quality feed, kept under good conditions (protected from light, heat and humidity) and use up rapidly (in less than 2 months) (p. 139).
6. In outdoor ponds (p. 126):
   —Disinfect the ponds with quicklime and use clean water for filling, if possible filter through a gravel bed, to limit contamination with predators or the intermediate hosts of parasites.
   —Ensure that appropriate organic or mineral fertiliser is applied and destroy predatory plankton several days before stocking the fry into the ponds (p. 130).
   —Use high-quality artificial food where there is a deficit in the food web in the ponds (p. 139)
   —Ensure the turnover of the water, an optimal level of oxygen (6 to 8 mg/l) and a minimal level of NH_3 throughout the day.
7. The biomass in the tanks or ponds must always be compatible with the size of the fry and the quantity of food available.
8. A complete examination for parasites, bacteria and viruses is needed during the rearing stage and before transfer to the on-growing systems in order to avoid contamination and to carry out any treatments needed.

Disinfection

This concerns personnel, equipment, rearing structures and sometimes the water itself. It should be remembered that disinfection is only effective if clean material is in

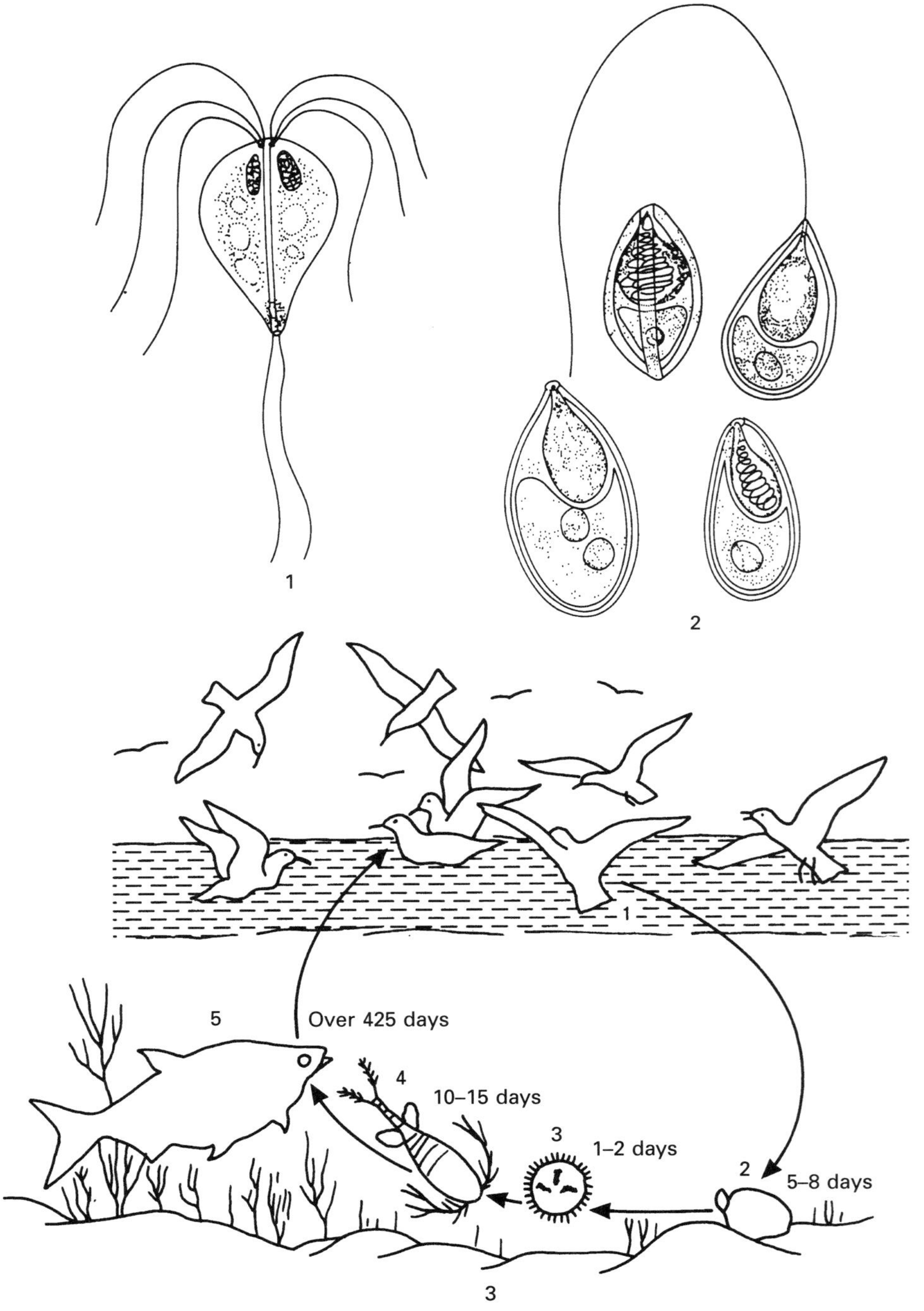
1
2
1
5
Over 425 days
4
10–15 days
3
1–2 days
2
5–8 days
3

contact with a newly prepared product for sufficient length of time. Doses are expressed here in mg/l (equivalent to ppm, parts per million).

Personnel. Measures to be taken are chiefly concerned with hands, boots and clothes. Before any contact, the hands of operatives (with or without gloves) should be disinfected with:

• iodophores	100–200 mg/l or
• quaternary ammonium compounds	1000 mg/l

These two products are highly effective disinfectants which do not damage hands; the former are effective against Infectious Pancreatic Necrosis (IPN) which may be carried with no symptoms but they are toxic for fish.

The best way to disinfect boots is to install a footbath at the farm entrance and even a bath for disinfecting the wheels of lorries and other vehicles that come onto the farm or move between rearing units. The disinfectant solution should be renewed often; the frequency being a function of the location and of the usage. The footbath should be protected from rain; everyone should be forced to pass through it to enter the part of the farm to be protected. Products suitable for use on clean boots are:

• quaternary ammonium compounds:	2000 to 5000 mg/l for 1 min
• calcium hypochlorite:	120 mg/l of chlorine/20 s or 0.06 degree on the chlorometric scale
• sodium hypochlorite:	180 mg/l of chlorine/20 s or 0.06 degree on the chlorometric scale

These three products are highly active but the two hypochlorites break down spontaneously especially in the presence of organic matter. It is good practice to renew solutions every 4 or 5 days; hypochlorites are dangerous to fish.

For clothes, it is advisable to keep special clothes for sensitive areas such as the hatchery and the nursery to avoid the risk of contamination.

Fig. 5.5. Internal parasites of cyprinids.

1. *Hexamita* sp. (7 to 12 μm, after Paperna, 1982).
2. *Thelohanellus catlae* (spore: 18–25 × 10–13 μm, after Bykhovsjaya-Pavlovskaya *et al.*, 1962).
3. *Ligula intestinalis*: life cycle (simplified, after Dubinina, 1953); (1) development, in birds of the adult ligula in 2–5 days after ingestion; (2) capped egg eliminated via the faeces; (3) free-swimming coracidium stage, released from the egg in 5–8 days; (4) coracidium ingested by an intermediate host (planktonic copepod) developing into a procercoid in 10–15 days; (5) a second intermediate host (fish) ingests the copepod with the procercoid which is transformed into a plerocercoid stage which develops into an adult in the bird that eats the fish.

Equipment. In order not to maintain and spread pathogens within the rearing unit and from one rearing unit to another, all equipment must be disinfected (nets, boots and other small items). These may often have metallic parts that may oxidise and any active product (mainly hypochlorites) which cannot be used frequently and must always be rinsed away with fresh water. This must not be allowed to run off into stocked ponds, as such products are toxic to fish:

- quaternary ammonium compounds: 2000 mg/l for 1 min
- calcium hypochlorite: 180–200 mg/l of chlorine/20 s; or 0.06 degrees on the chlorometric scale;
- formalin: in some circumstances this can be used as a vapour; this requires an airtight vessel.

Rearing equipment. Zoug bottles and other structures for egg incubation and larval rearing, walls and pipes should be disinfected by running disinfectant through in the absence of fish: any disinfectant compatible with the material can therefore be used:

—steam under pressure;
—gaseous disinfectants: chlorine, formalin;
—liquid disinfectants: aldehydes, iodophores, quaternary ammonium compounds.

It should always be remembered that disinfectant is only effective on clean surfaces and that discharges containing the products must be treated.
Ponds should be cleaned out and disinfected with lime before juvenile fish are introduced. In some places, molluscicides and compounds that ensure the destruction of intermediate parasite hosts or other unwanted creatures should be applied. Rotenone can be used a few weeks before the juveniles are stocked.

Water. In some places the water supplying either open- or closed-circuit farms must be treated. In open systems the problem is linked to the volume to be treated and the clarity of the water. In both open and closed systems ultra-violet light is the only treatment available at present; this is effective even on viruses. Chemical products can only reduce the bacterial load at the water inflow, but the active dose must not be toxic to eggs, larvae or juveniles. In a closed-circuit system, chemicals are dangerous to the active organisms in the biological filter.

Treatments

Treatments can be given systematically throughout the rearing period or as curatives when mortalities occur. It is important to keep in mind that the use of any treatment is a sign of technological failure and that treated individuals will be weakened. In economic terms it is difficult to treat and remain profitable.

The main principle is to instigate systematic treatments, called preventative treatments, which have been designed after analysis of the risks associated with the system. These will take into account the origin of the water and its treatment prior to arrival in the hatchery, the mix of species and age classes, the quality of the broodfish, etc.

Another principle to be taken into account when calculating therapeutic doses is that the efficacy and toxicity of products used varies according to the quality of the water (temperature, pH, hardness, etc). Doses indicated are average doses and trials should be carried out to define, for a given farm, the optimum dose to be used against various diseases. This dose will vary according to the season and the species of fish being farmed. It is essential that the dose and duration indicated are only taken as a guideline; a bath treatment can be used for fry but if there has been a diagnostic error or the fish are particularly fragile, the mortality rate may be high. Consequently, the behaviour of individuals should be closely observed so that immediate intervention can be made if the fish show any signs of distress.

Treatments lasting only a few minutes can be given by shutting off the water at the inflow; in other treatments it is necessary either to provide a supply of oxygen for the duration of the treatment, or to use a dose adapted to the flow of water (a running bath).

Systematic preventative treatments. Such measures can be taken from the incubation of eggs and during the first stages of rearing the larvae; they are simple to apply in closed circuits, open circuits and natural-type ponds.

These systematic treatments are directed against ectoparasites including fungi and use various disinfectants and antiseptics; the ease of application varies according to the nature of the rearing system.

Curative treatments. When disease breaks out in juveniles it is a sign of a failure of preventative measures or a technological breakdown. Once mortalities have begun to occur it is likely that the overall survival rate of the batch will be low and that the survivors cannot be reared profitably. In order for proper treatment to begin, it is essential that an accurate diagnosis be made before any therapeutic intervention is carried out. This should be achieved by examining both sick and healthy individuals. At the same time as a specific treatment is initiated, an epidemiological study should be performed to determine the predisposing causes which have favoured the development of the disease organism.

The methods used to cure diseases utilise the same substances as those used for their prevention, at the same dose rates but at a higher frequency, according to the severity of the outbreak and the state of the fry. The frequency can be daily on two or three consecutive occasions. To destroy crustacean parasites (*Argulus*, *Ergasilus*, etc.) various highly effective insecticides can be used. The treatments have to be repeated two or three times and some of the plankton will be destroyed (p. 130). The treatments must be used with care. When the presence of a bacterial disease has been demonstrated to be present various antibiotics are available for use:

—either as a bath (but they are dangerous for the environment);

Tanks and raceways

- Anti-fungal

—eggs	malachite green	5 mg/l/1 h
—larvae/fry	malachite green	1 mg/l/1 h

- Anti-protozoan

—formalin (30–40%)	150 mg/l/15 min
—formalin (1 l) + malachite green (4 g)	25 mg/l/5–6 h
—chloramine T	10 mg/l/1 h
—sodium chloride (kitchen salt)	20 g/l/15–20 min
—in association chloramine T	10 mg/l/1 h
+ malachite green	0.5 mg/l/1 h

Chloramine T is broken down by organic matter and, in some situations, doses should be multiplied by 2 to 5.

- Anti-flavobacterium of gills and skin (gill-rot and fin-rot)
  —specific measures:

—chloramine T:	10–50 mg/l/1 h
—quaternary ammonium compounds	2 mg/l/15–30 min

  —complementary measures: improvement of the quality of feed (for example adding vitamins), improvement of rearing structures, reducing density, self-cleaning tanks, etc.

Earth ponds

In such rearing systems, systematic treatments are much more difficult because of the volume to be treated and the low turnover rate of the water. It is therefore necessary to use products that are easy to spread at an acceptable cost: in practice these do not exist. It may be possible to use the products listed above if the level of water can be lowered to decrease the volume to be treated and then allowing fresh water back in. This is unfortunately not always possible.

- Anti-fungal

—malachite green	1 mg/l/1 h

- Anti-protozoa: in combination, once a week
  —malachite green 4 g + 1 l formalin (see above)

In practice, this type of rearing pond (used for the first 4–5 weeks) should only be supplied with bioaggressor-free water because of the difficulties involved in applying treatments.

—or in feed granules which will need to be given even if larvae are being reared on plankton alone.

CONCLUSION

Fingerlings and fry are very fragile and, in intensive rearing systems, are extremely sensitive to the least stress and the smallest quantity of pathogenic organisms. Successful rearing therefore depends on the quality of the rearing system and on rigorous husbandry. This can prevent disease and should be applied scrupulously after all of the risks associated with the rearing system have been identified.

BIBLIOGRAPHY

Bykhovskaya-Pavlovskaya I.E. *et al.*, 1962. Key to parasites of freshwater fish of the USSR. *Israel prog. Scient. Transl.*, 919 p. Transl. from Russian.

Bykoskii B.E., 1957. In Bykhovskaya-Pavlovskaya I.E. *et al.*, 1962.

Dubinina M.N., 1953. In Bykhovskaya-Pavlovskaya I.E. *et al.*, 1962.

Charlon N., 1990. Technologie: Automatisation de l'élevage larvaire en eau douce. *Aqua Revue*, **33**, 19–24.

Charlon N., Bergot P., Escaffre A.M., 1986. Alimentation artificielle des larves de carpe (*Cyprinus carpio* L.) *Aquaculture*, **54**, 83–88.

Paperna I., 1982. *Parasites, infections et maladies du poisson en Afrique*. FAO, CPCA, Doc. Tech. (7), 202 pp.

6

On-growing in ponds

6.1 TRADITIONAL EUROPEAN METHODS FOR REARING FISH IN PONDS[1]

The methods of rearing and the types of ponds used for rearing fish for human consumption or for restocking have been described in many publications on fish culture (listed at the end of this book in the General bibliography). The system of pond rearing described below is based on the culture of the common carp, because this fish is most widespread and best understood in Europe.

6.1.1 Different types of pond

In the oldest production systems, the phases of spawning, growth and overwintering take place naturally in the same ponds but in different physical zones (Figure 6.1). In more modern farms these zones have been recreated in separate ponds; the spawning pond is shallow and rich in macrophytes (p. 78), the overwintering pond is the deepest and the ponds used for on-growing are of intermediate depth. Rationalisation of production has also led to the division of growth into different phases and the separation of fish in the first, second and third years of growth. The idea of grouping the fish by age is ancient (in Europe this was pioneered by the Czech, J. Skala (Dubravius) 1488–1553) and is traditional in China with the "multigrade conveyor system". The succession of ponds used in the different growth phases is shown in Figure 6.2. The management of fry ponds is described on pp. 126-135.

The production cycle in on-growing ponds extends over 3 years: the ponds stocked with C_1 and C_2 carp are sometimes referred to as "grow-out ponds" but their characteristics are not very different from the fattening ponds where the fish are grown to commercial size. Ponds have very different characteristics especially with regard to their surface areas which may vary from a few to several hundred hectares.

---

(1) R. Billard.

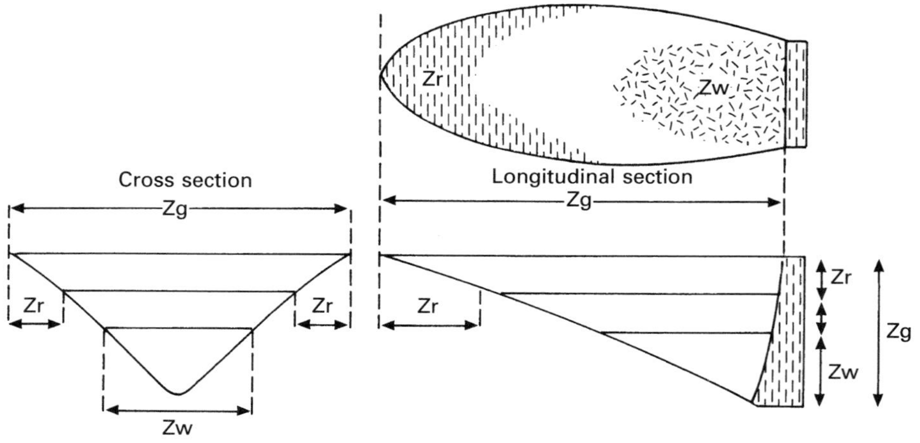

Fig. 6.1. Areas of the pond used for reproduction (Zr), growth (Zg) and overwintering (Zw) (after Voican *et al.*, 1975).

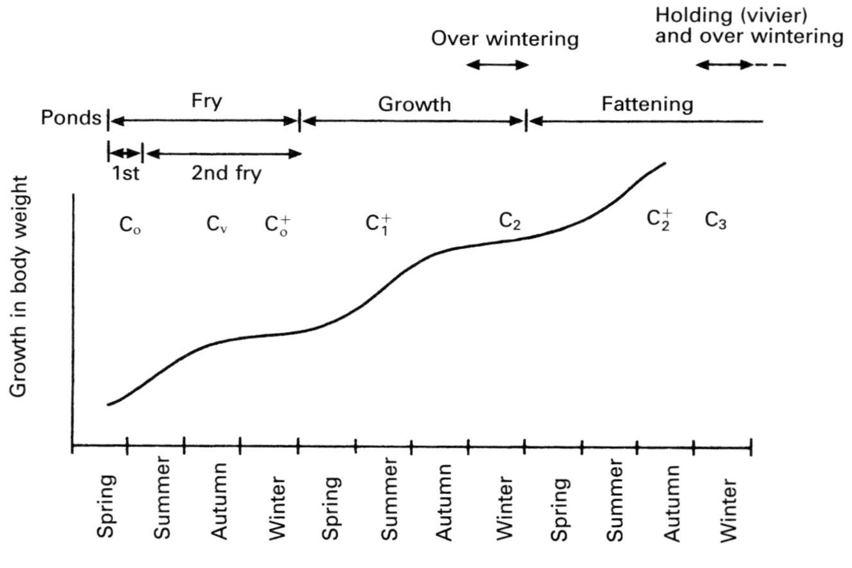

Fig. 6.2. Diagram of growth curves and types of rearing pond (3-year cycle). The abbreviations C_o–C_3 are derived from German terminology; C_o–C_v: yolk-sac fry, C_1–C_2–C_3: 1, 2 and 3-year-old fish; C_o^+ C_1^+ C_2^+ designates the fish at the end of the growth season in autumn (summer for C_o^+). The letter C is used for carp, T for tench, P for pike, etc.

Depth also varies but the best production comes from ponds with a mean depth of 1.5 m. The stocking rate of fish depends on the richness of the water in terms of nutrients, the method of management, the degree of fertilisation and the stocking formula used to achieve the expected production. This presupposes a thorough

Table 6.1. Example of the formula used for stocking carp in extensive rearing (2nd and 3rd year): $N = (R/p') + m$; $p' = f - p$.

Year	2nd	3rd
Predicted production (R) g/ha/year	600 000	600 000
Weight of individuals		
Initial (p) g	20	250
Final (f) g	220	1250
Gain per animal (p') g	200	1000
Mortality (m) %	10	5
Number to be stocked (N)	3300 C_1/ha	630 C_2/ha

knowledge of the productivity of each pond. An example of a stocking formula is given in Table 6.1.

Overwintering and holding ponds are further categories of pond type. The wintering ponds are only used in cold, snowy places or where the water is covered with ice which is several tens of centimetres thick in winter. The C_1–C_2, C_2–C_3 fish and broodfish are stocked here but not the C_0–C_1 fish which are more often left in second-fry ponds which have a strong flow of water passing through. The overwintering ponds are deep (1.8–2.5 m) and have a high water turnover rate to ensure oxygenation. The ice which forms on the surface prevents the exchange of gases with the atmosphere. Holding ponds have the same characteristics and are used to keep fish between harvesting and marketing. They should be readily accessible and are generally situated alongside a grading unit and a station for weighing or processing. Another distinctive feature of these ponds is their small area and the high concentration of fish. These fish are not stressed: in large lakes free-living carp tend to concentrate in favoured spots in winter when temperatures fall below 8°C.

In the most common type of rearing system used for cyprinid culture each farm has its own hatchery to supply juveniles. This is likely to change in the future with the development of schemes for genetic improvement and will have a similar pattern to the poultry industry, with the separate management of the selection of broodfish, production of juveniles and on-growing (p. 101). The size of the hatchery depends on the duration of the production cycle which is generally 3 years in temperate regions, although this may be different if carp are marketed at a weight of 1 kg after a rearing cycle of 2 years. For example, in a 100 ha farm:

—10 ha are given over to the production of 1-year-old juveniles, of which:
- 3 ha are used for the first-fry stage (two ponds at 1 ha and five small ponds of 2000 m^2 which also serve for overwintering)
- 7 ha for the second-fry stage (two ponds at 2 ha and one at 1 ha, for example)

—20 ha for 2-year-old fish
—70 ha for the final on-growing phase.

6.1.2 Fertilisation of the ponds with minerals

Several observations and experiments have shown that adding fertilisers increases the productivity of ponds; this has a beneficial effect on all trophic levels. It has been shown that the limiting factor for primary production, particularly the chlorophaecae (which are most important for the production of daphnia), is nitrogen and that a N/P ratio >4 or 5 should be maintained (Sevrin-Reyssac and Pletikosic, 1990). Values as high as 8 and even 11 have been suggested (Opuszynski in Michael, 1987) (see General bibliography). Levels of nitrogen in the water decrease more rapidly than those of phosphorus as the production season progresses. In some examples, phosphorus may even show a tendency to increase (Figure 6.3). Many observations have shown that the levels of nitrogen have a direct influence on fish production. It appears that algae exhaust the nitrogen in the water before the phosphorus. It should also be remembered that phosphorus can also be released into the environment

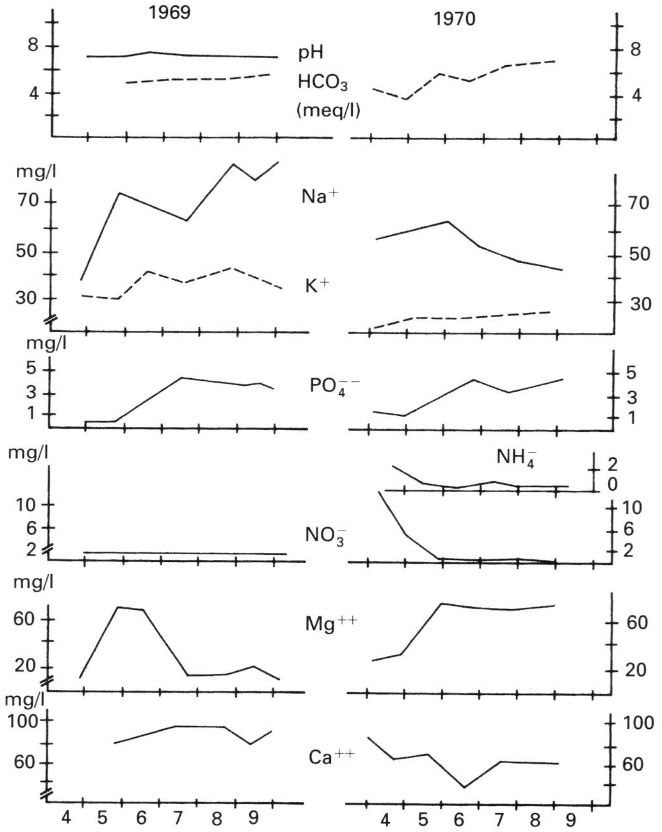

Fig. 6.3. Change and level of ions in the waters of the Moravian pond over 2 consecutive years. The pond was dried out between the autumn of 1969 and spring 1970 (after Ulehlova and Pribil 1978, reprinted in Billard 1980).

under various influences (e.g. changes in pH and water movements of water under the influence of wind-generated currents). These patterns are not always identified by taking spot measurements of phosphorus concentration in the water. The decision to add fertiliser and the application rate depend on a range of factors, in particular the richness of the environment, the method of management, the state of the community and the anticipated level of production. The fish farmer should first assess the existing N and P levels in the water; this will give an indication of the optimum rate of fertiliser application. Minimum values, above which grow-out ponds should be fertilised are 0.2 to 0.3 mg/l for PO_4^{3-} (inorganic phosphorus) and 1.5 to 2 mg/l for nitrogen. In more intensive production, using polyculture, levels between 3 and 5 mg/l for nitrogen are accepted (total inorganic nitrogen).

Once these figures (particularly those for phosphorus) have been taken into account, other factors should be considered. One of these, the Secchi disc, gives information not only on the productivity of the environment (by assessing algal density) but also the state of the whole community. Measurements should be taken when the water is low in minerals or in suspended organic matter or detritus (in calm weather). A high concentration of algae (transparency <25 cm as measured by the disc) shows that the water is eutrophic and any fertiliser applied would be wasted and might even be dangerous. This also indicates whether cladocerans, particularly the large daphnia, are low in numbers (or even absent) and that predation on them by fish such as carp is too high. At transparencies between 25 and 35 cm applying fertiliser may be beneficial and above 35 cm it is essential to stimulate production and to maintain a suitable level of dissolved oxygen. Water temperature must also be taken into account; mineral fertiliser has little effect in winter and is only used at temperatures between 15 and 25°C (taken at midday). Above 25°C, applications should be divided into several doses and the development of the water quality should be monitored. Fertiliser is not advised when the pH of the water is greater than 9. Other parameters such as the nature of the soil and its fertility must be considered. In addition, the behaviour of the fish and their growth should be assessed.

Application rates are thus extremely variable and depend on a wide range of parameters which must be brought together by the fish farmer. The farmer will take into account the history of the pond and its previous performance and refer to records of significant events and the production from each pond. Application rates of fertiliser can reach 200 to 400 kg ammonium/nitrates (30–33% nitrogen) and 100 to 300 kg superphosphates (16% P_2O_5) for intensive production. Table 6.2 shows the rate of application used in the former Czechoslovakia: fertiliser is spread twice and urea, which has a different effect, prolongs the effect of the treatment. In Hungary, Horvath *et al.* (1984) (see General bibliography) spread fertiliser according to the type of soil (Table 6.3). In France, Martin (1987) used 30 to 60 kg of nitrogen in solid form (90 to 160 kg/ha ammonium nitrate 33%) and 30 to 40 kg/ha nitrogen in liquid form (50 to 80 l of 39/0/0); for P_2O_5 doses are 40 to 60 kg/ha in a solid form (200 to 250 kg of Super 18) and 40 to 50 kg/ha in liquid form and should be divided and spread throughout the season and adjusted in relation to demand: spreading twice a month is common practice. Distributions of liquid fertiliser can be continuous, making use of reservoirs located around the edge or upstream of the pond. Where

Table 6.2. Levels of nitrogenous and phosphate fertilisers to be applied to on-growing ponds (from Billard and Marcel 1980).

Level of the element in the water (mg/l)	Application rates for a depth of 1 m* (kg/ha/application)					
	High rate at the beginning of the season (April–May)			Low rate at the end of the season (June–July)		
Inorganic phosphorus	Units of P	Superphosphates†		Units of P	Superphosphates†	
0	3	42		2	28	
0.1	2	28		1	14	
0.2	1	14		0	0	
0.3	0	0		0	0	
Inorganic nitrogen‡	Units of N	Amm nitrate. 30% N	Urea 46%	Units of N	Amm nitrate. 30% N	Urea 46%
0 (0)	15	50	33	10	33	22
0.5 (0.25)	10	33	22	5	17	11
1.0 (0.5)	5	17	11	0	0	0
1.5 (0.75)	0	0	0	0	0	0

* For deeper ponds application rates are calculated using a linear coefficient up to a depth of 1.5 m (coef. 1.1 for 1.1 m, 1.2 for 1.2 m, etc.) above 1.5 m it is considered that photosynthesis does not convert minerals into organic matter efficiently.

† Superphosphate at 7% P (16% P_2O_5).

‡ In brackets, level of ammoniacal nitrogen in the water.

Table 6.3. Annual dose rates of fertiliser used in Hungary for ponds excavated in different types of soil. (Horvath *et al.* 1984)*

Types of soil	Nitrogen fertiliser (kg/ha) Ammonium nitrate (25% N)	Nitrogen fertiliser (kg/ha) Urea (carbamide; 46% N)	Superphosphate (kg/ha)
Boggy	150	70	200
Sandy	150	70	150
Sodium rich	152	100–200	300
Compact, poor	300	150	400
Compact, fertile	350	180	500

* Cf. General bibliography.

liquid fertiliser is applied over a long period it is good practice to release it when winds are causing water movement in the pond. Methods of spreading fertiliser and the equipment used are described on p. 233.

Aside from the limitations imposed by temperature, pH and environmental richness, mineral fertilisers are not recommended when production is highly intensive (i.e. using artificial diet and high water-turnover rate) unless it is necessary to stimulate the production of phytoplankton to limit the levels of NH_4^+ (p. 177). It is essential to monitor the status of the pond and the community and, based on the results, adjust the rate of fertiliser application and if necessary use complementary feeding.

We have already described how organic manure increases the productivity at all trophic levels. To gain maximum benefit from the application of fertilisers it is necessary to exploit them by using a combination of species; this is polyculture and will be described later. To the problem of species diversification can be added that of the optimum predator–prey ratio. In operational pond conditions there is a major increase in zooplankton in the spring but this is of short duration and the biomass of zooplankton decreases in summer. This decrease is associated with a high degree of predation by fish on the zooplankton (particularly the largest forms). The zooplankton is thus unable to make best use of the phytoplankton, the production of which increases because of the application of the manure. In order to maintain satisfactory fish growth, feed in the form of cereals and granules must be given (Figure 6.4).

6.1.3 Organic fertilisers

Whether it is organic or mineral, the fertilisation of the ponds used for fish culture is part of a system of aquaculture which is similar to the farming of mammals on pasture; the animals consume production which is generated within the system (endogenous production). The application of organic fertiliser is more complex than

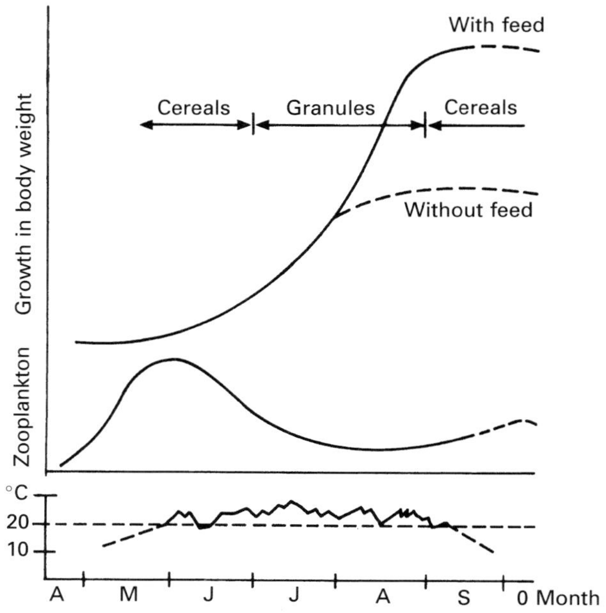

Fig. 6.4. Growth of carp in a pond and availability of food in relation to water temperature (from Billard and Marcel, 1976).

that of inorganic fertiliser but the basic materials are cheaper and the effectiveness several times better (see Figure 2.14). It also generates a more complex trophic web which Chinese fish farmers have sought to optimise through polyculture, the stocking of several species of fish into a pond, each of which feeds on a different level of the trophic web. Traditionally, these are phytoplankton consumers (silver carp), zooplankton consumers (common carp or bighead), benthic feeders (common carp or black carp which feeds selectively on molluscs). The performance of organic fertiliser is considerably improved by polyculture which can only be justified when this type of fertiliser is used (Table 6.4).

Organic fertilisers and polyculture are strongly based on the wider concept of the integration of agriculture and aquaculture. Aquaculture activities make use of land which is low in productivity or not utilised, share equipment and recycle wastes from agriculture as food for the fish (cereals, oil-producing crops). This concept is particularly widespread in developing countries such as China but shows great promise in richer countries which are faced with problems of a decline in agriculture and its negative environmental impacts. However, there are other problems, such as the difficulty of establishing the system and the management of the environment, particularly within large bodies of water. In addition, there are difficulties with the health status of the product from highly fertilised waters and of setting up complex associations in developed countries (availability of juveniles in the required numbers and problems of grading at harvest). In practice, the expertise required to manage

Table 6.4. Comparison of the performances of polyculture systems in either ponds treated with organic fertiliser (Chinese method) or unfertilised ponds (Taiwanese climatic conditions, from Billard 1991).

		Production (kg/ha/yr)	
Fish	Principal diet	Fertilised pond	Unfertilised pond
Perch	Fish	170	13
Common and black carp	Benthos, molluscs	910	34
Grass carp	Macrophytes	263	73
Bighead carp	Zooplankton	736	25
Mullet	Zooplankton	2502	17
Silver carp	Microzooplankton + phytoplankton	2706	262
Total production		7287	424

such systems can only be found in Asian countries, particularly China. Here there are adapted structures such as small ponds, between 3000 and 10 000 m^2, equipped with aerators, whose banks are accessible to vehicles.

Chinese techniques were copied in Central Europe in the 1950s and then in Israel in the 1960s where they were rationalised. From there they were reintroduced and spread through Europe. In Central Europe, France and Germany, a light application of organic fertiliser is traditionally given but because of the difficulty of spreading from inaccessible banks it is added when the ponds are dried out. Small heaps of composted cattle manure are deposited on the bed of the pond at a rate of a few tonnes per hectare per year.

Organic fertiliser is mainly spread in summer during the growth period but it can also be added in winter at a lower rate to maintain a small production of plankton: this appears to benefit the fish (p. 29). Numerous factors affect the summer application rate; these include climatic conditions (particularly temperature) and the degree of intensification which is itself linked to the size and structure of the ponds and the equipment in use, particularly aerators. In Israel, between 1974 and the beginning of the 1980s, a long series of experiments in small ponds (400 m^2) with polyculture systems including common, silver and herbivorous carp and sometimes tilapia and macrobrachium showed that high application rates of various types of manure from cattle and poultry can be added to ponds: for example 50 kg/ha/day dry matter (DM) or 1 m^3 of manure (5% DM) during the first weeks. Application rates can be increased by 12.5 kg per month, accompanying the increase in biomass up to a maximum of 175–200 kg DM/ha/day. Under these conditions, with a stocking rate of juvenile fish of around 10 000/ha, mean production is around 32 kg/ha/day (or around 10 tonnes for the growth period of 200 days per year) with a survival rate of 80–90%. Several literature reviews have shown a production of fish of around 30 kg/ha/day (which is the mean maximum observed) is obtained with levels of DM of the order of 100 kg/ha/day added every day and a stocking rate of 10 000 individual fish/ha. It has been demonstrated that production increases with the number of fish stocked between 2000 and 10 000 per hectare: above 10 000 there are no further improvements. It is also apparent that it is unnecessary to put together complex associations of species and that excellent performance can be obtained using only two species, for example common carp and silver carp. Production levels of 30 kg fish/ha/day have been achieved in temperate zones as well as in subtropical regions although the number of days over which the production is measured is different.

In ponds of over 300 m^2 in China observations made by Zhu *et al.* (1990) have shown that a daily application of pig manure at a rate of 30–50 kg DM/ha in a polyculture system of 7000 individual fish/ha 15% carp at 120 g, 50% silver carp at 60 g, 10% bighead at 150 g and 2.5% Crucian carp at 12 g gives a mean fish production of 10 kg/ha/day or 1 kg fish (live weight) for 8 kg (DM) manure spread. Daily distribution increases production by 38% in comparison with weekly distribution.

These levels of performance are subject to a certain number of constraints which are acceptable at an experimental or pilot level but not always on a larger scale. The results are obtained in small ponds with aerators and a daily application of manure as well as environmental monitoring.

In large ponds of over 5–10 ha which are common in Europe and generally used for the culture of carp, roach and tench, levels of organic fertiliser applied can be of the order of 10–30 kg DM/ha/day with weekly applications (1.4 to 4.2 m^3 per week per hectare with manure, 5% DM). There are various methods of spreading: very dilute manure (5 or even 1% DM) can be sprayed from the banks using a laterally directed jet. Composted or fresh manure or poultry wastes (in which dry weight can reach 25%) can be spread either from a boat or simply from heaps distributed around the edge of a pond. Dehydrated poultry waste is also used and is spread from a boat. The dry weight content of manure is highly variable and should be measured using a densitometer before spreading, in order to determine the correct quantities required for distribution. Even at low rates of application, the environment should be monitored periodically, chiefly by taking readings with a Secchi disc. The application rates and frequency should be altered in relation to the transparency of the water. When this reaches 25 cm or less, the application of fertiliser should cease, between 25 and 50 cm it should be continued at the normal rate and above 50 cm, the rate should be increased. Monitoring should be particularly rigorous in cloudy or stormy weather, when the levels of dissolved oxygen should be checked and aerators used if necessary. The use of organic fertilisers in ponds has only recently been quantified by Schroeder *et al.* (1990), who examined its effectiveness and showed that most of the nutrients taken up by the fish come from the algal trophic level. However, their conclusions only apply to naturally rich waters; organic fertilisation remains an economically viable way of increasing the production of earth ponds which are acidic or poor in bicarbonates and organic matter or fed by water with a low mineral content and (as a way of disposing of excess manure) (Wohlfarth and Schroeder, 1991). Additional information can be found in general works (Hepher and Pruginin, 1981*; Billard and Marcel, 1986a*; Moriarty and Pullin, 1987*).

6.1.4 Complementary feeding[(2)]

The culture of fish in ponds is based on the principle of optimising the natural production of the pond. However, it may be possible to justify the addition of feedstuffs either technically (e.g. cereals which can complete the natural balanced diet of carp where there is an excess of protein) or economically (where production must be intensified to justify investments which have been made). Sometimes the food naturally available is not enough for the biomass of fish present leading to a destabilisation of the trophic web (p. 25). The fish farmer may desire a faster growth rate so that the fish reach market size earlier. In order to decide on the appropriate amount of supplementary food to be given it is essential to know the availability of food in the water (see above) and the biomass of fish present. It is difficult to measure the biomass of fish; experimental farms sample periodically using various techniques but information on individual weights and growth rates only gives an indirect indication of the total biomass by modifying the initial number of fish stocked using the coefficient of mortality. It is difficult to see how major improvements can be made in this method.

Traditionally, food is distributed daily by hand using a small boat, at points marked out precisely by posts (five to seven distribution points per hectare). It is common practice to check whether food has been eaten by sampling the bottom of the ponds near to the posts using a type of hoe whose blade is made of a fine-mesh metal grid. All food should have been consumed within 2 or 3 h of distribution. If this has happened, the next day's rations can be increased slightly, in line with the mean projected growth rate of the fish stock. The amount of food given should be recorded in the farm logbook.

One method of improving the yields from fish ponds is the use of complementary feeding as a way of adding to the natural production of the environment, either continuously or during periods which are thought to be critical, but without the aim of (artificial feeding) covering all of the fishes' nutritional requirements. This is only applicable in extensive or semi-extensive systems: it is not appropriate in intensively farmed ponds. For more detailed information see: Cowey *et al.* (1985), Hepher and Pruginin (1981)*, Hepher (1988)*, Jauncey (1982), Neuhars and Halver (1969), Takeuchi and Watanabe (1977, 1982).

Nutritional requirements of freshwater fish

Protein

In all of the species which have been studied in detail it appears that, whatever the dietary regime, proteins form a significant proportion of their intake (32% of the intake of adult carp, 24% for catfish, 40% for trout). The remainder of the ration is

(2) A. Demaël.

* See General bibliography.

composed of lipids, carbohydrates in variable quantities, minerals and vitamins. The high protein requirement is the result of several factors of which the three most significant are:

—not all of the proteins ingested have the same coefficient of digestibility. This coefficient is higher for proteins of animal origin in comparison with plant protein, but even within the category of animal proteins, some are more easily assimilated than others. Thus for carp, the coefficient of digestibility of maize protein is around 70%, that of casein (milk protein) 75% and that of anchovy meal is 80%. It should be noted that the simple addition of amino acids to a protein increases the coefficient of digestibility. The digestibility of pure casein is 75% but if a mixture of amino acids is added to the casein (methionine, leucine, lysine, valine and threonine), digestibility rises to over 80% (see below).

—a significant part of the amino acids absorbed by the intestine after digestion of protein is oxidised in the body tissues and cannot therefore be used for the synthesis of protein tissue. Complete oxidation of these amino acids supplies energy which is used directly by the fish. This leads to the excretion of nitrogenous wastes as NH_4^+.

—fish have a high protein requirement because growth is slow and thus protein synthesis is low. It takes much longer to obtain large animals than in other types of animal rearing systems.

In extensive rearing systems that depend on the richness of phyto- and zooplankton, fish have a nitrogen balance which may be positive or negative. The farmer aims to maintain a positive balance or at least an equilibrium for as long as possible to minimise the retarding of the growth of the fish, even when the climatic conditions are unfavourable. The nitrogen balance can be expressed using a simple equation; at equilibrium:

$$\text{Nitrogen supply} = \text{Losses (intestinal + metabolic)} + \text{Nitrogen retained}$$

The goal of all rearing operations is to maximise the retention of nitrogen, i.e. protein synthesis in different tissues, particularly muscle, which represents more than 60% of body weight. In order for fish to grow, it is necessary that the supply of nitrogen is greater than the total nitrogen demand. Whenever supply fails to exceed demand there will be a breakdown of reserves, particularly protein and thus a loss of weight.

Losses of nitrogen which result from elimination through faeces and through oxidative metabolic pathways cannot be completely prevented even if the fish is starved or given a protein-deficient diet. Intestinal losses occur through digestive secretions (e.g. enzymes) and the renewal of the intestinal mucus. Metabolic losses are still high, representing 48% of the ingested nitrogen according to Fauconneau (1983). This loss of a significant part of the ingested nitrogen ensures that the coefficient of nitrogen retention remains modest for all species of fish, generally of the order of 30% ingested proteins.

For a simple method of determination, calculate:

—either the protein efficiency coefficient, which is the relationship:

$$\frac{\text{Gain in weight by the fish}}{\text{Quantity of protein ingested}}$$

—or the apparent coefficient of protein utilisation which corresponds to the relationship:

$$\frac{\text{Protein gain by the fish}}{\text{Quantity of protein ingested}}$$

With a diet containing 33% of protein of animal origin fed at a rate of 3% body weight/day, the daily retention of nitrogen in carp is 80.3 ± 4.7 mg/100 g body weight (Ogino, 1980).

Different proteins have different amino acid compositions. Not all proteins in the diet have the same quantities of essential amino acids which are needed for satisfactory growth. Although there are slight differences between species, it appears that essential amino acid requirements are similar for all species of freshwater fish. Ogino (1980) showed that there are no major differences between carp and trout, two species at the opposite ends of the dietary spectrum. The amino acids essential for growth are leucine, isoleucine, valine, threonine, phenylalanine, tyrosine, methionine, cystine, tryptophan, arginine, histidine and lysine. In the protein part of the diet, lysine should represent 5.3% of the amino acids, leucine 4.1%, arginine 3.8%, aromatic amino acids 6% and the sulphur amino acids at least 3%.

These requirements change in relation to temperature and increase as it increases. Thus, at 20°C with a daily feed input of 3% body weight, carp require 5.3% lysine, as opposed to 8.4% at 25°C (Viola and Arieli, 1989), an increase of around 60%. These experiments demonstrate that fish have both quantitative and qualitative requirements in relation to nitrogen.

Lipids

The diet of fish must also contain a significant proportion of lipids: this varies between species. For salmonids, lipids should make up at least 10% of the diet (10–20% or even more) but cyprinids require only 7%. Lipids are mainly in the triglyceride or neutral form which is one molecule of glycerol attached to three molecules of fatty acids.

The coefficient of apparent digestive utilisation of triglycerides is very high in fresh-water fish, reaching around 90%. The quality of fatty acids is strongly linked to their origin. If the triglycerides come from beef they contain around 50% of saturated fatty acids while the polyunsaturates represent no more than 12% of the total fatty acids. In contrast, when the dietary fatty acids are derived from fish oils, saturated fatty acids represent no more than 25% of total fatty acids; polyunsaturated ones make up around 50%.

After ingestion, some of the lipids are oxidised; the rate of oxidation increases at high temperatures. The fatty acids which are not degraded are hardly modified by the fish before being incorporated into cells. The fish has no adipose tissue where lipid reserves can be stored. Deposits are made in the liver, muscle and, in some species, the peritoneal cavity.

The growth of trout can be improved, without increasing the level of protein in the diet, by increasing the lipid level. Polyunsaturated fatty acids ($n = 3$) are required for this: these have three double bonds and must make up at least 10% of the lipid intake. Carp also show an improvement in growth if the dietary lipids contain more than 14% polyunsaturated fatty acids ($n = 3$): this cannot be provided by beef fat.

Carbohydrates

The third major component of the diet of fish is carbohydrate, essentially starch in the alpha form and dextrines which are short-chain carbohydrates. These make up around 30% of the diet of salmonids and 50% for cyprinids. The assimilation coefficient for carbohydrates is low for all fish: only 60% for starch in carp. This low assimilation coefficient is associated with the fact that carbohydrates, in comparison with fatty acids, are poorly oxidized by fish, especially at high temperatures, and therefore play a relatively modest role in the growth of fishes. At high water temperatures fish tend to accumulate carbohydrate reserves in liver and muscle in the form of glycogen. All fish have, in their intestines, the enzyme system which allows the complete degradation of starch and absorption of simple sugars.

As in mammals, the pancreatic amylase of fish hydrolyses cooked starch much faster than raw starch. In the natural environment fish only ingest raw starch, which explains the low digestibility of this substance, but on farms cooked starch may be supplied, which is far more readily digested. Glucose, which results from the complete hydrolysis of starch, is absorbed by the intestinal mucosa. As it is only partially oxidized at mid-range temperatures there is a tendency for it to be stored in the form of glycogen in the liver and muscle and also as triglycerides in the liver, muscle and, possibly, the peritoneal cavity. Glycogen is only used in emergencies such as during a long period of starvation and periods of hypoxia and even anoxia which may occur towards dawn in summer.

Digestion is possible in the intestine of herbivorous fish through the bacterial flora, but at a low rate. The benefits are divided between the fish and the bacteria.

Vitamins and minerals

Fish require vitamins for growth; some of these are lipid-soluble and are thus absorbed in the intestine because of the presence of lipids. Vitamin C is essential for the normal development of all fish. A supply of minerals in the diet is also essential.

Potential for the improvement of production in extensive rearing ponds

In extensive rearing systems, where no supplementary feeding is given, production can be described as modest, especially in temperate regions. In the absence of large zooplankton, fish will stop feeding; the energy expenditure in the hunt for food is greater than the gain achieved. In addition, the energy requirements of the fish increase as the water temperature rises. In spring, the water warms up and the requirements of the fish increase, but there is a delay in the rise of phyto- and zooplankton. This creates a deficit between the energetic requirements of the animal and the amount of energy available; the fish must draw on its reserves.

At low temperatures, below 7°C, the fishes' metabolism slows down and the limited metabolic requirements are covered either through the diet (if the environment is productive enough and the intestine is still actively digesting) or through the use of tissue reserves (glycogen from the liver and lipids from liver and muscle). These reserves are laid down by the end of autumn and are proportional to the quantity of energy assimilated.

Finally, by spring, the animals are often in a poor nutritional state and thus extremely vulnerable to a range of pathogens (viruses, bacteria, fungi) (pp. 143 and 195), all the more so as their immune system is practically non-functional below 18°C.

During summer, when water temperatures are high, there is often a scarcity of zooplankton (medium and large sizes) which creates a new problem of partial starvation, causing a marked growth check in individual fish, particularly where the stock is subject to episodes of fairly severe hypoxia. It is therefore beneficial to limit these periods of starvation and their consequences by giving supplementary feeding that adjusts the energy supply to the current needs. It is not necessary to give all of the dietary intake, merely a proportion to make up for the deficiencies of the environment.

In winter, for any species, a supplement equivalent to 1.3–1.5% body weight per day is sufficient. At the beginning of spring and in summer a daily ration of 1.7 to 2% is enough to prevent periods of starvation.

Types of feed which can be given as supplements

Supplementation can be made entirely with a commercially available diet, but this is likely to be expensive which will significantly increase the final product price. However, it is possible to reserve these commercial diets for periods when the fish are capable of assimilating them at maximum efficiency.

In cold weather it is unnecessary to give a protein-rich feed as a major proportion of the proteins will not be assimilated; it has been shown that if carp are given a supplement (1.7% body weight per day) of granules with a 41% protein content or cereals (10% protein) after 3 months there is no difference between the batches in terms of weight or mortality rate. In both examples the carp remained the same weight throughout the winter as at the start of the experiment.

Such results have shown that giving diet supplements is chiefly aimed at maintaining the weight of the fish and the stability of the energy reserves in the tissues so as to encourage the onset of the growth phase in the spring. This experiment also demonstrates the value of cereals in the diet; these provide a relatively cheap means of giving significant quantities of carbohydrates (60–70%) (Table 6.5) that can be digested and assimilated reasonably effectively by the fish at low temperatures.

Table 6.5. Percentage composition of the major cereal types. (A.E.C. 1978)

	Oats	Wheat	Maize	Barley	Rye
Dry matter	87	87	87	87	87
Proteins	10.5	11	9	10	10
Lipids	4.3	2	4	2	1.5
Cellulose	10	2.5	2.5	2	1.5
Carbohydrates	59.2	70	70.2	67.5	71
Minerals	3	1.5	1.3	2.5	1.5
Amino acids					
Lysine	0.42	0.31	0.25	0.37	0.4
Methionine	0.17	0.17	0.19	0.17	0.18
Methionine + cysteine	0.47	0.42	0.39	0.4	0.45
Arginine	0.65	0.53	0.4	0.51	0.55
Glycine + serine	0.98	0.93	0.77	0.84	0.89
Leucine	0.74	0.72	1.17	0.7	0.63
Phenylalanine + tyrosine	0.86	0.78	0.81	0.83	0.73

Cereals can beneficially be complemented by amino acids. Carp fed on a daily ration of 3% body weight with soya flour (30% protein) have been compared with carp which received soya flour enriched with L-methionine (0.39%). Fish in the second batch had, after 3 weeks, a weight gain 25% higher than the first batch. After 6 weeks the difference was 60% in favour of the second batch (Murai *et al.* 1981). Comparable results have been achieved in trout fed on casein (35%), alone or enriched with methionine and arginine. In an experiment performed in small ponds in the Dombes region of France between November and October a batch of carp of around 10 g were given crushed barley with the addition of DL-methionine (2.6 g/kg) and cysteine (4 g/kg) and a control batch was given standard granules (feeding rate 1.7% body weight per day). Growth remained poor and comparable between the two batches during the winter period but was higher later in the study for the carp given cereal and amino acids (Figure 6.5).

Levels of triglycerides in the muscles are higher and more variable (25.9 ± 13.02 mg/g) in carp given supplemented cereals than those given granules (8.65 ± 1.5 mg/g) although differences were not significant. The levels of phospholipids were similar but the relationship between ω^3 fatty acids and ω^6 fatty acids was higher in the fish given cereals (Table 6.6). Other experiments have all shown that the addition of essential amino acids (arginine, methionine, lysine) improve nitrogen retention.

Examination of such results confirms that the addition of one or more essential amino acids to the protein ration of fish can have a positive effect on their weight gain. Because of their high level of amino acids, oats may be the most appropriate cereal for fish feed, but production is low and other cereals may be used with the addition of one or more essential amino acids. This may provide a means of utilising excess grain production. Amino acids should be added to the cereal ration so as to

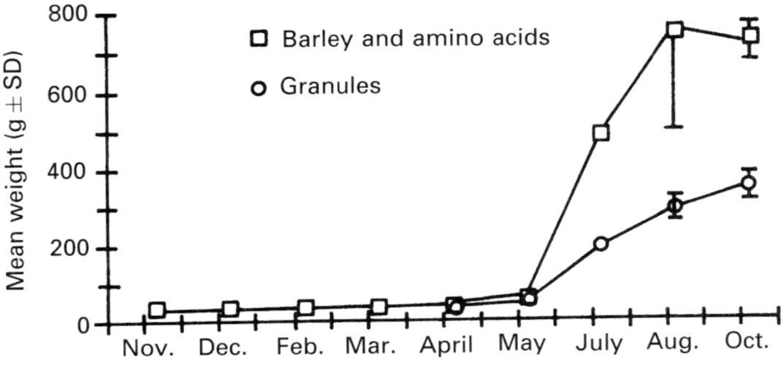

Fig. 6.5. Growth of two batches of carp, one fed on standard UCAAB granules with a 31% protein and 8% lipid content, the other on cereal (barley) supplemented with methionine (2.6 g/kg) and cysteine (4 g/kg). Feeding rate 1.7% body weight per day for both groups (Demaël *et al.*, 1991. *Ichthyophysiol. Acta*, **14**: 57–70). Weight mg ± standard deviation (SD).

mimic the concentrations found in fishmeal. The quantities of methionine needed are around 2 g per kg cereals; for arginine the figure is 4 g.

The role of amino acids is to increase the digestibility of plant proteins but they may have a more complex action. In fish they may have a stimulating effect on the olfactory and gustatory system which increases appetite and the search for food. In their presence the animals are able to ingest greater quantities of food.

If cereals are used to provide supplementary feed they should be crushed prior to distribution so as to give particles suitable for the size of fish to be fed. Only crushed grain should be used when mixing with amino acids and this should, if possible, be applied as a light coating in order to avoid losses when the feed is given. Gentle steaming of the cereals promotes the digestibility of the starch. The rations given are around 1.5% body weight in cold weather and 2% in warm weather (see Table 6.8).

Table 6.6. Percentage distribution of principal fatty acids in the triglyceride fraction of carp muscle coming from the experiments reported in Figure 6.5.

Fatty acids in the muscle	Carp diet		
	Barley + amino acids	(1)	Granules
C 16	20.71 ± 0.17	†	12.89 ± 0.21
C 16-1	11.81 ± 0.15	†	5.06 ± 0.22
C 18	5.78 ± 0.17	NS	5.51 ± 0.16
C 18-1	35.74 ± 0.19	*	30.37 ± 0.28
C 18-2-ω6	8.27 ± 0.17	†	22.82 ± 0.31
C 18-3-ω3	3.13 ± 0.12	NS	3.98 ± 1.35
C 18-3-ω6	0.53 ± 0.08	†	3.49 ± 1.21

(1) Statistical comparison: significant differences $^*P < 0.05$ and highly significant differences $\dagger P < 0.01$; NS = not significant.

The distribution of supplements

A single daily distribution by hand appears to be sufficient. If food is given less frequently, the mass to be ingested will be excessive and a large proportion will not be eaten by the fish. The supplement should be increased progressively to get the fish in the pond used to the diet. It is possible to use automatic feeders often operating under the control of photocells, to give continuous feeding but this does not always result in homogenous growth of the whole batch. In addition, when the animals feed continuously, insulin secretion is continuous but moderate while the fish require a strong hyperinsulinaemia for effective protein synthesis; ingestion of a large meal provokes a high secretion rate of insulin which improves the dietary efficiency of proteins.

6.2 INTENSIVE PRODUCTION OF CARP THROUGHOUT THE YEAR(3)

The traditional system of cyprinid culture in temperate regions is based on a pattern of 100 days of summer growth (temperature $\geq$18–20°C). Harvest is by emptying the ponds during autumn and winter. In France and other European countries this results in surpluses and difficulties of drainage in winter and lack of any production during summer. Markets for freshwater fish for consumption, whether traditional or new (widespread distribution, processing) or a range of products require regular supply to the markets or the factories. The development of catfish production in the USA (90% of these fish go to processors) and carp in Israel (sold fresh) are all based on the management of production techniques to allow a regular supply to the market in both quantitative and qualitative terms.

Production technologies are still being developed and one of the problems which has arisen in the first attempts at production in France has been the development of water quality in intensive systems and its means of management. These problems which differ from the management of the ecosystem in extensive production are described below.

6.2.1 Development and management of water quality in ntensive rearing systems(4)

When the production of fish in ponds intensifies and reaches over 3–4 t/ha/year through the supply of food from outside sources, the contribution of the natural ecosystem becomes negligible and water quality must be managed so as to achieve optimum rearing conditions both in terms of the requirements of the fish and the impact on the environment. The part of the natural food chain which remains essential in intensive production is the phytoplankton, because of its links to two

(3) J. Marcel.

(4) J. Marcel and Y. Racapé.

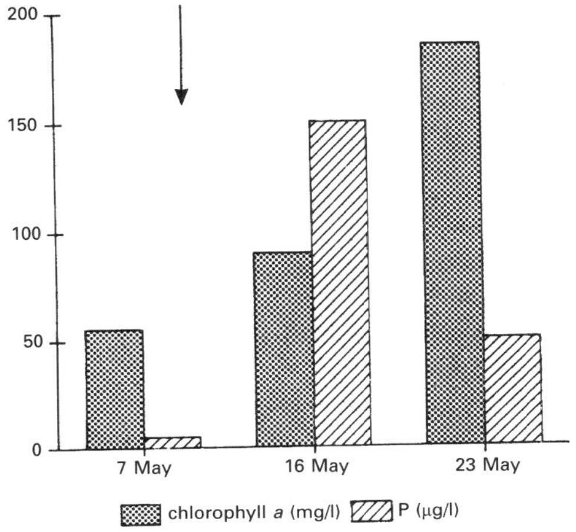

Fig. 6.6. Change in the biomass of chlorophyll *a* and phosphorus levels (expressed as PO_4^{-3}) after phosphate fertilisation. The arrow indicates an application of 80 kg/ha of superphosphates 14–48.

parameters: the supply of dissolved oxygen and the elimination of NH_4^+. The objective is therefore to maintain a high level of chlorophyll *a*, above 100 μg/l.

The following parameters have a direct influence on the development of the phytoplankton:

—the mineral elements nitrogen and phosphorus and the N/P ratio which should be 1/5–1/8. In the absence of nitrogen (N) and phosphorus (P) the phytoplankton will not develop. In an intensive system a deficit in mineral elements occurs, especially at the start of the rearing cycle, when there is little input of exogenous feed. After checking the levels of N and P, an adapted mineral fertiliser can be applied to initiate algal production (Figure 6.6).

—turbidity. This has a negative effect on the development of phytoplankton when it is caused by silt which is kept in suspension by the movements of the high numbers of fish. The level of suspended solids can reach 100 mg/l and even higher: this problem is worst at the start of the rearing cycle when photosynthetic activity is low.

A treatment based on aluminium sulphate, $Al_2(SO_4)_3 14H_2O$ (Boyd, 1979) (see General bibliography) causes the suspended particles to precipitate within a few hours. However, experiments have shown that a high rate of mineral fertilisation (80 kg/ha of superphosphate 14–48) induces a development of phytoplankton which is sustained in spite of high turbidity; 10–15 cm on the Secchi disc (see below).

When the turbidity is caused by phytoplankton which form the major part of the suspended particles (25 to 30 cm Secchi disc reading), the situation can be considered as ideal, both limiting the algal bloom and avoiding the appearance of filamentous algae.

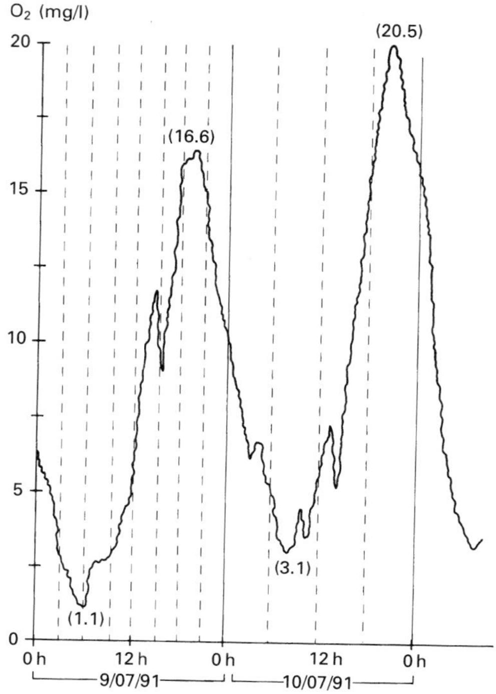

Fig. 6.7. Day/night cycle of dissolved oxygen in an intensive fish culture pond of 4 ha (4 t/ha of carp), without oxygenation. Note the times of the 10 lowest levels: 6 h on 9/07 and 8 h on the 10/07), while supersaturation occurs around 20 h (continuous measurement, SRAE, Châlon-sur-Marne, France).

Dissolved oxygen in intensive closed systems

The oxygen budget previously described (p. 40) is, in intensive systems, affected by the cumulative respiration of the fish–plankton–sediment complex except when production reaches significant levels.

The risks of a deficit in dissolved oxygen at dawn (5.00–6.00 h in mid-summer) are well known. Supersaturation occurs at the end of the afternoon (Figure 6.7). Levels of oxygen rise from daybreak and saturation will occur after a varying length of time, depending on the minimum value, intensity of photosynthesis and the biomass of the fish. Low oxygen levels continue until 8.00 h in feeding zones where large numbers of fish congregate; saturation may not be reached until 10.00 h (Figure 6.8). It is important to consider this when food is given in the morning (in general from 9.00 h) in the zones where fish concentrate. Low levels of dissolved oxygen have a strong limiting effect on food intake and conversion.

It is therefore essential to place an oxygenating device near to the feeding zone; this operates between 6.00 and 12.00 h. In summer (mean temperature $>22°C$) and when the biomass is above 5–7 t/ha oxygenators may function throughout the 24 h period for a short time.

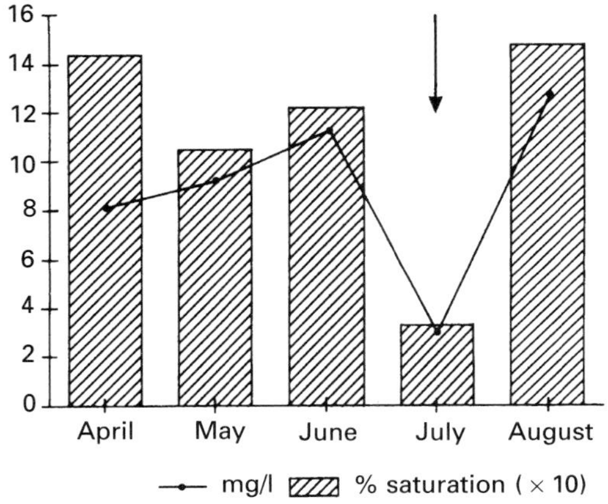

Fig. 6.8. Levels of dissolved oxygen determined from spot samples taken between 9 and 10am in an intensive fish culture pond (biomass 5 t/ha in April and around 6 t in July). In July, after the measurement was taken (arrow) and aerator was put into the 4 ha pond (samples SRAE, Châlon-sur-Marne).

Elimination of NH_4^+ in intensive closed systems

The quantity of ammonia (A) excreted by fish can be estimated from the protein utilisation (U_p) and the proportion of protein in the diet (T_p).

$$A = \frac{(1.0 - U_p)T_p \times 1000}{6.25}$$

where $A = NH_4^+$ excreted in g/kg feed
U_p = % protein utilisation
T_p = % level of protein in the feed
6.25 = coefficient representing the relationship $\frac{\text{Mass of protein}}{\text{Mass of nitrogen}}$

The coefficient 1000 makes it possible to obtain A in g/kg.

Thus for a carp diet with a 35% protein content and a protein utilisation of 58%, the quantity of ammonium (NH_4^+) excreted is 23.5 g/kg feed. NH_4^+ is thus converted and eliminated through a range of processes which limit and prevent any accumulation.

The loss of nitrogen to the atmosphere in the gaseous form of NH_3 is significant where the pH of the water is high. Nitrification transforms NH_4^+ into nitrite and then nitrate. Figure 6.9 shows the succession of the different forms of nitrogen over the course of the rearing cycle. The third process is the direct assimilation of NH_4^+ by various species of algae (p. 16). This phenomenon is demonstrated in Figure 6.10 which shows the succession of NH_4^+ in relation to the level of chlorophyll *a* over a production cycle.

Thus the major cause of the appearance of NH_4^+, which is very often linked to a deficit in O_2, is the low level of phytoplankton. There are many reasons: conditions

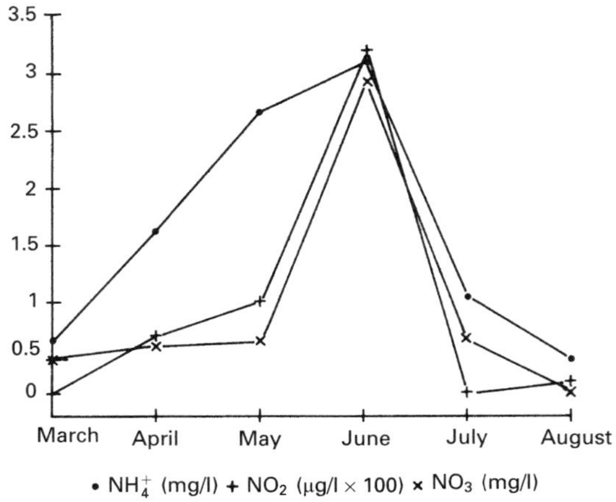

Fig. 6.9. Change in nitrogen compounds in a 4 ha intensive fish-culture pond (carp biomass 5 t/ha).

which are limiting for photosynthesis (light and temperature), mineral deficiency (particularly phosphorus), the use of minerals by higher plants (unusual in intensive systems) and excessive turbidity.

Thus the successful operation of an intensive rearing system (>5 t/ha), from the

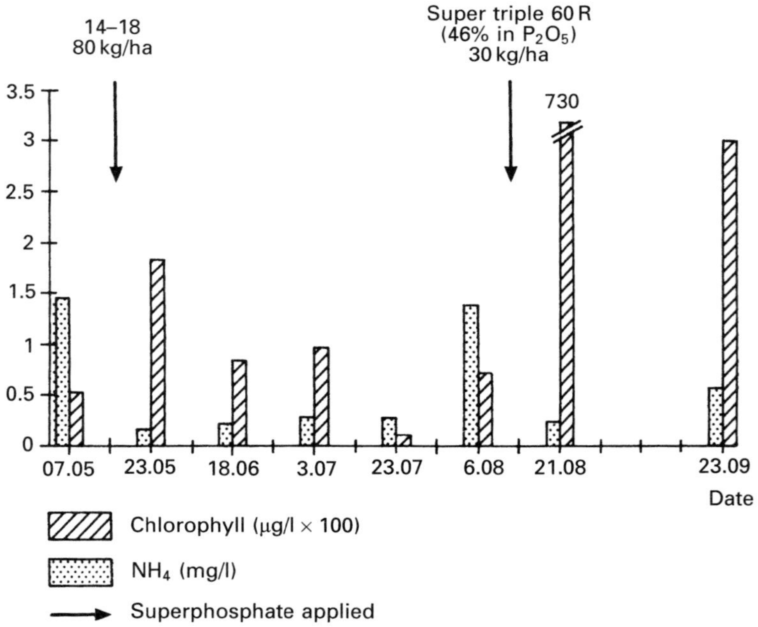

Fig. 6.10. Change in NH_4^+ in relation to the level of chlorophyll *a* during a cycle of intensive rearing of carp (final biomass of carp: t/ha).

point of view of ammoniacal nitrogen, entails the maintenance of a sufficiently high density of algae through the regular addition of mineral fertiliser and satisfactory levels of dissolved oxygen (> 5 mg/l).

The importance of the three factors (phytoplankton, dissolved oxygen and ammonia) is seen again in systems whose intensity is promoted to a greater or lesser degree by the addition of organic manure (p. 163). Balancing these in intensive systems is always complex and oxygenation using mechanical aeration is used universally as a means of ensuring supply.

Other adverse factors

In intensive systems where fish are reared for the table, the farmer must be on the lookout for substances which give an off-flavour to the flesh. The most serious problem is the muddy taste associated with cyanobacteria (p. 19). The fish can also take on other odours from substances such as petroleum products (in ponds alongside major roads) at concentrations of 0.02 to 0.1 mg/l of water in the rearing system or phenolic compounds at concentrations of 0.1 mg/l of monohydric phenols and 0.02 mg/l of chlorophenols.

6.2.2 A continuous production system

The scenario for continuous production combines three main procedures (Figure 6.11):

—the supply from traditional pond catch in the autumn and winter;
—the holding of fish over winter to supply the spring market;
—control of fish growth and continuous harvest during summer;

This last aspect is described in detail below.

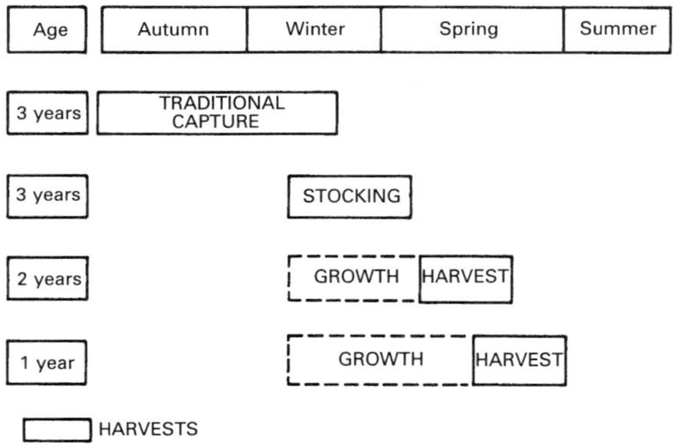

Fig. 6.11. Scenario for the continuous production of market-size carp. Traditional catches are in autumn and in winter by emptying the ponds, summer harvests are in open water.

Table 6.7. Weight of fillets and whole carp (in grams) in relation to their final destination (1991).

Market	Carp fillets (g)	Whole carp (g)
Restaurants	200	1200
Home consumption	100–200	600–1200
Processing	100–200	600–1200
Slicing	–	2000–2500
Smoking	250	1500

Definition of the commercial product (see p. 105)

The requirements of the market for aquatic products and the processing industry have a major impact on the characteristics of the farmed fish: the preferred product is a skinless, boneless fillet. The size of the fillet varies according to the intended destination and determines the harvest size of the fish, based on the information that the yield at filleting for carp is around 33% with the techniques currently in use. Table 6.7 shows the different uses of carp fillet; the range of weights of live fish needed is relatively large, between 600 g and 2.5 kg.

Continuous production through the control of growth

Principle

It is possible to obtain carp of market size in the summer (from the end of June or the beginning of July) after only 2 to 3 months growth if the animals are initially of an appropriate size that is adapted to the duration of the growth phase and the preferred market size. The growth curves for different size classes (between 100 g and 1000 g) are given in Figure 6.12 (p. 181). This develops slowly up to the end of May and increases during the summer months in relation to temperature, which regulates the feeding regime and growth.

Under a temperate thermal regime carp of 320, 213 and 142 g in March reached commercial size at the ends of June, July and August respectively. A management plan aimed at the continuous production of commercial-sized fish can be based on the rearing of different-sized juveniles in different ponds, to be harvested simultaneously. Stocking with part-grown animals results in high biomasses (400 to 500 kg/ha for a net summer production of 600 to 750 kg/ha).

Harvest of fish which have reached market size can either be after the pond has been completely emptied or through the operation of a fishery within the pond. Emptying the pond in summer has several drawbacks (authorisation, loss of water, putting a pond out of use during the best part of the growing season); fishing within the pond is preferable but may be difficult to initiate, particularly in traditional ponds which were not designed for this technique.

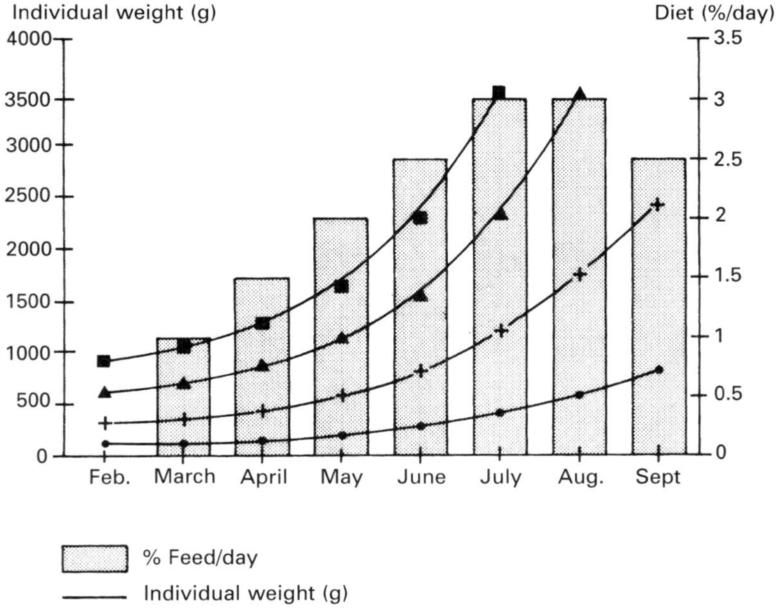

Fig. 6.12. Theoretical growth curves for carp above 100 g in relation to a feeding rate based on temperature and with a conversion coefficient of 2.

Production of juveniles of a range of sizes

In a temperate climate it is possible to produce juvenile carp of 70 to 100–120 g in 1 year by integrating a range of operations to control reproduction, advancing development in hatcheries, larval and fry rearing using artificial diets (p. 139) and using densities adapted to obtain the desired weight of individuals. In Israel, schemes for the production of carp of a range of sizes have been developed (Figure 6.13). The rearing phase is roughly between May and November, from an input of 4-week-old carp (1–1.5 g). The objective is to obtain batches of juveniles of 50–250 g by varying the rearing density (100 000 to 15 000 individuals per hectare) for productions of 3 to 5 t/ha. The rearing scheme starts with batches of individuals of identical weight.

In order to optimise the use of the fry-rearing system during the growing season (temperature >20°C) all of the ponds are stocked with 100 000 fry/ha from the end of the first-fry stage. Then, once or twice during the growing season (Figure 6.13), the fish are removed and counted, and the density of animals is reduced in relation to the desired final density.

Feeding is based on a manufactured diet containing 25% protein of which 15% is fishmeal (plus an addition of methionine, cystine and lysine). The daily distribution rate can reach 3 to 4% body weight during the hottest period. Juveniles stocked into wintering ponds will be given a daily maintenance ration when the water temperature is above 10°C (0.5% body weight/day).

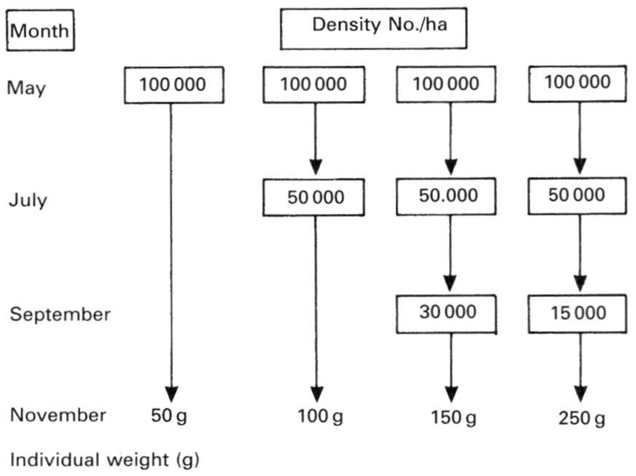

Fig. 6.13. Diagram showing production of different sized juveniles used in Israel. The rearing tanks are stocked initially at the same density (100 000 4-week-old fry/ha), this is reduced in order to achieve individuals of the desired size.

Production of commercial sized carp

A management plan has been devised, based on the definition of the product (with regard to the final weight of the individual fish), the frequency of supply and the production system.

Semi-intensive production in traditional ponds (p. 167). Young carp (1 or 2 years old), weighing 300 to 350 g are stocked from March onwards at a density of 450–500 animals/ha (production of 450–500 kg/ha). These are given a complementary ration of cereals (barley or wheat) (50%) and soya flour for the first 4 weeks (50%) then cereals alone, when the water temperature exceeds 10–12°C, or at the end of March. A simplified table for the weekly cereal ration is given in Table 6.8. The weight of individuals increases by a factor of 2.5 by the end of June and 3 by the end of July

Table 6.8. Daily distribution rate of cereals and mean monthly growth % in relation to temperature (based on a mean conversion index of 2) (carp of 400 to 1200 g).

Temperature (°C)	Feeding rate (%/day)	Growth (%/month)
10–12	1	14.6
14	1.5	22.6
16	2	30.9
18	2.5	39.8
20	3	49.0
22	3.25	53.8
24	3.5	58.6

Month				
February	250 g	150 g	100 g	50 g
March				
April				
May				
June	*			
July	*	*		
August	*	*		
September		*	*	
October			*	*
November			*	*
December				*
January			■	■

* Partial harvest ■ Harvest through emptying the pond

Fig. 6.14. Diagram of continuous production as applied in Israel. Each rearing pond is stocked with juveniles of a mean weight to allow the production of 1.2 kg carp throughout the year.

but this depends mainly on the temperature of the water and the availability of natural food. The net conversion coefficient (weight of food given/gain in weight) is 2 to 2.5. This tends to increase with the increase in the initial size of the animals.

Intensive production in ponds (e.g. Israeli fish farms). When the aim is to produce fish of market size as quickly as possible, on-growing ponds of around 10 ha each are stocked from March at a density of 10 000 juveniles per hectare. Each pond has a single weight category: variations in size are established between different ponds. The final net production varies between 7 and 9 tonnes/ha with a mean individual weight of 1.2 kg. Figure 6.14 shows the different production scenarios designed to provide a constant supply of fish of market size over about 8 months (June to January). After this the market is supplied from holding ponds (February to mid-June).

All of the fishes' nutritional requirements are provided by externally supplied feed. This feed is made up of 25% protein (fish meal and soya meal). The rate of distribution is related to water temperature but never exceeds 3% body weight per day. The quantity of feed given is adjusted periodically after sampling the weight of individual fish. The feed is distributed at a single feeding point from a silo which is installed on the bank (15–20 m perforated metal through which food is moved by a worm shaft-screw device) so as to distribute feed almost continuously throughout the day (except between 5.00 h and 7.00 h in the morning and 12.00 h and 16.00 h in the afternoon) (p. 243).

It may be possible to transfer this practice to temperate areas (France and Central European countries) (see growth curves, Figure 6.12). To meet the demands of the French market, the fish should have an initial weight of between 100 and 300 g at the end of June.

6.2.3 Open-water harvest system

Partial harvests by net

There have been several trials of this technique, particularly in Israel in polyculture units where carp is the dominant species and in catfish farms in the USA. The results owe more to local know-how and practical expertise than to established standard techniques.

In Israel, the ponds used for fish culture have a maximum width of 100 m which makes it possible to fish the whole pond with a net pulled along its length. The operation begins at the deepest part of the pond and the fish are drawn towards the most accessible part where they can be removed (by elevator, fish pump, worm screw, hand net). The depth and length of the net are respectively 1.5 times the depth and breadth of the pond but the net never exceeds 150 m in length. The central part of the net is expanded to form a pocket (10 m). In practice, several sheets of 50 m long net are used; these are usually of different mesh sizes and depths. Wooden rods are fixed between the float and lead lines at the two extremities of the net. These hold the net open and in contact with the bed of the pond while it is being pulled. To avoid the lead line becoming buried in the silt, bands or clumps of synthetic material are attached. This helps the net to glide over the bed of the pond. This fishing procedure can be used for the whole pond or for part of it such as the feeding area, where partial removal is required, for example to provide samples for market.

Jensen (1981) described an enclosure for capturing catfish in open water in reservoirs in Alabama. A 60 m long net is installed permanently 15 m out from the bank (close to the feeding point), a 30 m length parallel to the bank and a 15 m length taken back to the bank at both ends (Figure 6.15). During fishing operations the two ends are pulled quickly with the help of ropes but the central part of the net remains in place. The net is thus closed near to the bank for the removal of the fish. This enclosure makes it possible to capture 12% of a catfish population in a 9 ha pond in

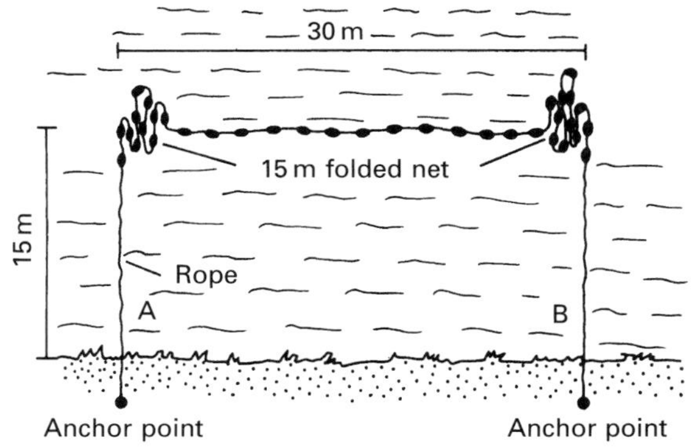

Fig. 6.15. Diagram of a trap system for periodical sampling of catfish in open water. A 60 m net is installed and partially deployed (over a length of 30 m). During fishing the net is closed by pulling the two ropes A and B (after Jensen, 1981).

a single operation. In a 0.5 ha pond the rate of capture can reach 70% of the biomass.

Both carp and catfish become extremely suspicious after such operations and take 7 to 14 days to return to normal feeding behaviour (Busch, 1985). Because of this, harvests (or sampling) should be at intervals of at least 3 weeks (Figure 6.15).

Partial harvests using fixed fishing gear (e.g. in Israel)

These are operated in open water when sampling shows that 20 to 30% of the animals have reached market size. Even though carp may be stocked initially as a homogenously sized batch, the extremely heterogeneous growth pattern results in some individuals growing much more rapidly than others. This characteristic is exploited in partial harvests. On-growing ponds have fixed apparatus close to the feeding and aeration points. These consist of four posts, at the extremities of which are ropes which can be used to lift up a square net (50 × 50 m). The net is submerged before fishing and raised rapidly with the help of four electric winches fixed to the posts, when large numbers of fish have accumulated above it to feed. The fish are concentrated at the base of the net into a large detachable bag (Figure 6.16) which is removed and taken to the bank for the fish to be collected. The carp are graded by size and either returned to the water or to a holding facility before marketing.

Such harvests are carried out at dawn in ponds of around 10 ha with high stocking densities and are repeated so long as the proportion of fish of market size is acceptable. After one pond has been harvested, attention turns to another, returning to the first, 4 to 6 weeks later.

Fig. 6.16. Apparatus used to capture pond fish in open water using a net which is submerged at the feeding station and lifted suddenly; the bag at the base can be removed and taken to the bank with the fish concentrated in it (photo J. Marcel, ITAVI).

6.2.4 Holding ponds

These are earth ponds, a few hundred square metres in area and 1.5 to 2 m in depth, where fish are held for a few weeks only. It is essential that water flows into the ponds constantly; depending on the flow rate biomass can reach 15 kg/m^3 (Hepher and Pruginin, 1981) (see General bibliography).

It may be possible to hold carp for 1 or 2 months (April to June) in these ponds. They are small and can be emptied and refilled during the course of the summer. Around 500 kg carp/ha can be stocked and food given to satisfy nutritional requirements and avoid weight loss; the environment is unlikely to supply much food at this time. Fish kept at high density (30–40 t/ha) in ponds at summer temperatures (>25°C) for a few weeks require a constant inflow of water, a system of aeration and a supply of feed which meets the fishes' maintenance requirements. This system is currently used in Israel.

6.2.5 Conclusion

The assurance of continuous production from a farm is essential for the development of a market system for freshwater fish in the same way as any other form of animal production. Techniques now available make it possible to meet market demands throughout the year, even under temperate climatic conditions such as are found in Europe. However, temperature remains of fundamental importance and probably limits the development of the system, in terms of intensification, to a short period of rearing in the spring (growth and water quality). Methods of removing fish from open water must still be regarded as experimental (or empirical). However, much interest is being shown in improving them: this should soon lead to an improvement in the technology.

6.3 MODELLING THE GROWTH OF CARP IN PONDS[(5)]

6.3.1 Construction of the model

An equation modelling the growth in body weight of carp (crosses between Polish and Hungarian strains) through a cycle of 3 years of rearing has been established for the climatic conditions of southern Poland at the Golysz Station (longitude 18° 48′ E, latitude 48° 52′ N, altitude 275 m) (Augustyn and Szumiec, 1985), taking account of weight, the cumulative temperature at which effective growth can take place, the density of fish in the rearing system and the quality of the diet. The model has been tested using multiple regression analysis; verifications and validations give a good correlation between theoretical growth and measurements (Szumiec and Szumiec 1985, Szumiec, 1990). The procedure is described below and shown as an example that can be used, after adaptation to other climatic and rearing conditions.

At the end of the season the body weight (*G*) or weight of an individual fish (a

[(5)] Maria V.A. Szumiec.

Table 6.9. Characteristics of experiments on carp growth (Szumiec, 1988; Kolasa-Jaminska, 1988).

Age	1st year (C_{0-1}) exp *A*	1st year (C_{0-1}) exp *B*	2nd year (C_{1-2})	3rd year (C_{2-3})
Initial density/ha × 1000	90–150*	90–120*	6.0–24.0	2.4–4
Final density/ha × 1000	20–95*	16–95*	2.8–23.3	2.1–4
Fertilisation:				
none, $v = 10$	+			
organic, $v = 20$	+			
mineral, $v = 30$	+		+	+
organic + mineral, $v = 40$	+	+		
Diet; % proteins:				
10		+	+†	+†
20		+		+‡
30	+	+	+	+
40		+	+	+
Feed conversion	1.6–2.3	3.2–4.6	2.5–5.3	2–4.3
pH, extreme values	6.8–9.7	6.9–10.3	6.8–10.6	7.2–9.5
O_2, mg/l extremes	0.9–16.6	0.9–18.5	1.3–20.2	0.9–18

* Ponds with the highest density of carp are aerated.
† Wheat distributed.
‡ Granules, 25% protein content.

variable) is dependent on a series of other variables (called explicatives) such as the initial individual weight (G_0), sum of the temperatures $\sum \theta_e$ (>14°C), the final density, *d* and the dietary value *v* of the feed which may be natural or complementary. The following relationship has been established:

$$G = f(G_0)f\left(\sum \theta_e\right)f(d)f(v)$$

The water temperature is the only variable which cannot be managed by the fish farmer. $\sum \theta_e$ is established by taking into account the mean daily temperature (measurements taken 3 times per day) above 14°C (generally between the beginning of May and the end of September). The model can be simplified for the 1st year of growth by subtracting the period over which mortalities occur during the first weeks after stocking the fry into the second-fry ponds; this corresponds to a total of 200°C. This has been verified for 70% of the examples used to construct this model.

The dietary value *v* is affected by a coefficient varying between 10 and 40, which corresponds to the percentage protein content of the diet (Table 6.9). Where fertiliser is applied the coefficients which apply to the different regimes are as follows:

10: without fertiliser,

20: organic fertiliser,
30: mineral fertiliser (N,P),
40: combined mineral and organic fertiliser.

The environmental conditions are given for extremes of pH and dissolved oxygen (Table 6-9). The model is calibrated by using the best possible available information but this does not necessarily guarantee the consistency of the result. Of the various regressions tested those with the highest level of significance and the best correlation between the theoretically derived masses and those measured have been retained.

6.3.2 Growth during the first year of culture (C_{0-1})

The best agreement between the calculated weights of individuals and those measured (Figures 6.17 and 18) is given by the following equations:

—where fertiliser is applied:

$$G_{1A} = 7.054 \times 10^{-9}\left(\ln \sum \theta_e\right)^{11.7145} \exp(0.000010d)\ v^{0.2330} \qquad (1)$$

—where complementary feeding is used:

$$G_{1B} = 3.723 \times 10^{-9}\left(\ln \sum \theta_e\right)^{12.3488} \exp(-0.000012d)\ \ln(v)^{0.5695} \qquad (1')$$

These two equations have the same form and differ only in the value of the parameters. The application of the model shows that growth is highly dependent on the cumulative temperature (degrees Celsius): husbandry practices such as fertilisation and additional feeding have little effect in cold water ($\sum = 500°C$) while the effects are amplified considerably when the water is warmer ($\sum = 1000°C$) (Figure 6.19).

The flexibility of the model has also been confirmed by the good correlation between the calculated weight of individuals in comparison with actual measured weights (other than those used to calibrate the model), when the ponds are stocked at their highest density (Figure 6.20). When mortality rates are high, it is likely that they will not occur immediately after stocking but will be progressive; this confirms the close correlation between the observed weights and those simulated from the observed density at final harvest. When mortalities occur early in July in warm years (Figure 6.20, pond 6), the gain in individual weight observed corresponds closely to that given by the model. When mortalities occur later (or during a cold year) the true weight of individuals is considerably lower than that calculated because the density in the pond has remained high (Figure 6.20, pond 7).

Fig. 6.17 and 6.18. Theoretical measurements (lines) and observations (points) of the weight of individual carp during a growth season (temperature >14°C) in relation to the cumulative temperature ($\sum \theta_e$) for different final fish densities (d in ind/ha) and v (see Tables 6–9) 6–17: fertilised pond; $v = 10\%$; without fertiliser $v = 20\%$: organic fertiliser, $v = 30\%$: mineral fertiliser NP, $v = 40\%$: mineral and organic fertiliser, 6–18: fish given feed from external sources, v corresponds to the level of protein in the diet.

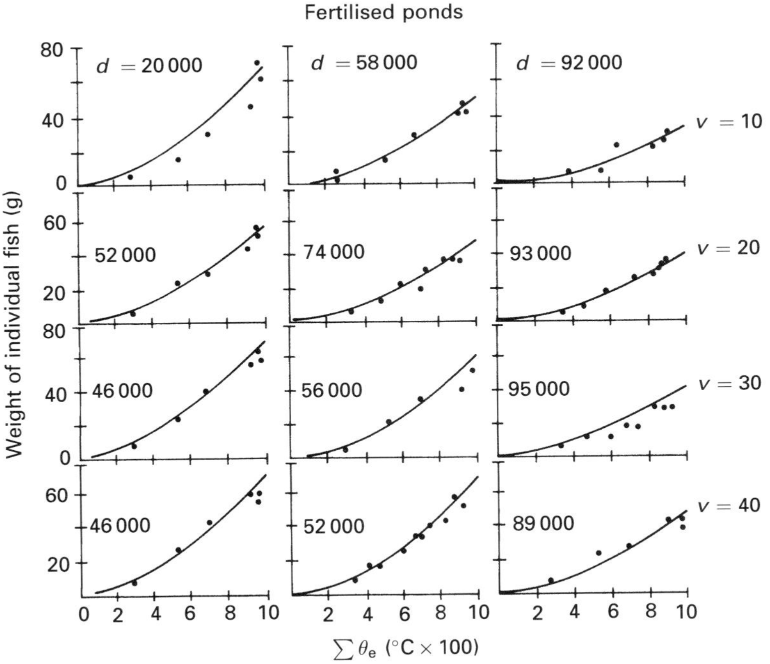
Fertilised ponds
Weight of individual fish (g)
d = 20 000
d = 58 000
d = 92 000
v = 10
52 000
74 000
93 000
v = 20
46 000
56 000
95 000
v = 30
46 000
52 000
89 000
v = 40
80
40
0
60
20
80
40
0
60
20
0 2 4 6 8 10
2 4 6 8 10
2 4 6 8 10
∑θe (°C × 100)

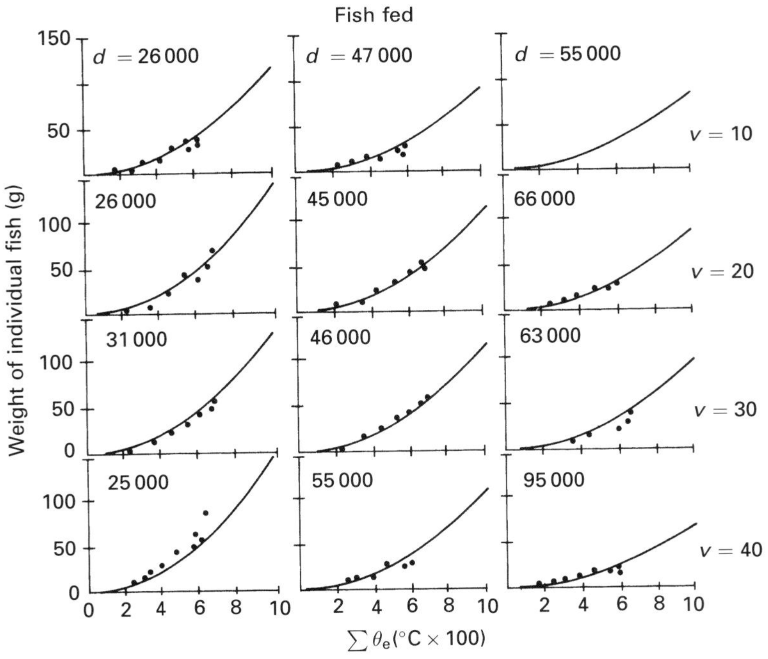
Fish fed
Weight of individual fish (g)
d = 26 000
d = 47 000
d = 55 000
v = 10
26 000
45 000
66 000
v = 20
31 000
46 000
63 000
v = 30
25 000
55 000
95 000
v = 40
150
50
100
50
100
50
0
100
50
0
0 2 4 6 8 10
2 4 6 8 10
2 4 6 8 10
∑θe(°C × 100)

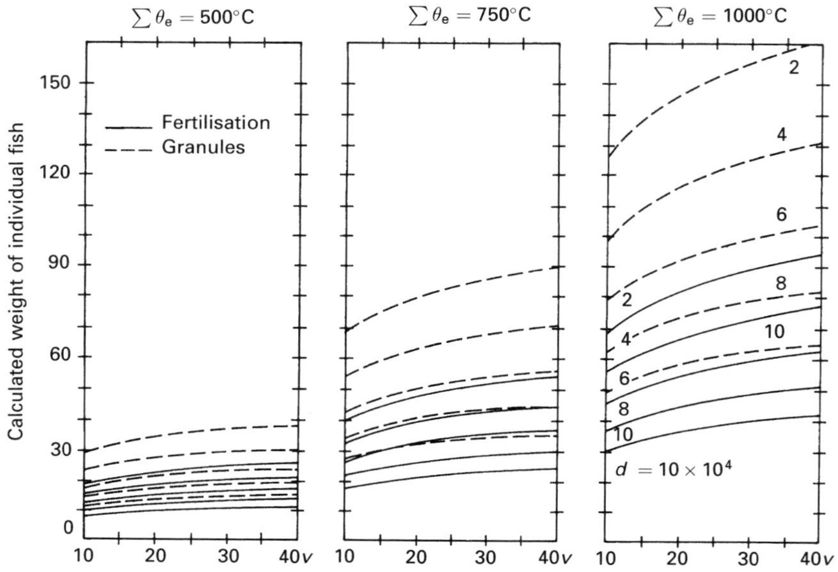

Fig. 6.19. Change in body weight of individual carp during the 1st year calculated for three different degree sums over the season and four values of v in the case of fertiliser dominating (lines) and artificial feed dominating (dashes). The final densities (d) of fish (shown in the figure on the right) vary between 20 000 and 100 000 individuals/ha/season. The total production for the season will be 3.2–4.8 t/ha at 1000°C and 0.8–1.1 t/ha with 500°C.

In equations (1) and (1′), the expression n^x assumes that when n is large, the accuracy will also be high. It should be remembered that when such models are used certain parameters should be measured to a high degree of precision; for others such precision is superfluous. For example, it is essential to keep 0.000 012 in G_{1B} as a multiplication factor taking into account that d varies from 20 000 to 100 000 while 12 in place of 12.3488 hardly changes the precision of the equation.

6.3.3 Growth of carp during the second year of culture (C_{1-2})

A good estimate of the theoretical change in the weight of individuals (G) is obtained from the equation:

$$G_2 = G_0 + 1.53810 \times \left(\sum \theta_e\right)^{1.14239} d^{-0.55190} v^{0.27330} G_0^{0.51793} \qquad (2)$$

The agreement with observed values is shown in Figure 6.21. The correlation coefficients are slightly below those obtained for the 1st year. Here, a fourth variable is included; this is G_0, the initial size of the fish which has a major impact on the final weight. However, in the final protocol d (which has a negative effect on the growth of the individual) and v, the food variable (which has a positive effect) vary at the same time so that they are difficult to quantify individually.

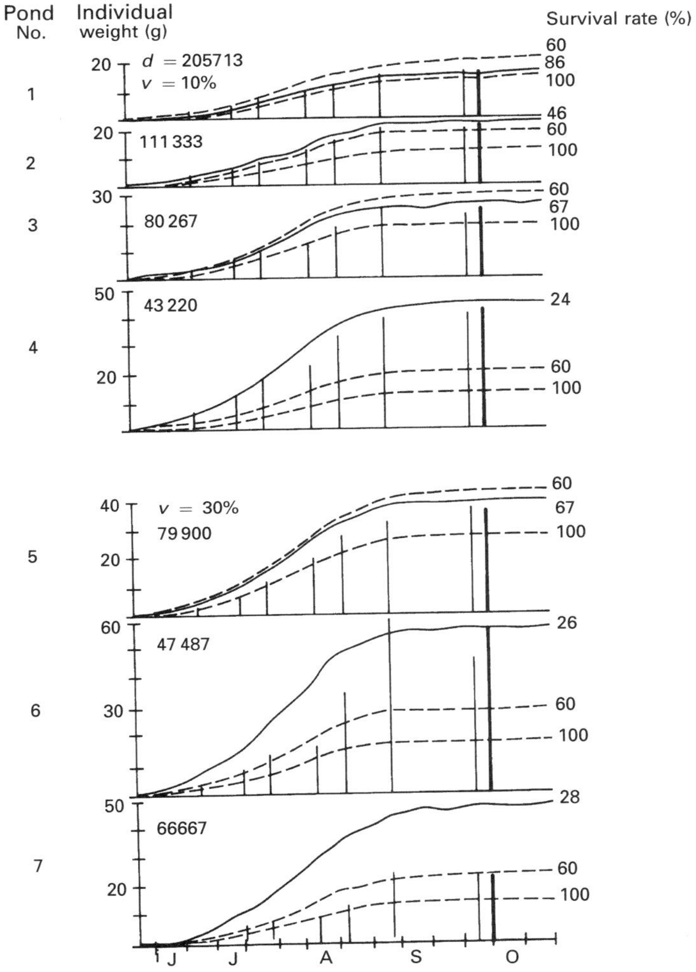

Fig. 6.20. Validation of the growth model; growth of C_{0-1} in a series of seven ponds from values observed during sampling (small lines) and the final harvest (thick line to the right) and from calculations (curves) from different values of (individuals/ha) and v as well as survival rate; the unbroken lines correspond to information from sample fishing, dashed lines from simulations.

The sum of the temperatures has, as during the 1st year, a crucial importance for the final yield. During the warm season the mass G varies from around 170 g to 1000 g per individual per season. This is valid when the lowest weight G_0 is a 20 g individual, $v = 10$ and $d = 20\,000$ fish/ha/season and the highest value is obtained when the weight reaches 80 g/individual, $v = 40$ and $d = 5000$ fish/ha/season. During the cold season, G reaches 90, around half of the growth rate during the warm season (Figure 6.22).

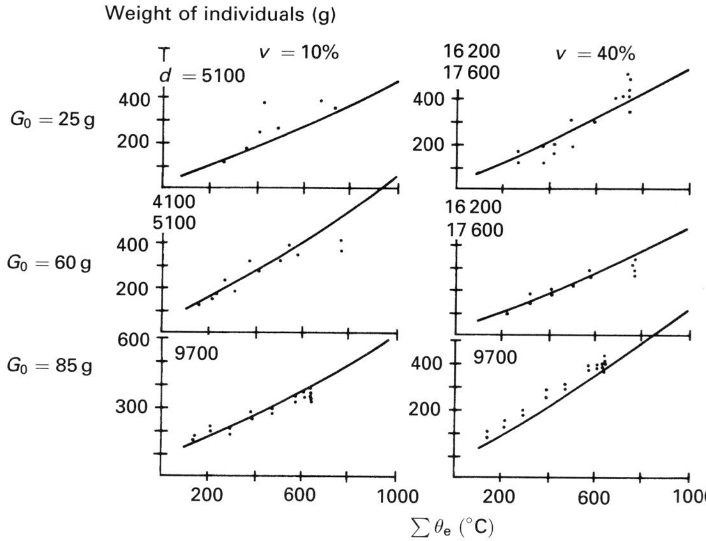

Fig. 6.21. Verification of the growth model for carp during their 2nd year for different values of d (individuals/ha/season for two ponds); v (percentage protein in the diet); G_0 (individual weight at the time of stocking); dots: actual measurements; lines; theoretical calculation.

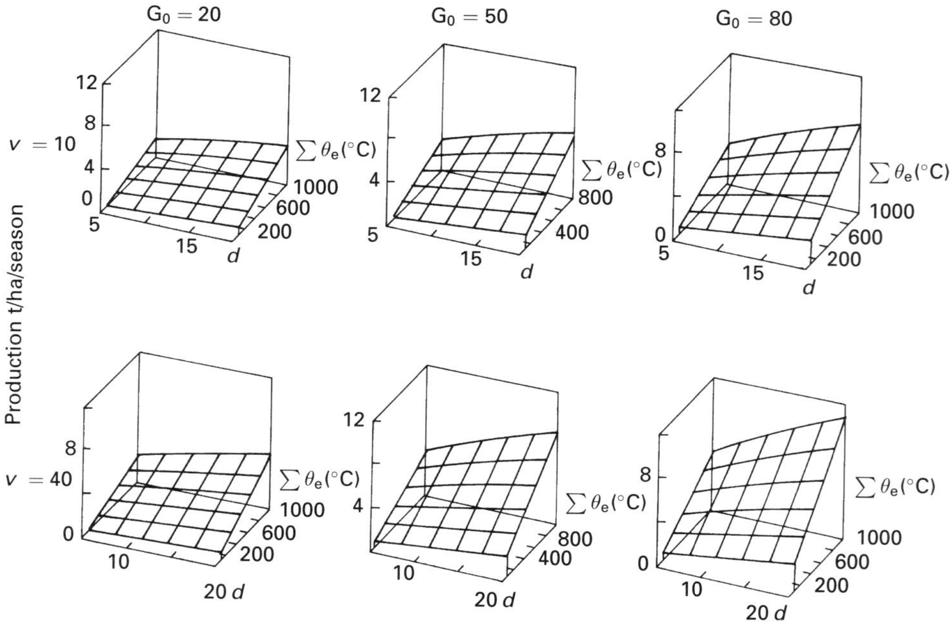

Fig. 6.22. Three-dimensional representation of theoretical yields based on the initial weight G_0 (g/ind), d (ind./ha × 10^3), v and $\sum \theta_e$ (°C).

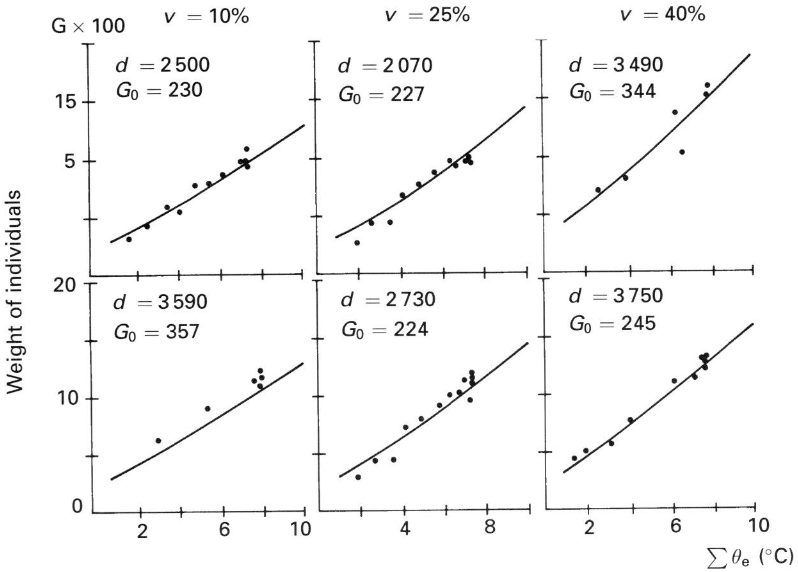

Fig. 6.23. Verification of the model of the growth of carp during their 2nd year; d, density in numbers of ind./ha/season, G_0, initial weight in g/individual, v, protein level in the diet; points—actual measurements; lines, from calculations.

For the 1st and 2nd years of growth, the model is valid for the climatic and operating conditions prevailing in Poland.

6.3.4 Growth of carp during the 3rd year of culture (C_{2-3})

The information given in Table 6.9 does not allow density to be manipulated because of the low variation (2100 to 4000 individuals/ha). The best equation takes the following form:

$$G_3 = G_0 + 0.018\,258 \times \sum \theta_e^{1.16864} 1.007\,98^v \ln(G_0)^{1.67437} \qquad (3)$$

The calculations of individual growth agree closely with those measured (Figure 6.23) although, for high initial weights (e.g. 357 g for $v = 10\%$) the theoretical curve is below the points observed. The simulated growth values for three temperature regimes (500, 750, 1000°C) show again the major role played by temperature in the determination of the final weight (Figure 6.24). At 1000°C and even at 750°C, it can be seen that it is possible to adjust the initial size and quality of the feed to obtain a final size between 600 and 1100 g at 500°C, 800 to 1500 g at 750°C and 1200 to 2000 g at 1000°C. Increasing all of these variables leads to a sharp increase in final production (Figure 6.25).

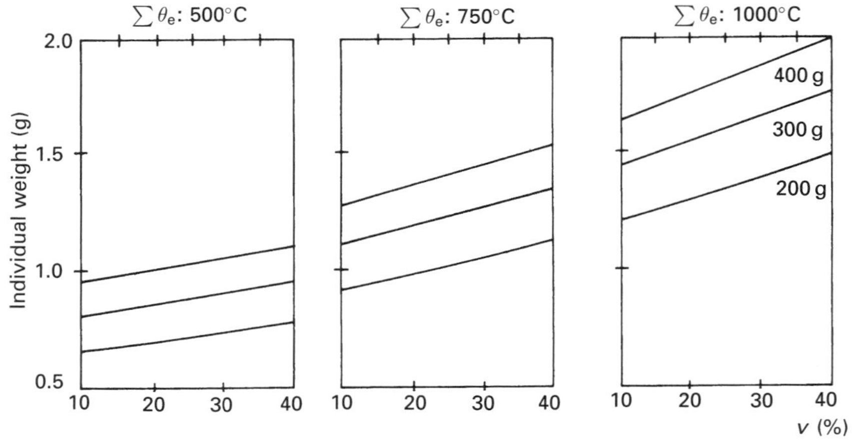

Fig. 6.24. Final calculated weight of carp in their 3rd year based on the sum of the temperatures ($\sum\theta_e$) during the growth season, the level in proteins (v) and the initial weight G_0 (in g).

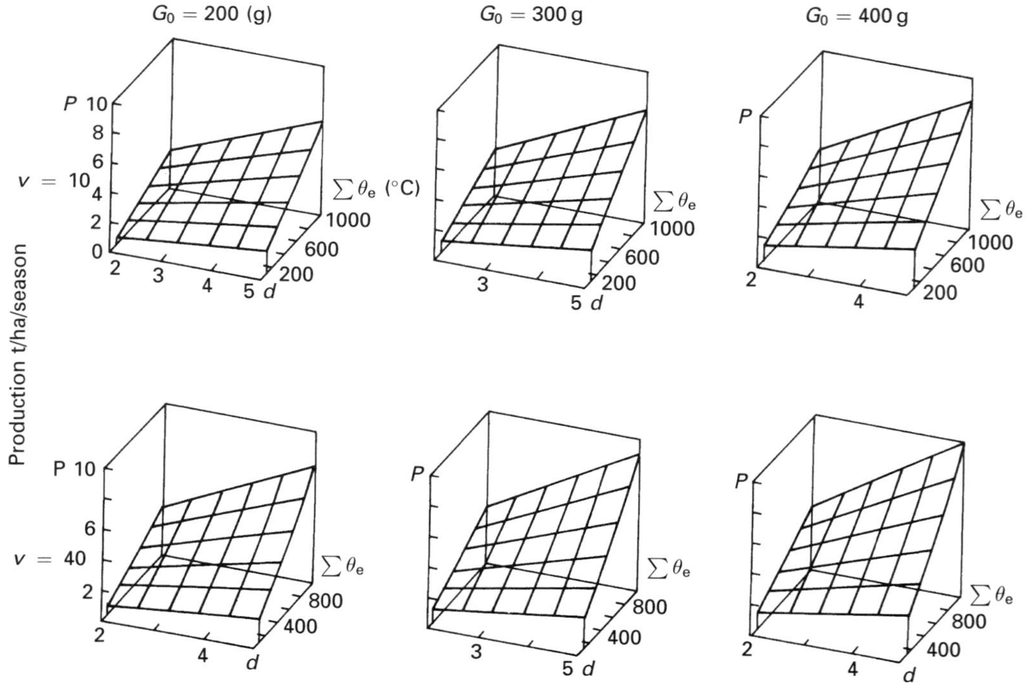

Fig. 6.25. Calculation of the productivity P (in t/ha/year) for carp during their 3rd year based on individual weig (G_0 in g), their density d (ind/ha × 10^3), the level of protein in the diet v (%) and the temperature sums $\sum\theta_e$(°C).

6.3.5 Conclusion

This example of the modelling of carp growth shows that it is possible to make a number of predictions relating to the growth of individuals and final yield, to simulate development taking into account temperature and to instigate a shift during the course of rearing by altering a range of variables such as the quality of the feed (quantity can also be considered) and the density of the fish. There is further interest in the comparison between the information from sampling (which is good practice in rearing operations) and the values given by the model. However, a model is only valid in the rearing conditions under which it has been established. Predictions are highly likely to be false when rearing conditions are altered. Models of this type can be established for each climatic region and for operations with similar characteristics. A model can also take economic variables into account to allow more effective management. Models are relatively easy to establish so long as sufficient basic information, measured with the same precision, is available, as demonstrated in the example above.

Information kept by the fish farmer in his own records can be a good starting point for a specialist. Once established, the model can be operated by the fish farmer himself with relatively simple computers which are now accessible and are becoming widespread in fish farms (p. 259).

In summary, this type of model is capable of explaining a major part of the information relating to growth, observed through the explicative variables utilised; it can predict growth (but only within its validated range) and gives an idea how to express variables within the range of values encountered; this may guide the choice of more biologically valid expressions in the final model. However, it should be noted that the expressions obtained have no theoretical justification in as much as this model does not permit an understanding of what has happened in the past and even less of what will happen if there is a departure (even a small one) from the range of variables used in calibration.

6.4 CAUSES OF MORTALITY AND THEIR TREATMENT[6]

The causes of mortality in the first 4–5 weeks after hatching have been reviewed earlier in this book (p. 142) and it is clear that changes in the rearing system as on-growing starts bring new risks which add to or replace those there previously. It is essential to see this in context; fry of around 4 to 5 weeks old are transferred from nursery tanks to ponds or on-growing tanks: the next stage extends to the point when the fish are ready for market which is 2 or 3 years later depending on the system, technology and region. Two main types of technology should be distinguished; each has different associated risks:

—intensive rearing with artificial diet;
—extensive or semi-extensive rearing where feed comes from the natural food

[6] M. Morand.

chain which may or may not be stimulated by mineral and/or organic fertiliser and/or supplementary feeding.

In the first case the risks are those associated with industrial farming and the water is no more than a support medium which contributes a little to the purification of wastes. In the second case, the water provides at least part of the nutrients through primary production.

6.4.1 Extensive rearing

The survival of fish in extensive rearing systems is a reflection of the equilibrium between the bioaggressors and the environment with its chemical, physical and biological elements. If equilibrium is maintained, the well-being of the fish is preserved and survival is satisfactory. If the equilibrium is disrupted, conditions become unfavourable and mortality begins. The serch for causes of mortality begins by reviewing the causes of imbalance between the fish, the bioagressors and the environment.

Factors relating to the fish

In a healthy state, the fish is capable of withstanding a range of threats posed by changes in the external environment or following a proliferation of bioaggressors. When the alarm and defence functions are sufficiently strong, equilibrium is maintained and stress can be withstood and overcome. In contrast, when the physiological state is poor, the fish is incapable of reacting to stress and mortalities occur with varying degrees of rapidity. In ponds it is not always possible to appreciate the consequences of such accidents and it is never possible to put them right after the event. It is important to establish which fish are most susceptible to stress, either due to the environmental conditions or to technological aspects of culture (fishing, transport, treatment, etc). In the former, stress can result from meteorological factors and is difficult to predict, in the latter the stress factors can be predicted and managed. In both types, fish should always be capable of adapting to changes in the environment and be in a good physiological state. This comes through the provision of nutritional requirements (carbohydrates, proteins, lipids, oxygen, minerals, trace elements, etc.) in suitable environmental conditions (temperature, dissolved oxygen, NH_3, absence of pollution, etc.).

In favourable physiological conditions, the functions of the fish respond normally to a range of "aggressors" which are met during the rearing cycle; conversely, individuals which are deficient in some way are incapable of withstanding changes and mortality occurs, either because of pathogens, other biological factors or poor environmental conditions.

Bioagressors

Ponds represent a natural environment where technology has little influence and where a wide range of bioaggressors, parasites, bacteria or viruses come together either through the water or through animals such as birds, mammals, terrapins and crustaceans which live in and around the pond. It is important to recognise that these bioagressors are part of the fishes' environment in the pond and that long distance transfers have introduced new bioagressors, either with the species of fish which are already present or with new species. The equilibrium between bioagressor and fish is thus precarious and all efforts should be made to avoid its disruption. Apart from measures taken to ensure the optimum conditions for the fish, it is essential to limit the introduction, spread and proliferation of bioagressors by choosing methods to counteract them as appropriate for each type (parasites, bacteria, viruses) under the conditions of the particular farm.

Parasites (Figure 6.26)

Parasites are plants or animals which live at the expense of their host. They are localised either on the skin, fins or gills (external or ectoparasites) or inside the organism (internal or endoparasites). Whether external or internal, their pathogenicity varies in relation to their number and the organs concerned and also the species of parasite. Some of them show a distinctive pathology; others are more opportunistic and therefore more variable. The list of possible parasites is long and lengthening by the day because of extensive rearing in open systems with all of the other migratory or non-migratory species which are sometimes introduced; these frequently act as parasite vectors, either as a definitive or an intermediate host.

External parasites (Figures 5.3 and 5.4). Some of these are specific to the gills and the lesions that they induce may cause problems with feeding and then with oxygen uptake. Among the most common are the monogeneans (e.g. the *Dactylogyrids*) and sometimes crustaceans such as *Ergasilus*. Others may be found indiscriminately on the gills, skin and fins and their pathology varies with the number and age of the parasites. Some are visible to the naked eye, others with a hand lens ($> \times 10$ magnification) and others with a microscope. It is not possible to define a clear relationship between size and pathogenic potential buta specific fungus *Branchiomyces* can be seen on the gills.

The most visible of the external parasites are the fungi, and some crustaceans; the fungi appear as cottonwool-like tufts which are invasive to varying degrees and can sometimes develop very rapidly but are most frequently a secondary infection of an existing lesion resulting from a wound or other cause. Other visible external parasites include leeches and crustaceans such as *Argulus* (the fish louse), *Lernea*, etc.; parasites which break the skin or the branchial epithelium, opening up the fish to infections through entry of viruses or bacteria.

To the naked eye it is easy to see the lesions caused by the presence of microscopic parasites, either as the consequence of reactions of host tissue or as one of the developmental stages of the parasite itself. The host reactions may be:

—melanin patches (accumulation of pigments) around nematode larvae; these parasites have their adult stages in birds and are not particularly damaging to fish in spite of the unpleasant appearance they confer;

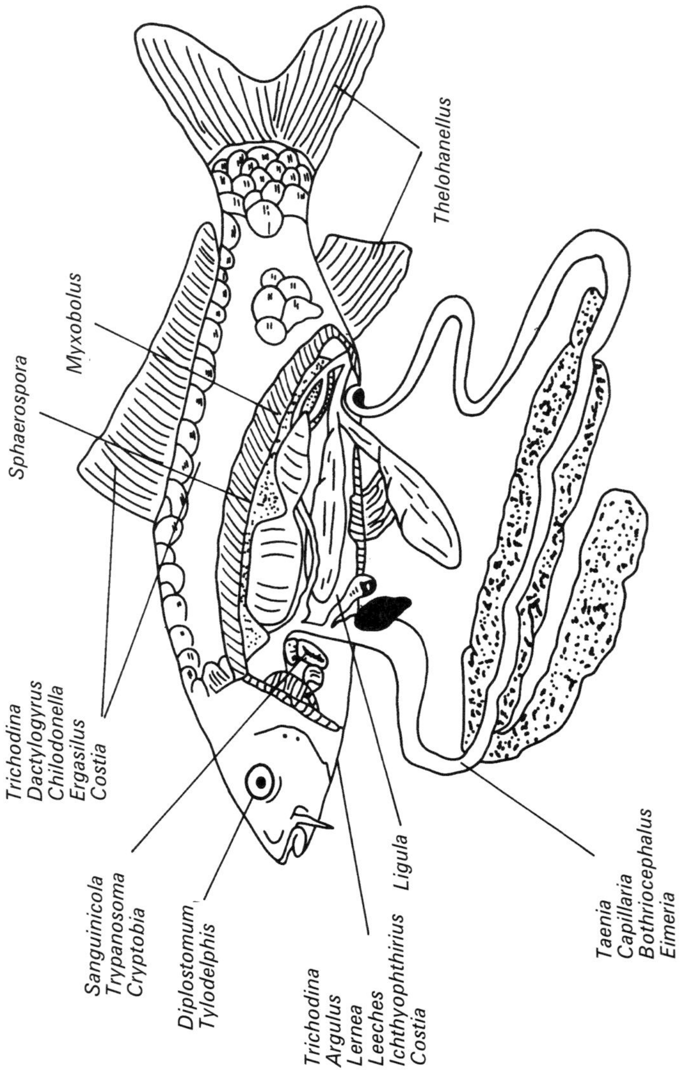

Fig. 6.26. Location of the most-common fish parasites.

—cellular proliferations around parasites such as white spot, caused by *Ichthyophthirius multifilii* (Figure 5.4) which has a very rapid life cycle when the temperature exceeds 20°C and is highly pathogenic for juveniles, lodging under the fishes' epidermis. This makes the parasite inaccessible to treatment of the fish: treatment must be through the water against the free larval stage.

—whitish or yellowish cysts, external or under the skin, due to the presence of myxosporidians including *Henneguya* sp., *Myxobolus* sp., *Thelohanellus* sp., including *T. nikolskii* which has been recently introduced to France. Depending on their location on the fish, the pathogenic effects may be significant (on the gills) or insignificant (on the fins).

—hypersecretion of mucus makes the fish sticky and sometimes nervous; they have a tendency to rub themselves and in doing so may become damaged. Such observations suggest the presence of small protozoans such as *Costia*, *Trichodina*, etc. When associated with fin erosion, the lesions are likely to be caused by *Gyrodactylus* sp.

Internal parasites (see Figure 5.5). All of the organs of fish in ponds are capable of harbouring parasites but only some of these are visible to the naked eye. The most spectacular of the endoparasites is *Ligula intestinalis*, the larvae of a tapeworm of piscivorous birds which can exceed 20 cm in length. These are sometimes referred to as solitary worms and they can exist in the abdominal cavity of the fish, sometimes from a very young age, because the fry feed on plankton. Apart from the repugnant appearance of the fish, this parasite may cause the death of its host when it perforates the abdominal wall in the perianal region.

The digestive tract can harbour numerous macroscopic parasites or smaller ones of uncertain pathogenicity which can perforate the intestinal wall and encyst in the abdominal tissues, as in the example of the acanthocephalans. It is not uncommon to find various tape worms such as *Bothriocephalus acheilognathus*, introduced with Chinese carp species. Using a hand lens or a microscope it is possible to observe the presence of various trematodes, small tapeworms, various nematodes and some protozoans such as *Hexamita* sp. It seems that only *Bothriocephalus* has a real pathogenic effect, increasing the mortality rate and slowing the growth of young carp.

The heart and blood vessels provide sites for the development of several protozoan parasites with low pathogenicity such as *Trypanosoma* sp., *Trypanoplasma* sp. and *Sphaerospora renicola* and also the trematode *Sanguinicola* sp., the adult form of which becomes lodged in the heart and releases large eggs into the blood circulation. The eggs block the capillaries in the organs, causing circulatory problems and lesions in the gills that provide a route for the release of larvae into the water. There they find their intermediate host, the pond snail *Lymnea*: the infective larval stage leaves the snail and penetrates the fish through the gills.

Many different parasites colonise other organs including the eyes (trematode larvae), gall bladder (bladder worms), or other organs indiscriminately (Myxosporidians, etc.). Some myxosporidians affect the nervous system causing nervous disorders and/or feeding problems together with malformations and abnormal coloration.

In summary, parasites which damage the skin, either directly or indirectly, and especially, the gills, can be a serious problem. On the other hand, many others are present and readily tolerated so long as they are in numbers compatible with the fishes' defence capabilities.

Bacteria

In extensive rearing systems, specific bacteria are unusual with the exception of *Aeromonas salmonicida novi*, the causative agent of erythrodermatitis in cyprinids, but some strains are confined to one species: pike, perch or roach. This bacterium causes a range of haemorrhagic septicaemias with mortality or cutaneous problems, the development of ulcers and sometimes healing and recovery.

In addition there are highly opportunistic water-borne bacteria: among these are *Aeromonas hydrophila, Pseudomonas* sp., etc., as well as flavobacteriaceae (previously referred to as Cytophagales or myxobacteria) which secondarily affect primary lesions. They are all equally likely to cause lethal septicaemias where the fish are stressed through lack of oxygen, effects of transport, etc.

These bacteria are sensitive to most of the antimicrobial agents used in fish culture but the symptoms are usually delayed, by which time the fish are unable to take in medicated or non-medicated feed. Because of this, fish of unknown health status should not be brought in and parasitic diseases which are often associated with lesions and provide a route for the entry of opportunistic pathogens should be prevented.

Viruses

Viruses affecting pond fish are poorly understood: spring viraemia of carp (SVC) is the most studied viral disease, but there are others. In practice, extensive rearing does not cause the outbreak of viruses and disease seem to happen by chance. These include perch rhabdovirus, carp rhabdovirus and iridovirus of different catfish species.

In carp, SVC is manifested by a short period where mortalities occur without lesions or symptoms, then the mortality increases to reach a ceiling and declines gradually; any lesions that appear are usually melanosis, exophthalmia and haemorrhages on the fins, gills, eyes and internal organs including the liver, swimbladder, visceral fat, intestine, kidney, etc. In the terminal phase, ulcers may appear on the skin, ascites develops and there are behavioural changes such as lethargy. The fish varies its behaviour between drifting in the current and phases of hyperexcitation. Fish which survive SVC may appear normal but be contagious carriers of the virus.

Stress often causes the manifestation of the clinical form of the disease; many mortalities in the days following stocking pass unnoticed and are the consequence of transport or the change of water. However, not all significant mortality is caused by the virus. It is therefore necessary to improve management by setting up routines for determining the cause of death.

In practice, fish reared in ponds are subjected to the pressures of pathogens and other adverse biological factors linked to primary production in the water which is itself controlled by climate. The presence of associated populations which may be directly harmful through predation or indirectly as carriers of parasites or pathogens must also be taken into account. The farmer who chooses to farm extensively must understand and be familiar with these constraints. Some of these are manageable, for example by disinfecting the pond, spreading organic manure and mineral fertiliser, choosing fish in good health and treating external parasites when the fish are held in tanks prior to stocking. In addition, it is important to study the behaviour of the fish in the weeks following the fry stage. This can be done by placing a few tens of fish in cages with easy access: if there is any mortality, it is easy to take a sample for laboratory examination.

Environmental factors

In order to survive, grow and reproduce normally, fish need comfortable conditions. These result from controlling technological factors such as handling, harvesting and transporting the fish, and the physical, chemical and biological factors described elsewhere.

Oxygen and food must be available throughout the year, particularly during warm weather when requirements are greatest. The younger the fish, the greater their requirements, chiefly in the need for a regular supply of suitable food which is rarely present in natural conditions in ponds. It is essential to satisfy the needs and also to suppress, or at least discourage, unfavourable factors such as toxins and irritants (chemical or biological). For example, badly managed fertiliser applications can lead to an oxygen deficit, an excess of nitrogen or an algal bloom that causes asphyxia or poisoning with toxins from cyanobacteria (*Mycrocystis* sp., *Anabaena* etc.)

6.4.2 Intensive rearing

In intensive rearing, the water only serves as a physical and chemical support for the fish. The biological function is largely replaced by the distribution of artificial diet. In terms of bioagressors, the high density of fish is more difficult to manage and the farmer must keep strict control of hygiene and the health of the fish introduced to the system, the quality of the water and the cleanliness of the pond banks and bed.

Factors relating to the fish

The constraints linked to the fish are similar to those described above. In addition, there are social factors, principally in relation to the supply of feed to fish at high densities. Intensification also favours the spread of pathogens and other bioaggressors.

Factors relating to bioagressors

Intensification of rearing in tanks or small ponds is likely to reduce the incidence of diseases caused by organisms with life cycles involving other species such as plankton, molluscs, mammals or birds. However, any infectious diseases in the general sense will find ideal conditions for their development and spread when a large, receptive population is present. This is true whether the disease agent is a parasite, bacteria or virus.

Parasites

The most significant parasites in intensive fish culture are those with a direct life cycle and those whose life cycle includes a maturation stage in a substrate which is rich in organic matter. These conditions favour ectoparasites such as *Costia* in young individuals, then *Gyrodactylus*, an obligate parasite then *Dactylogyrus*, *Ichthyophthirius*, crustaceans and other flagellates. Given the infesting potential of these parasites and their ability to induce pathogenic effects during massive infestations it is essential that they should be excluded from rearing systems. Prophylactic measures such as bath treatments should be rigorously applied before stocking. Unfortunately these are largely ineffective for *Ichthyophthirius* because its location under the epidermis gives it protection against bath treatments. This parasite can be devastating in intensive systems; when temperatures increase and the progress of the life cycle accelerates, huge, lethal infections occur. Prophylaxis for this parasite entails quarantine and hygiene control. The risks are lower for internal parasites other than *Hexamita* which has a direct life cycle in juveniles; most internal parasites, whatever their location in the fish, require an intermediate host.

Bacteria

The most recognised bacterial disease in cyprinids is carp erythrodermatitis, associated with *Aeromonas salmonicida* (atypical) which is now referred to as *A. salmonicida novi*. This has a characteristic pathology which appears as mortality with haemorrhages, and cutaneous ulcers associated with exophthalmia and melanosis. This disease appears after stress either from technological factors (grading, transport, etc.) or natural factors (change of water temperature, increase of suspended matter in the water, etc.). In addition to this specific disease, poor conditions in the rearing unit can favour the development of septicaemias caused by opportunist bacteria. Bacteria which are normally present in the water may be associated with mortalities, either when the fish are weakened or when there are abnormally high numbers of them.

The main prophylactic measure against bacterial diseases is, above all, to maintain the health and wellbeing of the fish by ensuring their needs are met with a balanced diet, an oxygen-rich and an ammonia-poor environment and preventing the occurrence of external lesions by taking measures against ectoparasites and by stocking with high-quality individuals. It is possible to treat erythrodermatitis with a diet containing antibiotic. In addition to these well known diseases, others may occur as new rearing conditions are introduced; amongst these are gill disease and septicaemias caused by flavobacteriaceae.

Viruses

One virus is currently dominant in the rearing of carp in ponds–the rhabdovirus. Each fish species seems to have its own distinct rhabdovirus; for carp this is the agent of SVC, a disease accompanied by high rates of mortality in its acute form or extended losses in its chronic form. It is well known that intensification reveals new viruses, as has happened in other species. The only possible prophylaxis is that of health management based on the rearing of juveniles from hatcheries which are certified disease-free.

6.4.3 Conclusions

The success of rearing depends on the management of technology in order to respect the well-being of the fish. Techniques must consider the "disease" in the design stage and include appropriate prophylactic measures.

- disinfection of tanks and ponds;
- control of the health of introduced fish together with treatment and quarantine;
- following introduced fish by installing some cages in ponds;
- use of a balanced diet, distributed in a suitable quantity for the numbers of fish present.

Mortalities may be caused by the presence of infective agents but they often reflect a technical mistake or the omission of a prophylactic measure. Specific treatments, given after accurate diagnosis, are only effective if the technical defects of the rearing system are rectified at the same time.

6.5 PREDATION BY PISCIVOROUS BIRDS AND METHODS OF PROTECTION[7]

The establishment of a fish farm entails changes to the environment of a varying degree of significance; this has an effect on the native species of animals. The farm increases the amount of food available and therefore the species associated with production. These have included rats and mice, which were associated with the first agricultural farms, and sparrows and starlings, after the intensification of cereal production.

The bird species which pose problems for the pond fish farmer throughout Europe are, in decreasing order of importance (Figure 6.27):

(7) H. Le Louarn.

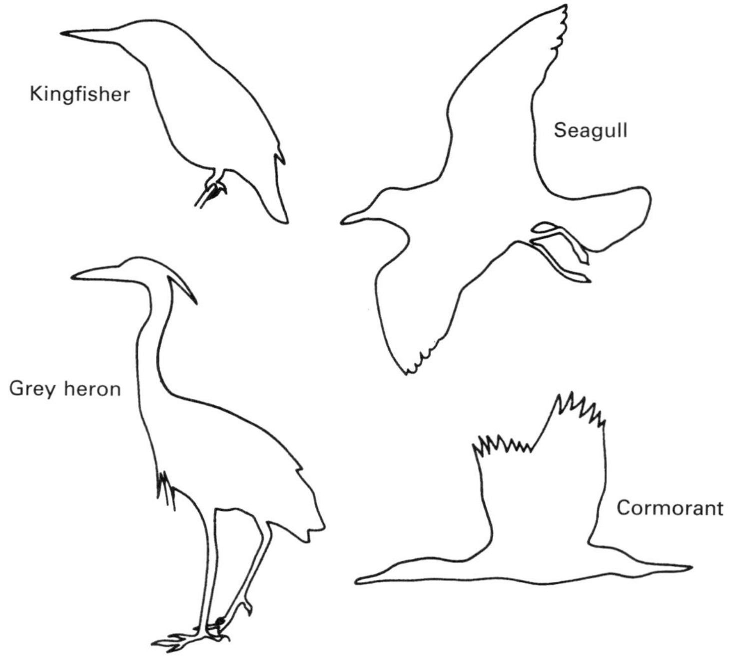

Fig. 6.27. Silhouettes of piscivorous birds.

- the cormorant, formerly confined to coastal zones (The Netherlands, Camargue and Atlantic marshes in France) but now spreading inland;
- the grey heron in Germany, the Benelux countries and northern France;

and, to a lesser degree:

- seagulls can be a problem in some areas when ponds are emptied;
- kingfishers in intensive farms.

Great-crested and little grebes are noted piscivores but do not pose a major problem because they are present in small numbers and do not overwinter in large flocks.

6.5.1 Species of predator

The cormorant (see Box)

This species has been the subject of extermination campaigns and for several years was threatened with extinction in some countries (The Netherlands, Australia). It

Losses caused by cormorants and possible means of prevention
(by C. Ferra)

The number of breeding pairs of cormorants (*Phalacrocorax carbo sinensis*) in Northern Europe has increased from 9000 in 1980 to over 100 000 in 1993. In France there were 15 000 overwintering birds in 1982–1983, in contrast to 60 000 in 1993, with a trend towards increased numbers being found inland where they arrive in August and September and remain for around 6 months, leaving in March or April. Peaks can be observed in November and December (Brenne) or in February (Dombes). Each individual consumes a mean of 400–500 g of fish from a wide range of species each day (reviewed by Staub and Ball, 1994). The size of the fish consumed ranges between 5 and 30 cm but the preference is for fish between 13 and 17 cm in length (Adamek, 1991). It is difficult to estimate total losses. A theoretical calculation suggests that 60 000 individuals overwintering in France for 120 days would consume 3500 tonnes of fish. For regions with many ponds, calculations made by Marcel (1994) based on counts of birds by FRAPNA and the ONC suggest losses of around 500 tonnes, including around 150 in Dombes, 80 in Brenne and 80 in Sologne. In the USA, a different species of cormorant, (*P. auritus*) causes severe losses on catfish farms with a mean hourly consumption of 4.75 catfish of 14 cm length (Stickley, 1991). Again, it is difficult to quantify true losses. There is the autumn removal of stocked or small fish, the spring removal of fish which are just about to be stocked into grow-out ponds (which leads to a considerable production deficit in the following year) and losses caused by wounds to bigger fish which lead to a reduction in health; these are estimated at 15% of the overall losses. Several species of cormorants are carriers and vectors of parasites including *Ligula intestinalis* and *Diphyllobothrium*. The cormorant is also the final host for other parasites such as *Diplostomum* sp. and *Posthodiplostomum cuticula*. Overall, losses appear to be considerable for what is, in France, a fragile industry. Marcel (1994) estimated direct losses at 7.2 million French francs but they should be considered as much higher than this because of the subsequent loss of fish for restocking, production deficits and losses to the market.

The Bonn and Bern conventions (23rd June and 19th September 1979) do not place the cormorant on their list of species to be protected. In France, the decree of 17th April 1981 ratifies the European Directive but Article 9 of this directive suggests the possibility of derogation "In the interests of health and public safety: to prevent significant damage to livestock cultures, to forests, fisheries and to waters". The application of this article rests with each member state: Article 2 allows that "To ensure the maintenance of biological equilibrium, the Minister charged with the protection of nature can establish, in the case of necessity, methods for the destruction of species listed in article 2 . . . and can proceed to the destruction of eggs and nests of these species". It is therefore possible for member states to take action. In Germany, Denmark, Italy and Ireland there is authorisation to capture, kill, scare and to destroy nests if damage to fisheries can be shown. In France, the decree of 2nd November 1992 authorised shooting under certain conditions in the following departments: Ain, Indre, Loir et Cher and Moselle in 1992/93. Instruction number 94/3 of 6th June 1994 of the Direction de la Nature et des Paysages (DNP) modifies and replaces the preceding one and defines the conditions for granting authorisation for destruction: establishment of a departmental committee for monitoring and authorisation of destruction, delivered by the Ministry of the Environment after advice from the Comité National de la Protection de la Nature. In the Dombes region, such measures have been found to be insufficient.

has been protected in France since 1972 and recognised by an EC directive since 1979 as a species where some protective measures are needed.

Many studies on the diet of cormorants have been carried out in Europe and have shown that they feed almost exclusively on fish. In general, the diet in freshwaters is made up of the most available fish and, in the sea, pelagic fish.

In the Camargue, where a quarter of cormorants overwinter in France, the daily individual intake is between 340 and 540 g, depending on the weight of the bird (2 kg for females, 2.7 kg for males). Prey are very variable in size, from 6 to 27 cm. Food intake is very rapid and cormorants can satisfy their daily requirements within 10 min. Carp, which are not very active in winter, are extremely vulnerable in contrast to non-commercial species (sunfish, catfish, bream). Predation is greatest on young carp in their first year weighing between 60 and 80 g. Carp of 200–250 g are particularly prone to attack (see below). The situation is becoming catastrophic for pond farmers.

The impact of this species is increasing and some farmers find themselves with a dilemma. Damage has occurred in France, in the Arcachon Basin (cormorants overwinter around the coast), Dombes Region, Central Region where several thousand birds frequent the Loire and feed on the fish farms of the Sologne, parts of Germany, Switzerland and, particularly, The Netherlands where the large fish farm at Lelystad has been forced to cease operation. This farm was situated close to the largest breeding area in Europe for cormorants (6000 pairs) and was visited continually by several thousand birds.

The grey heron

As with the cormorant, this species has been protected in France since 1975 and since 1979 in the rest of the EU. In Bavaria, shooting the grey heron is authorised for a radius of 200 m around fish farms from 16th September to 31st October each year. In France the population is estimated to be 13 000 breeding pairs which are closely associated with watercourses. There are two main populations: the Atlantic coast from the south of Brittany to the North of Aquitaine (5000 pairs in 40 colonies) which feed on the 200 000 hectares of the Marais in the west; a third in the north-east of France with a more widely dispersed population (over 100 colonies) with the same number of breeding pairs. Growth in population size has been noted, especially in an isolated population in the Camargue. This growth in numbers is accompanied by the establishment of new colonies, often temporary, and a greater migratory tendency of juveniles in summer following the overcrowding of feeding grounds. The effect of the migrants and the overwintering birds found mainly in the north-east is not known with any degree of precision. A significant population increase has been recorded for all of Europe with colonisation (sometimes temporary) of new environments. Populations have probably doubled but the new colonies are somewhat fragile, particularly in Great Britain.

The diet is very eclectic, ranging from shrimps and dragonflies to fish, voles and frogs. In salt marshes, the area with the richest fauna, the average composition of the biomass consumed is as follows: 85% fish of 18 species (44% eels), 10% other vertebrates (6% rodents). The daily dietary requirement is between 270 and 370 g, depending on the weight of the predator. Herons mainly take prey of 5 to 15 cm, the size which they find most frequently. They also wound larger fish. In Britain, it has been shown that grey herons consume between 3 and 8% of the annual production of a pond used for the polyculture of roach, rudd, tench and pike and sometimes up to 35% of the fish in holding ponds (unpublished information by H. Le Louarn).

Other species

Black-headed gulls. Breeding populations are widely dispersed and mainly associated with the sea. However, migrants and overwintering gulls cause problems for the farmer when the pond is being emptied. The gulls, although generally only occasionally piscivorous, find sources of food which are absent elsewhere at this time of year.

The kingfisher. The distribution of pairs and the general low density reduces the impact of this species. Predation is generally on fish of 4 to 7 cm (10 cm maximum). Impact is sometimes significant in intensive culture tanks.

6.5.2 Estimation of losses

Information on this subject is almost non-existent except for a 1990 study in Germany which estimated losses of 5000 ecus to an intensive carp farm where a mean number of 15 grey herons were permanently resident.

For practical guidance the following advice can be given:

Intensive fish farms

Losses are variable but can be very high (cormorant, grey heron, kingfisher). Removals depend on the size of the prey and the layout of the tanks or ponds. Apart from a few examples of fish of certain sizes where the mortality rate is low, the impact is high (up to 1800 trout in 6 months by one heron). The market value is also relatively high whether the fish are trout from a fry unit or cyprinids destined for fishing (unit price 2 to 6 French francs (FF)). Whether rearing is in tanks or raceways, protection is often but not always achievable.

Medium-size waterbodies

Carp produced in grow-out ponds are subject to predation during their first 2 years. From the second summer onwards the fish are too big. However, fish which are wounded and unsaleable (200 to 250 g) must be added to the direct losses; the rates of such damage can reach 30 to 40%. Finally, major losses may occur through flocks of gulls when ponds are being emptied.

Larger waterbodies (including salt marshes)

Because of the size of the predator population, the biomass of fish consumed is sometimes significant: up to 10 tonnes for 700 ha of marsh by cormorants; 0.5–3.5% of the biomass of commercial fish in a lake by the same species in Switzerland. Production cannot be studied precisely where ponds cannot be emptied and accurate counting techniques used to quantify the impact. It is certainly lower than in production ponds as it is mainly based on species of little commercial importance.

6.5.3 Possible methods of protection

Several methods have been tried for different bird species.

Nets

It is clear that this is the only method which can give complete protection but it is only practical in a limited number of situations, such as channels or small tanks and ponds. If handling is not necessary a net may be stretched at ground level with several support posts driven into the pond. However, if it is necessary to move around the pond, overhead nets must be stretched across a frame. Prices for net systems vary from 0.38 to 0.77 €/m^2.

Partial protection

To counteract the grey heron, it is possible to partially enclose the ponds: nets around the sides, wires stretched overhead, or, more simply, wires stretched in one or two directions. Protection is variable and wires have no effect on kingfishers. Against the cormorant, reliance is placed on the fact that the bird needs a distance of 8 to 10 m to take off and this is more than tripled after the bird has fed. In the Camargue, polyamide wires were stretched every 20 m, 1.5 m from the surface, in two directions. The protection was effective for ponds of over 1 ha. (Figure 6.28).

Bands of plastic or lines of floats situated 1 m from the bank have also been shown to be effective against herons.

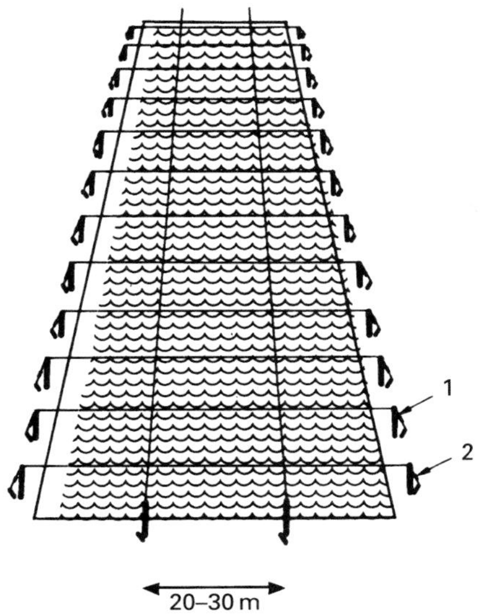

Fig. 6.28. System for the protection of a 1.2 ha pond. 1.5 m high stakes are placed at 20–30 m intervals and polyamide wire is stretched between them. This is held rigid by an iron wire (1) and a tightening device (2) (from Im and Hafner, 1985).

Management of the bank

It has been observed that herons are unable to fish when the depth of the pond is greater than 50 to 70 cm. A steep bank thus causes difficulty, as does the absence of shelter since this species likes some protection (Figure 6.29).

Scaring systems (sound, vision or a combination of the two)

In the Camargue, all types of scaring systems have been tried against cormorants: canons, cartridges, detonators and lights and silhouettes of hunters or birds in the alert position. They show some effectiveness but are not compatible with the usual operating practices on most bodies of water. Distress calls have been shown to be ineffective for cormorants. However, herons are deterred by the sound of distress calls or sounds of juveniles flapping their wings. Frightening birds at roosts may also be possible: this has been successful for crows and starlings. Cormorants would probably abandon their roost and regroup some distance away. These roosts support birds from a range of at least 30 km but results are of doubtful value as the birds will return to the same place or to a neighbouring farm. This will always be a problem with the various scaring methods.

In conclusion, reducing the impact of piscivorous birds on fish farms has no simple solution even if only two species (grey heron and cormorant) cause major problems. The combined use of different deterrent techniques would probably lead to an improvement: nets over intensive fish-culture units, partial protection (and

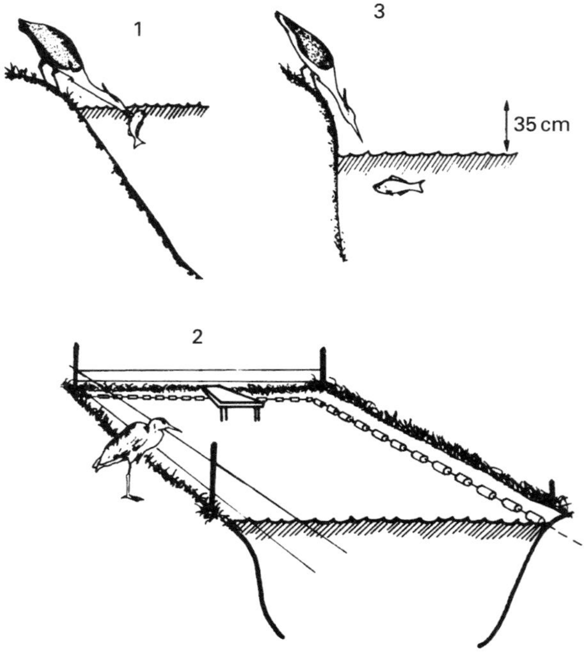

Fig. 6.29. Partial protection against herons: relatively easy removal of fish (slope of bank $<60°$) (1); fishing hindered by wires and floats (2) or by steeper slope and lowering of the water level (3) (after Templeton, 1984).

management of banks for grey herons), completing harvests in big ponds at the end of October and beginning of November before the influx of large numbers of cormorants and keeping fish in small tanks or ponds which are easy to protect. If population control is planned, at least for cormorants, it must comply with European Commission Regulations because the breeding birds posing problems are the most northerly ones.

Since the above was written, the population of cormorants has continued to increase; control measures including shooting have been insufficient to such a degree that the existence of many fish farms in the pond region is threatened.

6.6 PREVENTION OF ADVERSE IMPACTS OF POND-BASED FARMS ON THE ENVIRONMENT[(8)]

6.6.1 Possibilities for the reduction of environmental impacts

Rearing fish or higher vertebrates often has an adverse impact on the aquatic environment. Fish farms can have a range of impacts. Salmonid-rearing units

---

[(8)] Parts furnished by P. Cerda i Mulet and C. Salomoni.

produce organic wastes similar to those produced by intensive agriculture. In ponds, even where production is extensive, the quality of the effluent water is altered (in particular there is an increase in the water temperature as it passes through in summer) and suspended waste matter is lost when ponds are emptied. These factors can all cause deterioration in water quality downstream. Effects can be reduced by management of the water, by reducing the frequency with which ponds are emptied and using settling ponds to receive the waste-laden water coming from the rearing units (as has been successful at M. Dannancier's fish farm at Cormoz, Ain, France). In fish-farming ponds, the wastes produced by the fish themselves are not harmful as they are recycled *in situ* and act as fertilisers. This property of recycling the effluents produced in rearing within the ponds is well known and has been applied for many years in Asia (p. 297) with the management of the trophic web in the pond where microalgae, zooplankton and fish are living together. More elaborate systems are being developed in lagoon systems in Europe where algae, plankton and, in some places, fish and higher plants are in separate ponds. Living matter can be extracted from these ponds and exploited. The harvesting of algae is not yet economically viable; that of zooplankton is beginning to be considered for direct distribution to fish and for drying and incorporation into feeds for feeding to other domestic animals. Two examples of recycling are given below, one from Spain based on the recycling of effluents from an intensive carp unit (that could be applied to salmonid culture); the other, from Northern Italy, is a pilot plant for recycling manure from piggeries which produces zooplankton and fish.

6.6.2 Recycling systems and the purification of water in intensive fish culture

Systems for recycling and/or purification are still little developed in fish rearing, except in hatcheries (p. 95). However, it appears to be increasingly necessary to recycle waste to a varying degree as intensification of rearing increases, partly for reasons of water economy and partly to minimise the release of organic matter into the natural environment. There are no standard systems, a range of types is in use. In Majorca (Spain) a small carp farm (Carpeix Pollenca) which has a seasonal production of 9 tonnes per annum (summer sales of 1-year-old fish from 250–400 g) uses water which eventually goes for irrigation in summer. The water is drawn from a 30 m well at a rate of 20 m^3/h, feeding 1500 m^3 of tanks (one-third of the water is replaced each day) and is then used to irrigate lemon and orange trees. Outside the irrigation periods, the water is recycled to avoid permanent pumping of groundwater and a purification system is used. This is made up of two ponds which form a type of lagoon system.

The first pond was designed as an overflow ditch with a butyl liner and a volume of 1000 m^3 (330 $m^2 \times 3$ m deep) from which water flows into a second pond (300 m^2), no more than 1 m deep. Before being fed into the large overflow channel all of the water from the fish farm goes through a 3-m-deep collection channel (volume 10 m^3). Suspended organic matter is decanted into this collecting channel and then into a bigger settling channel; it is removed by pumping, six times each year (four times in spring and summer and twice during the rest of the year) and is spread on the orchards. The wastes are removed when they reach a depth of 40 cm in the bottom of the channel. By removing such large quantities of sediment, the recirculation of organic matter is reduced and the outflow channel offers a degree of purification by producing algae, daphnia and even fish. The quantities of ammoniacal nitrogen, which reach 6 mg/l in the collecting channel are reduced to 2 mg/l in the large outflow channel. Daphnids are captured at night using a net after having been concentrated by a light source (100 W). In summer the daphnia are fed to young carp in the fry ponds but they also have other uses (animal feeds, aquarium food).

Complementary purification takes place in the second pond where algae, rotifers and daphnids develop together with several clumps of water hyacinth. This pond has the effect of slightly reducing the level of NH_4^+ from 2 to 1 or 0.5 mg/l. In this last pond, as in the 3 m channel, some black bass are kept in order to eliminate any juvenile carp which may have accidentally escaped from the rearing system. If these carp were to develop in the lagoon system they would predate on the large daphnia which would therefore not be available to consume the microalgae effectively. There is managed removal of daphnia using a net and also by cyprinids in a cage. Organic matter is thus exported from the system.

6.6.3 An example of manure recycling in Italian lagoon ponds with the production of daphnia and fish

In the Emilia Romagna region there are 4 500 piggeries and 2.6 million pigs. The local authorities have planned a system for processing the manure in lagoons and have developed a pilot scheme covering an area of 1 ha (2000 m^2 of water for algae, 900 for zooplankton and 3000 for macrophytes and fish) which is capable of treating part of the wastes from a piggery (Billard *et al.* 1990).

Wastes which have been digested anaerobically are pre-diluted in a small mixing pond before being spread in four ponds with a total area of 2000 m^2 for algal production. These are lined with butyl sheets and raised by 1.5 m to allow emptying by gravity into the downstream pond. The water is kept permanently in circulation (30 cm/sec), to avoid any sedimentation, by means of a central longitudinal wall and a paddle wheel. The depth of the water (60 cm in winter, 30 cm in summer) can be adjusted in order to change the retention time in relation to seasonal climatic conditions (12 days in winter, 2 days in summer). The manure is added to the system at a rate of 2 l/m^3 of culture in winter, 6 to 7 l in summer; the maximum application can reach 12 l. Intermediate levels are applied between seasons. The microalgae cultured are *Chlorella* sp., *Scenedesmus* sp. and *Scenedesmus acutus*. The cultures are harvested on a semi-continuous basis at the outflows of the ponds, the algae being fed once a day, 6 days a week, into a series of small earth ponds for zooplankton. The species of zooplankton produced are *Daphnia magna*, and occasionally *Moina brachyata* and *Brachionus calyciflorus*. Because of the high rate at which the manure is spread the level of NH_4^+ is still very high in the daphnia rearing units where it can exceed 5 mg/l. A final purification takes place downstream in a small pond of 3000 m^2 (earth pond, 1–2 m deep) by means of macrophytes (*Lemna* sp. and ferns such as *Azolla filicoides*). Levels of NH_4^+ are reduced to below 5 mg/l. It is possible to raise fish in the final pond which benefits from artificial aeration. The water is not filtered at the outflow of the zooplankton ponds, the organisms which pass out with the water become food for the fish. The retention time of the water is around 50 days. The outflow water is returned to dilute the manure at the beginning of the system. Floating plants are harvested at the outflow screens.

Production performance of the system

Taking account of the short summer season during which the system can absorb the highest rate of application of manure (12 l/m^2/day) the annual quantities recycled are around 20 000 m^3 manure at around 0.5% dry weight (after fermentation and extraction of biogas) or around 100 t dry weight per hectare of algal culture per year.

The highest levels of algal production (>20 g DW/m^2/day) is only achieved for around 100 days per year and the total production is no more than 30 t dry weight/ha/year. This production may be greatly reduced if rotifers develop in the system. This can be prevented by maintaining a high pH and by microfiltration.

The productivity of zooplankton depends on temperature, the quantity of microalgae given and the density of zooplankton in the culture. In terms of fresh weight per m^2 per week, it is possible to harvest 30 g in the cool season, 250 g under more favourable conditions, or a total of 80 t/ha over the course of a year (in a 1.2 m deep pond).

The productivity of the last pond (3000 m^2) is higher for plants; 5 g dry weight/m^2 in March and 15 g at the end of June or 9000 kg dry weight over the whole year. A first experiment in the rearing of fish consisted of stocking common carp (346 individuals), grass carp (142) and goldfish (182) in a 3000 m^2 pond in summer, a total of 2250 individuals per hectare. The final production, in autumn was more than 1 tonne/ha (common carp 800 g, Grass carp 250 g and goldfish 150 g), the survival rate was 99%. The quality of the water leaving the system was satisfactory with levels of NH_4^+ below 3 mg/l with a lagoon area of 12 m^2/pig.

A lagoon system for pig manure is currently being tested in France (Aisne), but the French system differs from the Italian pilot plant in its simplicity (two algae ponds instead of five) and by the fact that only microalgae are used in the purification process.

BIBLIOGRAPHY

Adamek Z., 1991. Food biology of cormorant (*Phalacrocora carbo* L.) in Nové Mlyny Reservoirs. *Bull. VURH Vodnany*, **4**: 105–111.

Augustyn D., Szumiec M.A., 1985. Studies on intensification of carp farming 3. Meteorological conditions, solar radiation and water temperature in ponds. *Acta Hydrobiol.*, **27**: 159–172.

Bérnard A., 1993. Effects d'une fertilisation riche en matiéres organiques azotées sur les relations trophiques (bactéries, phytoplancton, zooplancton) dans un étang de pisciculture. Thèse Doctorat, Muséum d'Histoire Naturelle, Paris, p. 215.

Billard R., 1991. Divers aspects de l'aquaculture dans le monde. *Pisc. Fr.*, **103**: 25–40.

Billard R., Marcel J., 1980. Quelques techniques de production de poissons d'étangs. *Pisc. Fr.*, **16**: 9–49.

Billard R., De Pauw N., Micha J.C., Salomoni C., Verreth J., 1990. The impact of aquaculture in rural management. In: N. De Pauw, R. Billard (eds), *Aquaculture and the Environment*, European Aquaculture Society, 12 Bredene, Belgium, pp. 57–91.

Busch R.L., 1985. Harvesting, grading, and transportation. In: C.S. Tucker (ed.), *Channel Catfish Culture*. Elsevier, Amsterdam, Oxford, New York, pp. 543–567.

Chang W.Y.B., 1991. Integrated lake farming for fish and aquatic plants. *Aquaculture Magazine*, **17**: 31–38.

De la Noue J., De Pauw N., 1988. The potential of microalgal biotechnology: a review of production and uses of microalgae. *Biotechnology Advances*, **6**: 725–770.

Im B.N., Hafner H., 1985. Impact des oiseaux piscivores et plus particulièrement du grand cormoran sur les exploitations piscicoles en Camargue. *Bull. O.N.N.C.*, **94**: 30–96.

Jauncey K., 1982. Carp (*Cyprinus carpio*) nutrition, a review. *Advances in Aquaculture*. Croom Helm, London, Canberra, 215–263.

Jensen J., 1981. *Corral Seine for Trapping Catfish*. Alabama Cooperative Extension Service, Auburn University, Alabama, Circular ANR-257.

Kolasa-Jaminska B., 1988. Investigations on investigation of carp fingerling production 5. Physical and chemical properties of water. *Acta Hydrobiol.*, **3**: 325–337.

Marcel J., 1994. Situation des populations de grands cormorans en France et leur impacts sur le milieu piscicole, *Echo-système, ITAVI*, **25**: 5–12.

Marcel J., Le Gouvello R., 1990. Production intensive de carpillons d'un été. *Echo-Système*, **15**: 7–10.

Martin J.F., 1987. La fertilisation en étang. *Aqua-revue*, **12**: 35–41.

Murai T., Ogata H., Nose T. 1982. Methionine coated with various materials supplemented to soybean meal diet for fingerling carp *Cyprinus carpio*, and channel catfish, *Ictalurus punctatus*. *Bull. Jap. Soc. Scient. Fish.*, **48**: 85–88.

Ogino C. 1980. Requirements of carp and rainbow trout for essential amino acids. *Bull. Jap. Soc. Scient. Fish.*, **46**: 171–174.

Schroeder G.L., Wohlfarth G., Alkon A., Halevy A., Krueger H., 1990. The dominance of algal-based food webs in fish ponds receiving chemical fertilizers plus organic manures. *Aquacultuer*, **86**: 219–229.

Sevrin-Reyssac J., Pletikosic M., 1990. Cyanobacteria in fish ponds. *Aquaculture*, **88**: 1–20.

Staub E., Ball R., 1994. Effects of cormorant predation on fish populations of inland waters. *EIFAC* XVIII/94/inf. **8**, 29 pp.

Stickley A.R., 1991. Cormorant feeding rates on commercially grown catfish. *Aquaculture Magazine*, **3**: 89–90.

Swingle H.S., 1947. Experiments on pond fertilization. *Alabama Polytech. Inst. Agr. Expt. Sta. Bull.*, **264**: 34 p.

Szumiec J., 1985a. Studies on intensification of carp farming. 1. Introduction and general programme. *Acta Hydrobiol.*, **2**: 131–145.

Szumiec M.A., Szumiec J., 1985. Studies on intensification of carp farming. 2. Effect of temperature on carp growth. *Acta Hydrobiol.*, **2**: 147–148.

Szumiec J., 1988. Investigation on intensification of carp fingerling productin. 1. Optimization of rearing biotechnics. *Acta Hydrobiol.*, **3**: 275–289.

Szumiec M., 1990. Stochastic model of carp fingerling growth. *Aquaculture*, **91**: 87–99.

Takeuchi T., Watanabe T., 1977. Requirement of carp for essential fatty acids. *Bull. Jap. Soc. Scient. Fish.*, **43**: 541–551.

Takeuchi T., Watanabe T., 1982. Effects of various polyunsaturated fatty acids on growth and fatty acid compositions of rainbow trout *Salmo gairdneri*, coho Salmon *Oncorbynchus kisutch* and chum Salmon *Oncorhynchus keta*. *Bull. Jap. Soc. Scient. Fish.*, **48**: 1745–1752.

Templeton R.G., 1984. *Freshwater Fisheries Management*. Fishing News Books Ltd., Farningham, England, 190 pp.

Viola S., Arieli Y., 1989. Changes in the lysine requirement of carp (*Cyprinus carpio*) as a function of growth rate and temperature. *Isr. J. Aquacult. Bamidgeh*, **41**: 147–158.

Wurtz-Arlet, J., 1980. La fertilisation en étangs. In: R. Billard (ed.), *La Pisciculture en étang*. INRA, Paris, 99–106.

Zhu Y., Yang Y., Wan J., Hua D., Mathias J.A., 1990. The effect of manure application rate and frequency upon fish yield in integrated fish farm ponds. *Aquaculture*, **91**: 253–251.

7

Creation of ponds, equipment and mechanisation

7.1 CREATION OF PONDS[1]

7.1.1 Various types of pond

The practice and success of carp culture depends largely on the availability of an adequate area for production and a sufficient water resource. There are several types of water resource: ponds, rivers and boreholes, and strict regulations on their use may be imposed. In all operations, the size of the fish-culture unit should be based on knowledge of the volumes of water available to refill the ponds after emptying in winter and to compensate for losses through evaporation and seepage in summer. The different operations concerned with the management of the ponds (filling, harvest) are other elements that must be taken into account in a fish-culture project.

In general, four main types of water body can be used for fish culture:

- reservoirs. These are traditional ponds in hilly terrain constructed by damming a river or stream.
- catchment ponds constructed to capture runoff water (ponds typical of regions where fresh-water fish culture is common such as Dombes, Sologne, Brenne).
- ponds created by excavation which are fed by underground springs which determine the water level (quarries, gravel pits)
- ponds created above ground by building dykes around the perimeter: these are generally not gravity fed.

[1] B. Lanoiselée.

Table 7.1. Some regulatory, technical and economic constraints associated with the construction and exploitation of various types of pond.

Constraints	Reservoir	Catchment pond	Excavation pond	Tanks
Regulation	Constraint	Medium	High	Low
Management of the water	Difficult (leaching)	Depends on the catchment basin	Depends on the level of the water table	Equipment for raising the water
Management of the environment	Low	Medium	Nil	High
Technical management	Low	Medium	Low	High
Mean cost	Low	Medium	Low	High

Fish can be reared in all of these types of ponds: however, consideration of management techniques and regulations limit the possibilities to certain types of pond (Table 7.1).

7.1.2 Technical aspects for the creation of ponds for fish culture

Water resource

In contrast to other types of fish culture, the main characteristic of pond culture is the use of water resources from a range of origins (rivers, catchment ponds, streams, wells or underground springs). Ponds are bodies of standing water; the quality of the water is managed to a varying degree depending on the objectives of production. The water serves as both a physical and biological support for rearing. Over the year, the minimum water requirement for the pond is the initial volume required for filling the pond (if it is emptied annually) and the amount needed to compensate for evaporation and seepage. The water balance for the pond can thus be established in terms of inputs and outputs.

Water inputs

Precipitation. The annual precipitation rate and its distribution throughout the year can be estimated from meteorological stations in the vicinity. These values are significant in terms of direct supplies to the surface of the pond and also in terms of the volume supplied by the catchment. There are significant variations between the statistical record and true precipitation. It is essential to have a margin of security that allows for wet and dry years. In the regions of France that are suitable for ponds the annual precipitation is between 600 and 800 mm.

Run-off surface water from the catchment area. This ensures that the canals or streams from which the ponds are filled are fed. However, only a proportion of the precipitation ends up as run-off; the percentage varies according to the time of year and the use of the catchment area (around 5% on average over the whole year, maximum 15%, minimum 0.5%. Calculations of the inflow from the catchment area should be based on an evaluation of the surface area feeding the site. The height at which the inflow is positioned can have a significant effect on the area as can the presence of tracks, paved roads or any other obstacle that is associated with a network of ditches.

The run-off from the catchment area should be calculated in terms of the maximum instantaneous flow during heavy rain in order to calculate the dimensions of the sluices needed.
Run-off also has an effect on water quality and the possibility of contamination with unwanted species should be evaluated together with the potential for contamination caused by activities upstream.

Watercourses. These can be used to top up ponds as necessary during the year. Analysis of hydrological records will give an indication of what is available. There are strict regulations controlling abstraction from rivers that are likely to cover the amounts which can be removed and the periods when it is prohibited.

Underground source of water (aquifer). It can be a key element in maintaining the level of water in the pond. Where rearing ponds are excavated in a water table itself, it is difficult to manage the quality of the water and the fish population, as it is impossible to empty the pond completely.

Boreholes. These may offer a solution in some places but there may be regulatory and financial constraints.

Losses from ponds

Evaporation. Evaporation from the surface of the water and transpiration by plants can reach high levels in warm and windy places. As a general rule there are few plants in the ponds used for fish culture and so only evaporation needs to be taken into account. This is mainly dependent on water temperature and on wind. In Europe, losses due to evaporation have been estimated to be of the order of 30 to 40 cm during summer or 0.5 to 0.8 cm per day during the period when evaporation is at its fastest. Ponds with a large amount of plants may lose water twice as fast.

Leakages. These are impossible to evaluate precisely *a priori*. A measure of the permeability of the ponds before construction (soil in place, soil present as mud) and the determination of the textural composition of the soil makes it possible to calculate an order of magnitude. As a general rule, the monthly loss through leakage will be between 1 and 5 cm.

Pond emptying. This represents a major loss to the system. In some farms the water can be reused by transfer to another pond or the creation of a storage pond. In the case of a pond collecting from a catchment area or a pond constructed on a watercourse there will be an excess of water during floods when the pond reaches its maximum level and cannot hold any more water. The water released must be considered as a loss to the system in the calculation of the annual water budget. The

Table 7.2. Hydraulic study (pond with a surface area of 20 ha in a catchment basin of 100 ha).

Hydraulic balance of ponds

The direct balance of the pond can be established by the water height (∂h in cm). The calculation allows the assessment of the balance in volume terms (∂V in 1000 m^3): the annual overall deficit is around 30 cm per year or around 60 000 m^3, over 20 h of ponds.

	Jan.	Feb.	March	April	May	June	July	August	Sept.	Oct.	Nov.	Dec.
Rainfall (cm)	9.5	8.0	6.5	5.5	5.5	5.0	4.0	5.5	5.5	6.0	7.0	8.5
Evaporation (cm)	1.0	2.0	3.0	5.6	7.6	10.1	11.1	10.6	8.6	6.1	2.5	1.5
Leakage (cm)	3.0	3.0	3.0	3.0	3.0	3.0	3.0	3.0	3.0	3.0	3.0	3.0
∂h (cm)	5.5	3.0	0.5	−3.1	−5.1	−8.1	−10.1	−8.1	−6.1	−3.1	1.5	4.0
∂V ($\times$ 1000 m^3)	11.0	5.9	0.9	−6.1	−10.1	−16.2	−20.2	−16.2	−12.2	−6.1	2.9	7.9

Supply from catchment basins: the distribution of run-off over the year varies between 0.5 and 15% depending on the time of year. The run-off is calculated by volume but converted to water depth in the ponds.

	Jan.	Feb.	March	April	May	June	July	August	Sept.	Oct.	Nov.	Dec.
Run-off ($\times$1000 m^3)	17.1	11.5	7.8	5.3	4.6	2.4	1.4	0.7	3.3	0.7	4.2	10.2
Run-off (cm)	34.2	23.0	15.6	10.6	9.2	4.8	2.9	1.3	6.6	1.4	8.4	20.4
Balance ($\times$100 m^3)	28.1	17.5	8.7	−0.8	−5.5	−13.8	−18.8	−15.5	−8.9	−5.4	7.1	18.1
Balance (cm)	14.0	8.7	4.4	−0.4	−2.8	−6.9	−9.4	−7.8	−4.4	−2.7	3.6	9.1

Overall hydraulic balance

Over the water surface of this example, it is estimated that all ponds were refilled, the deficit reaches around 68 000 m^3 over the summer (*drop in level 35 cm*) while overall it reaches 11 000 m^3 over the year. Where it is possible to recover run-off water in summer, the deficit for each pond represents a maximum drop of around 35 cm in October.

	Jan.	Feb.	March	April	May	June	July	August	Sept.	Oct.	Nov.	Dec.
Cumulative deficit ($\times 1000\ m^3$)				−0.8	−6.4	−20.1	−38.9	−54.4	−63.3	−68.7		
Cumulative deficit (cm)				−0.4	−3.2	−10.1	−19.5	−27.2	−31.6	−34.3		

To maintain the levels it will be necessary to use additional pumping between April and October for a total duration of around 1400 h at a flow of 50 m^3/h (wells, irrigation pipes).

	Jan.	Feb.	March	April	May	June	July	August	Sept.	Oct.	Nov.	Dec.
Monthly pumping (hours)				17	110	276	375	311	177	108		
Cumulative pumping (hours)				17	127	403	778	1089	1266	1374		

However, if a drop of 20 cm is accepted in all of the ponds the water requirements can be met up to the end of July and are no more that 28 000 m^3 or around 580 h pumping at 50 m^3/h.

	Jan.	Feb.	March	April	May	June	July	August	Sept.	Oct.	Nov.	Dec.
Cumulative deficit ($\times 1000\ m^3$)								−14.4	−23.3	−28.7		
Cumulative deficit (cm)								−7.2	−11.6	−14.3		
Monthly pumping (hours)								289	177	108		
Cumulative pumping (hours)								289	466	574		

balance between the resources available and the losses allows a calculation to be made of the risks of water shortage; from this the provision of supplementary systems such as recycling, boreholes and the emptying period can be organised.

Water balance. The balance of inflow and outflow allows the calculation of a water budget for the fish-culture unit and thus the determination of the risks likely to be associated with any deficit, which must be made good by a supplementary water source (Table 7.2).

Water quality

The quality of the water and the temperature of the pond determine the site's potential and the choice of species that can be farmed there. The quality of the water in a pond depends mainly on the interactions with the bed of the pond and changes which occur in relation to inputs such as fertilisers and feed (p. 47). However, the initial quality of the water may influence the success of the farming operation in the pond. As a general rule, run-off water can be used for fish-culture ponds, especially water coming from agricultural catchment areas. Water from forested catchment areas is generally more acidic and less fertile than that of farmed catchment areas which have organic residues and plant detritus. However, where there is intensification (>3 t fish/ha) the risks of plankton blooms following the spreading of agricultural wastes and animal faeces will be increased.

The bed of the pond

Fish farming has the advantage of allowing the use of large areas that are not exploited by traditional agriculture. It is possible to create ponds in wetland areas, whether or not the soil is clay-based (Table 7.3).

A *minimum* level of clay for the construction of a dyke is between 17 and 20% but should not exceed 60%. The optimum level for ease of construction is around 40%. If the level of clay decreases it should be compensated for by better compaction (compaction represents 90% of success).

The moisture content of the dykes should also be monitored closely in order to give the soil its maximum plasticity. Where there are difficulties, the optimum moisture content can be measured in a laboratory (Proctor test). The permeability of the substrate must be considered as part of the water management of the ponds. For example, the production of organic matter during the rearing cycle will contribute rapidly to inhibiting seepage. The process of blockage of the gaps in porous substrates can be improved by high application rates of manure when the water is first introduced (10 m^3/ha). Only rocky and gravelly ground is unsuitable for the construction of ponds although it is possible to use plastic liners or membranes. Places where the water table is close to the surface do not lend themselves to the straightforward construction of ponds and their successful management because it is impossible to empty them by gravity and the construction techniques in general use are not very adaptable.

Table 7.3. Different soil types and their characteristics for the construction of the banks of ponds

(a) Classification of soil texture:

Texture	Sand (%)	Alluvium (%)	Clay (%)
Sandy soils	70–100	0–30	0–15
Alluvial sand	50–70	0–50	0–20
Alluvium	20–50	70–85	5–25
Alluvial clay	20–45	15–50	25–40
Clay/alluvium	0–20	40–60	35–55
Sandy clay	45–65	0–20	35–55
Clay	0–20	0–40	40–100

(b) Quality of the materials for the construction of banks:

Texture	Permeability	Plasticity	Compaction	Overall suitability
Sand	− − −	0	+ +	− − −
Alluvial sand	+/−	+ +	+/−	− −
Alluvium	+/−	+ + +	+/−	+/−
Sandy clay	+ + +	−	+ +	+ +
Clay	+ + +	+/−	+	+ + +
Peat	− −	− −	− −	− − − −

(c) Permeability (cm/day):

Texture	Natural soil	Wetted soil
Sand	>25	
Alluvial sand	15	5
Alluvium	10	3
Alluvial clay	7	2
Clay/alluvium	2	≈1
Clay	3	≈1

The homogeneity of the substrate should be studied in order to assess the risks of leakage caused by the presence of grains of sand or gravels and the need for their removal. The presence and depth of any humus-rich soil should also be investigated. This soil may be utilised either behind the dyke to build up the back or laid over the front to encourage the growth of vegetation.

Topography

Modern fish farming is based on strict control of the stock. This requires the option of complete harvesting of the fish, which is greatly facilitated if the pond can be emptied completely, if possible by gravity. This means that the bottom of the pond

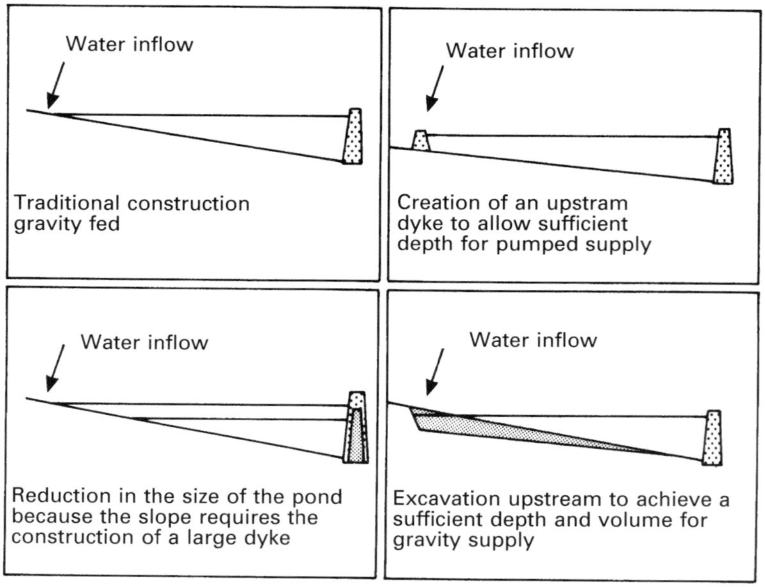

Fig. 7.1. Differences in water inflow to ponds.

should be above the water outflow point to the river, channel or return ditch (by pumping). Where complete emptying is impossible, special techniques for harvesting and stock management must be adopted. Not all of the ponds used in intensive fish culture are emptied annually; for example the ponds which are used for rearing channel catfish in the USA may only be emptied every 5–7 years.

The topography will also take into account the height of the water-inflow point which will determine how water is supplied (gravity or pumping) (Fig 7.1).

The optimum site for the construction of ponds has either a slight slope or is flat ($\approx$0.5 to 1%). Ponds can be laid out perpendicular to the contour lines. The slope of the bed of the pond follows that of the soil; this reduces the amount of earth needed to build the dykes. When the land has a greater slope the maximum depth of the pond should be calculated so that there is sufficient depth for fish culture at the shallowest point ($\approx$80–100 cm). This often requires the movement of significant quantities of soil.

In undulating, hilly land, high and costly dykes may need to be constructed. For economic reasons these should never be more than 3 m high; the natural slope of the land should be no more than 2%. Here the axis of the pond will be perpendicular to the contour lines. This is the most frequent layout of traditional ponds. The topography and the siting of the ponds must obviously allow access to the ponds in all weathers for all operations such as feeding, spreading fertiliser and harvesting.

7.1.3 Construction

Catchment-area ponds

There is a huge variation in the optimum size and shape of these ponds because of the important influence of topography, size of catchment area, and the characteristics of the substrate. Costs are largely dependent on topography and the transport of the necessary materials (1.8 to 2.7 €/m^3 based on the cost of removal, transport and construction).

Preparations

The choice of site for the dyke is controlled by matters relating to topography and to the size of the pond. A horizontal projection along the ground will allow the measurement of the total volume of the pond once filled. Determination of the size of a catchment area will allow an evaluation of the supplies and the risks during periods of maximum run-off which must be taken into account when calculating the size of the overflow which can be integrated into the emptying system or by a sluice built into the dyke. Where the catchment area is large, a safety margin of 50–60 cm may be needed to retain the large quantities of water after rain, while limiting the size of the overflow.

The dyke or dam

It is essential that a foundation trench is constructed to ensure the stability of the junction between the dam wall and the soil in which it will be constructed. A trench is hollowed out in the centre of the dam site. The depth will be determined by the nature of the ground and the width at the base should not be less than the width of the machinery used for compaction (2.5–3 m). Good-quality clay should be placed in the trench and compacted in the same way as for the superstructure of the dam. The height of the dam is determined as a function of the final depth of the pond, to which should be added 40–80 cm to allow for excessive run-off after heavy rain. For fish culture, the depth should be no more than 2 to 2.5 m. The top of the dam should be 4–5 m wide in order to allow the passage of machinery for compacting the dam and the equipment used during farm operations. The slopes of the dam wall on the interior and exterior of the dam will be 3 : 1 and 2 or 3 : 1 respectively. A slope of 1:3 at the outside of the pond, while increasing the volume of material required, is preferable because maintenance is easier. The volume of earth is determined by calculating the linear sections at points along the dam (it is beneficial to add 10% to allow for compaction). The volume of earth to be cleared can be calculated in the same way (a coefficient of 1.2–1.25 should be used to allow for expansion) (Fig 7.2).

The dam should be built in successive 20 to 30 cm thick layers depending on the material used. The layers are compacted and bonded together to ensure the watertightness and integrity of the dam (Fig 7-3).

The bottom of the pond

The contours of the pond are based on a minimum depth of 0.5 to 0.8 m throughout. Some parts of the periphery or the tail of the pond can be hollowed out and the

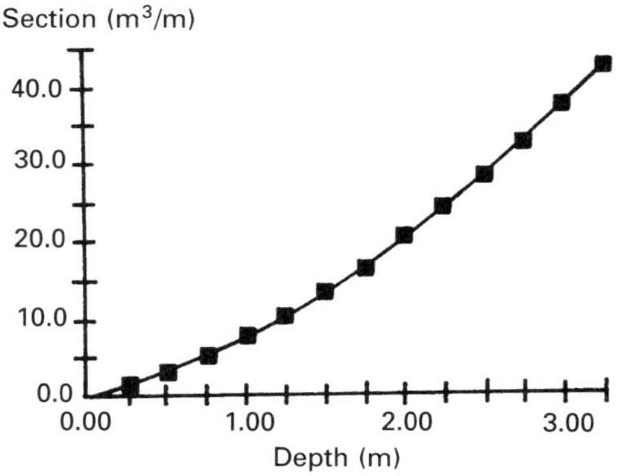

Fig. 7.2. Determination of the volume of earth in m^3 (y-axis) per linear metre of dyke in relation to depth (x-axis), dyke 4 m in height, slope 2 : 1 and 3 : 1, cf. Figure 7.3.

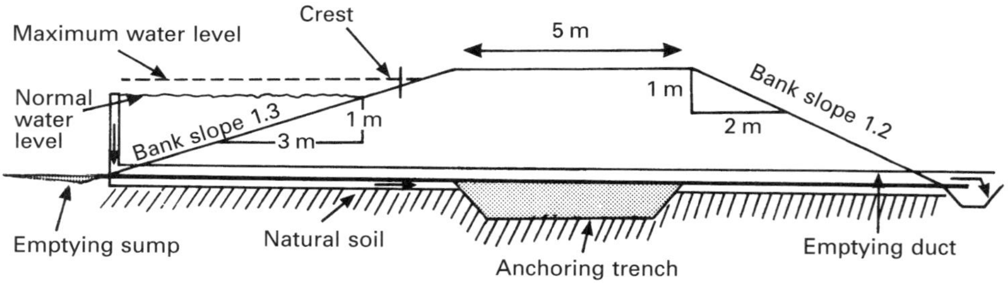

Fig. 7.3. Diagrammatic cross-section of a pond dyke.

material removed used in the construction of the dam. The slope of the bank of these hollowed-out sections should also conform to the 3 : 1 rule. All topographical irregularities in the ground should be removed or filled in to make sure that the pond can be completely emptied and also to avoid hindering open-water harvests with a seine net. If necessary, the deepest parts should be filled in to limit the maximum depth of the pond. Finally, a service path should be built around the pond if it is to be used for intensive or semi-intensive production.

The emptying system

The emptying system for these ponds is either a monk, which also acts as an overflow, or a system of sluice gates or pipes completed by an overflow system within the dam wall. There are different types of monk, which may be either a stack of planks that can be used to control the water level or a sluice gate on the bed of the pond. Each region has its own traditional system, with varying degrees of sophistication adapted to the type of management used in the ponds. Depending on the requirements, a simpler system made up of a movable elbow and vertical pipe can be installed. This system is widely used in the USA, even for very large ponds.

Ponds

The choice of the site and number of ponds to be constructed on an area of flat land depends largely on the amount of land available and its dimensions. In all cases, the costs of construction can be reduced if common dykes can be used for several ponds thus reducing the amount of material to be moved. The unit cost per cubic metre of dyke constructed depends essentially on the techniques and materials used.

Preparation

The main dam retaining the ponds should be orientated at an angle of 90° to the prevailing winds to reduce the effect of waves on the banks; the maximum depth should be 1.6 m and the minimum 1.2 m. The optimum shape of the pond is a rectangle with the length about twice the width. Ponds should be contiguous but each can be emptied individually either by gravity or by pumping. Water may come from a borehole by pumping or from a reservoir into which the outflow water can be fed and then recovered.

Construction of the dyke

The foundations of the dyke should be scoured out and cleaned to ensure the watertightness of the base of the structure. In some places it is necessary to build an anchoring trench. The dyke is constructed by placing the balance of the excavated material and hard core on the site (hollowing out 0.3 to 0.6 m across the whole area of the pond and raising the dyke to 1.5–1.7 m). This equilibrium can be calculated using a coefficient of expansion of 1.2 to 1.3. The width of the crest of the dykes should be between 4 and 5 m so that the machinery used for compaction and, at a later stage, equipment used to operate the farm can move around freely. The slopes of the interior and exterior embankments will be 3 or 4 : 1 and 2 or 3 : 1 respectively. Dykes, which are common to two ponds, will have slopes of 3 or 4 : 1 on both sides. From this it can be calculated that the mean volume of dykes is around 15–17 m^3/m for those around the periphery and 17–19 m^3/m for the dykes separating the two ponds. Dykes are constructed in 20 to 30 cm deep layers, depending on the material used. These layers will be compacted and bonded together to ensure the water-tightness and the integrity of the dyke.

Bottom of the pond

All topographical irregularities should be removed by excavation. A slope of 0.5–1% should be sufficient to ensure that the pond can be completely emptied.

7.1.4 Costs of construction (values are given in Euros, 1992)

There are few specialised contractors for the construction of ponds for fish culture in Europe. The work is generally carried out by contractors working in related public rural construction projects such as water treatment, irrigation and drainage who build

Table 7.4. Excavation costs in relation to the total investment for fish farms which are either new or under construction (€ × 1000).

Surface area (ha)	Investment total	Excavation costs	Total %
13	184	48	26
60	599	261	44
17	230	108	47
17	184	100	54
60	402	207	52
18	184	92	50
Mean cost/ha	10.9	4.8	45

These fish farms are made up of different types of ponds from 500 m^2 to 4 ha, five of which were less than 5000 m^3. The mean cost of excavation is 1.5€/m^3.

ponds for leisure or reservoir purposes as their principal activities. Such companies are likely to possess suitable equipment such as bulldozers, mechanical shovels and tipper trucks. This equipment has a large capacity over short distances (excavation–back-fill, <50 m) and often the compaction necessary for the long-term preservation of the dykes is neglected as only one specific type of equipment, a road roller is truly suitable. The cost of terracing varies from one region to another, one site to another and even from one farm to another (Table 7.4).

It is difficult to estimate costs whatever the size of the ponds; the unit cost per hectare of pond varies mainly in relation to area (Table 7.5). The impact of the cost of construction on the economic performance is high because of the great importance of depreciation and financial costs. Table 7.6 shows the impact on the overall performance of a variation of ±10–20% on the costs of construction. Thus in an example where terracing represents 45% of the construction costs, a reduction of between 10 and 20% on the investments gives an increase of 12 to 23% on the net result.

The machinery used to construct pond-based farms in France generally has the capacity to extract between 250 to 500 m^3/h with mechanical shovels and 200 to 400 m^3/h with bulldozers over distances less than 40 m. They are not very mobile; this is why the sequence of work commonly used for ponds over 2 ha is as follows:

Extraction and loading:	mechanical shovel (JCB)
Transport and backfilling:	lorry or tipper truck
Shaping	bulldozer
Completion of shape	bulldozer/shovel
Compaction	road roller
Grading of the pond bed	bulldozer

Table 7.5. Construction costs of ponds used for pisciculture (in € 1992).

(a) Examples in central France

Area and number of ponds:	7 000 m² ×1	1 ha ×2	4 ha ×2
Clearing	307		
Excavation	6 912	11 828	33 641
Pits	276	131	
Sluices	7 527	3 072	3 226
Emptying system	1 659	–	983
Pipes, etc. ⌀ 300	699	983	215
Installation, etc.	154	230	269
Total	17 535	16 244	38 333
Costs in €/ha	25 049	8 122	4 792

(b) Examples from the United States

Surface area of the water (ha)		Quantity (m^3*)	Total (€)	Total cost (€/ha)
8 (1.5)†	Clearing	43 000	44 585	
	Emptying (m)‡	4	461	
	Grass	4	384	
	Gravel	44	2 765	
			48 195	6 024
56 (7)†	Clearing	167 772	123 733	
	Emptying (m)‡	640	14 747	
	Grass	17	2 067	
	Gravel	766	6 169	
			146 724	2 620
224 (25.5)†	Clearing	638 862	483 871	
	Emptying (m)‡	2560	58 986	
	Grass	64	8 267	
	Gravel	3097	24 971	
			576 095	2 572

* Slope of the bank 3 : 1–4 : 1, crest 4–5 m.
† Area of the ponds in ha.
‡ Length of pipe 25 m, ⌀ = 400 m.

The same volume of soil is removed three or four times and should be deposited in piles rather than thin layers which result in poor homogeneity of the embankment, leading to poor compaction.

In the USA, most ponds have been constructed by a specialised technique using scrapers which are pulled by farm tractors of 150 to 250 hp (Figure 7.4). This equipment extracts, transports, shapes and compacts in the same sequence (Box 7.1). Construction costs are thus reduced considerably as is shown in Table 7.5.

Table 7.6. Simulation of the impact of construction costs on economic performance.

Cost changes, %	−20	−10	0	10	20
Excavation	1467.1	1650.5	1833.9	2017.3	2200.6
Depreciation	236.5	248.8	261	273.2	285.4
Financial costs	88.3	94.2	100	105.8	111.7
Net result	191.3	173.1	155.1	137.1	119
Variation (%)	23	12		−12	−23

Fig. 7.4. 'Scrapers' in operation for the construction of ponds in the United States (photo J. Marcel, ITAVI).

7.2 EQUIPMENT USED ON FARMS AND MECHANISATION OF PRODUCTION[(2)]

At present mechanisation of the production of cyprinids in ponds remains very modest in scale; any equipment used is relatively unsophisticated. This contrasts with the significance of their production on a worldwide scale but fits in with the traditional character of rearing techniques in ponds, the extensive nature of production and the need to keep production costs low. However, it is clear that increasingly complex equipment is being used in hatcheries on new farms where electricity is available and where smaller-sized ponds are used more intensively. This facilitates the use of equipment such as aerators and automatic feeders. Another form of increased sophistication is the use of microcomputers and kits for water analysis. Other information is available in Aqualog (1996), (Beveridge and Ross, (1990)*, Hepher and Pruginin (1981)*, Horvath *et al.* (1986)*, Itavi (1989) and Varadi (1983, 1986, 1990a,b).

(2) Information supplied by L. Varadi.

* See General bibliography.

Box 7.1
Construction of ponds for channel catfish using scrapers
(example from the USA described by J. Marcel)

Equipment

The equipment used is generally a farm tractor and two scrapers (Figure 7.14)

—135 hp farm tractor (minimum) attached to a double front-wheel axle unit and automatic gearbox;
—scraper with a minimum capacity of 8–9 m^3 (5 m^3 when compacted). Opening, closure and tipping of the truck is controlled hydraulically. John Deere makes purpose-built equipment for this use. The cost is estimated at 12 645 € new (3612–4515 € second hand)
—a control box which brings together all the hydraulic functions which is placed at the tractor coupling (3612 €);
—possibly a laser system for automatic assessment of levels.

Techniques

The earth is removed by the successive passage of the scrapers. After filling, the trucks are towed by the tractor to the site of the dykes, the soil is released and the bank is built up in a single operation.

The opening of the blade of the scrapers is regulated so as to remove a layer of soil of the required volume to fill up the truck without stopping the tractor and without using the maximum power. This is influenced by several factors such as the nature of the soil, its humidity, the slope, the power of the tractor and the speed at which it works.

For two coupled scrapers this involves filling the first (immediately behind the tractor), lifting the blade and then lowering the blade of the second scraper and filling. All of this can be controlled from the tractor. The automatic gearbox allows the driver to concentrate on controlling the scrapers. When a laser system is used to follow the contours it is connected over the opening of the blade of the second scraper.

Unloading material and construction of the dyke
When the two trucks have been filled, the tractor moves on to the site of the dyke. In a single passage, without stopping, the driver opens the first scraper; after emptying completely and spreading out with the blade the second scraper is opened. The front axle unit of the scrapers, together with that of the tractor, are sufficient to level and compact the volume of earth that has been transported. The dyke is thus built up in successive layers. Two set-ups have been used:

—a 185 hp tractor (4850 JD) and two 8-9 m^3 scrapers; mean volume of earth compacted, 150 m^3/h;

—a 350 hp tractor and two 14 m^3 scrapers; mean volume of earth compacted, 500 m^3/h.

The time taken crossing the pond, filling and spreading depends on the nature of the soil which determines the working speed and the distance. It is generally a few minutes and the set-up should be continuously going back and forwards. This technique of construction is used for fish-culture ponds (a series of 3000 m^2 ponds (Figure 7.5) or 3–4 ha ponds) as well as traditional reservoirs. This method is not as heavy as when using a bulldozer and works well in a wet environment.

Financial advantages in the use of classical agricultural equipment can be assessed in the same way as an agricultural unit. Investment is less than in Public Works material, operating costs, maintenance, distribution, fixed assets are also less heavy than Public Works. The cost per cubic metre of earth moved varies from 0.45 to 0.13 € (Table 7.5); 20 m^3 of earth is required per linear metre of 2 m high dyke (Figure 7.2).

To avoid delays two teams often work in parallel so that if one encounters any difficulties the second can come to its aid immediately. In addition, a small bulldozer may be used to shape the dyke, particularly in big ponds, but this is not essential.

Fig. 7.5. 3000 m^2 ponds, partially raised, constructed using scrapers (photo J. Marcel, ITAVI).

7.2.1 Spreading fertiliser

Organic fertilisers

There are several methods of spreading the various types of organic fertiliser (manure, poultry wastes) (Figure 7.6), including:

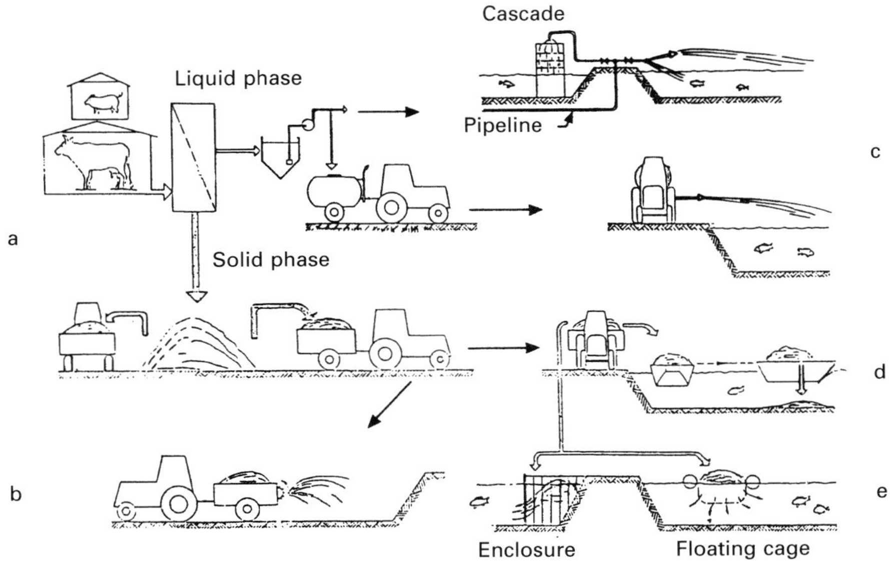

Fig. 7.6. Diagram showing different methods used for the distribution of organic fertilisers in ponds: (a) separation of liquid and solid phase, storage and spreading, (b) spreading on the bottom of the pond, (c) distribution of liquid manure by a muckspreader or a pipeline, (d) distribution by boat, (e) deposit in enclosures or floating cages.

- spreading directly on the bottom of the pond, after it has been emptied, using traditional agricultural machinery (e.g. muck spreaders);
- piling manure or poultry waste at the edge of a pond in an enclosure or on a submerged, floating platform where they dissolve progressively in the water;
- piling manure or wastes onto a boat for distribution around the pond either by hand or mechanically by either active or passive methods (Figure 7.7);
- distributing manure at low rates (in terms of dry weight) by spraying from the edge of the pond using agricultural equipment. This requires vehicle access to the periphery of the pond and a machine capable of lateral spraying. To avoid major losses by aerosol formation the slurry should be spread as close to the surface of the water as possible. One method uses a large diameter pipe, emptying just below the surface of the water.

Mineral fertilisers

In their solid form, fertilisers or supplements are spread in various ways, either manually or using a continuous screw system from a boat (Figure 7.8). The fertilisers may also be simply placed on a submerged platform (Figure 7.9) where they dissolve without coming into contact with the bed of the pond. In its liquid form the fertiliser can be distributed quite simply by opening a distribution pipe positioned close to the

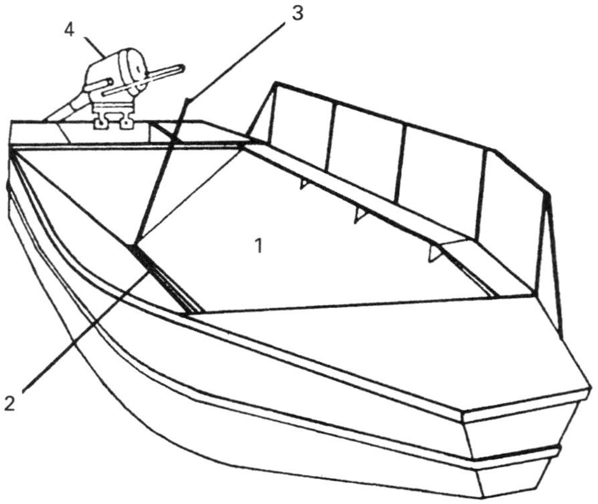

Fig. 7.7. Boat (Keszec type, used in Hungary) with a V-shaped hopper (1) which has, at the base, a cleft (2) which allows organic or mineral fertilisers or feed (granules, cereals) to be spread in the water; (3) lever which controls the opening.

Fig. 7.8. Boat (Veyssière type, France) with a hopper for mineral fertilisers and additional matter (e.g. granules and cereals) which it feeds to the pond by using a continuous-screw system (able to handle 10 t/day) (photo Etablissements Veyssière).

propeller of a boat moving through the pond. Spray bars can also be used to distribute liquid fertiliser (Figure 7.10), as can muck spreaders.

7.2.2 Aeration

Several approaches have been made to help maintain or restore levels of dissolved oxygen in the water. Some consist of introducing oxygen-rich water (this is only

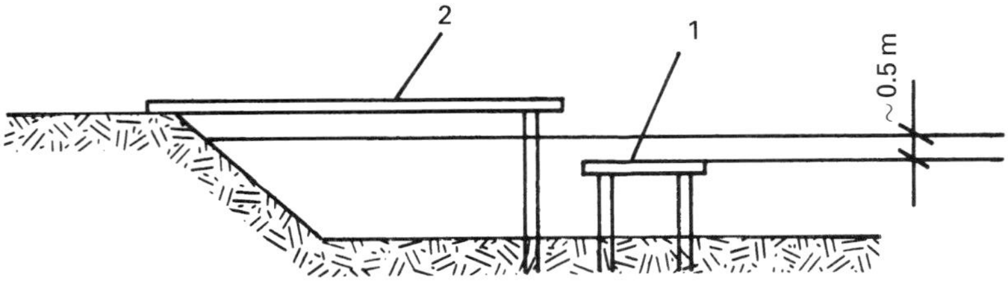

Fig. 7.9. Diagram of a submerged platform (1) designed to support piles of mineral fertilisers which will dissolve progressively; size 1.5 × 2.3 m; access ramp (2) 40 cm wide (after Boyd, in Billard and Marcel, 1986) (see General bibliography) (see p. 129).

Fig. 7.10. Boat used for speading liquid manure (Veyssière FxT-1) capacity of the tank: 400 l, spray bar 9 m long when unfolded (photo Establishments Veyssière).

possible and effective in a small pond in an emergency), others consist of pumping and spraying water into the pond or installing aerators. These systems introduce atmospheric oxygen, either directly by blowing air into the water or through mechanical agitation of the water, which increases the surface area of the water exposed to air and thus the rate at which oxygen from the air dissolves.

Systems for injecting air in the form of bubbles are made up of a means of supply, pipes and distribution; low-pressure air is injected in large quantities. Figure 7.11 shows a bubble aerator. The electric motor and compressor can be installed in the centre of the pond or at suitable points from which compressed air can be distributed through plastic pipes which either have perforated ends or airstones. This method is not very efficient, only 50–60% in comparison with surface aerators (Table 7.7).

Several types of surface aerator have been developed over the past few years (Table 7.7). An intermediate type is represented by inclined-axis aerators, which create water movement (AIRE O_2®, RSK9®, Figure 7.11). Their effectiveness, based on the efficiency with which oxygen is incorporated in the water rather than their capacity to move the mass of water is given in Tables 7.7 and 7.8. The Turbo Jet® works on the same principle: suction of air and forcing it back into the water.

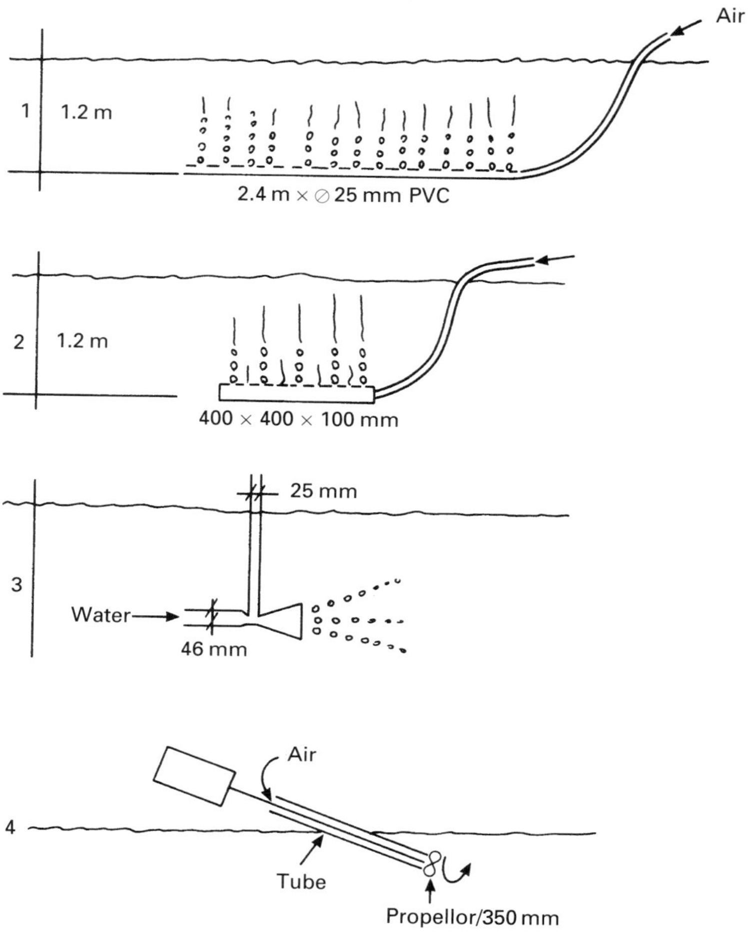

Fig. 7.11. Systems for air injection (1) through perforated PVC tube, (2) by porous diffuser, (3) by venturi, (4) by inclined shaft surface aerator (AIRE O_2® RSKG®...) type (cf. performance Tables 7.7 and 7.8, Figures 7.13).

Another type (hydroejector, Figure 7.12) is made up of a submerged motor and a vertical transmission tube and operates by projecting water into the air, ensuring oxygenation in a circle of no more than about 10–15 m; water oxygenated in this way is not displaced to other parts of the pond. A recent model has a floating hydroejector with a submerged motor (Flopusle®, Faivre) and this gives performances similar to those of AIRE O_2®. Other models are based on a paddle-wheel system (Figure 7.13 a-b) which agitates the water and induces a current that moves aerated water over 50–100 m.

Table 7.7. Comparative trials of the effectiveness of several aerators used in Hungary (tests in a clear water pond, initial dissolved oxygen level = 0) (after Kuli, 1982, MEM Muszaki Intezet Gödöllö 29, in Hungarian) cf. Figure 7.11.

Type	O_2 injected (kg/h)	kW/h consumed*	Efficiency (kg/kWh)
Air injectors (large bubbles, Figure 7.11)			
—with perforated pipes(1)	3	5.9	0.51
—with porous diffusers(2)	2.1	1.9	1.1
Aerator type			
—Venturi(3)	0.63	2.16	0.29
—Inclined shaft (RSK 9)(4)	2.1	3.84	0.55
Paddle wheel†			
—Bikal model(5)	3.1	2.56	1.21
—standard model (Halinno B)(6)	2.2	1.56	1.41

* True capacity of electric motors measured during the trials.
† Chinese paddle-wheel (blades on arms) (Figure 7.13).
(1) PVC pipe, 2.5 m long and 25 mm diameter placed at a depth of 1.2 m with 230 holes with a diameter of 1.5 mm; air pressure: 135 mmHg; flow 312 m^3 air/h.
(2) Diffusers 40 × 40 × 10 cm placed at −1.2 m; air pressure: 120 mmHg; flow 312 m^3 air/h.
(3) Venturi placed at a depth of 1.2 m; air flows in through a 25 mm PVC pipe; air flows from the venturi through a 46 mm pipe at a rate of 290 l/min.
(4) Speed of the shaft: 1440 rpm; diameter of the screw: 350 mm.
(5) Total diameter of the wheel: 650 mm; penetration of blades into the water 210 mm; rotation speeds: 126 rpm.
(6) Total diameter of the wheel: 1 m; penetration of blades into the water: 90 mm; rotation speeds: 90 rpm.

Table 7.8. Information on the performance of several surface aerators.

Type of aerator	Power (kW)	kg O_2h	kg O_2/h/kW
Air injection aerators:			
Vertical shaft			
Aqua Pilz	0.4	0.6	1.5
Aqua Jet	0.37	0.36	0.97
Inclined shaft			
AIRE O_2 (USA)(1)	1	1.1	1.1
Turbojet	0.35	0.5	1.43
	1.1	1.6	1.45
Paddle wheel			
House Manufacturing (USA)(2)	9.4	24	2.54
AEMCO USA(3)	9.4	14.2	1.4

(1) AIRE O_2 offers models whose power varies between 1 and 7.5 HP (1 HP = 0.94 kW)
(2) Paddles on a cylinder (cf. Figure 7.13) with four rows of 18 paddles, triangular plates (angle 180°) arranged in a spiral pattern on the cylinder; length 3.6 m ⌀: 20 cm; penetration of paddles into the water: 96 cm; rotation 84 rpm.
(3) Paddles on a cylinder with four rows of 16/17 paddles in 8.7 mm diameter tubes, 7.5 cm long cut longitudinally every 3.7 cm inserted in the cylinder (length 3 m, diameter 25 cm) with an angle of 120°; penetration into the water: 20 cm; rotation 105 rpm (cf Boyd and Martinson, 1984).

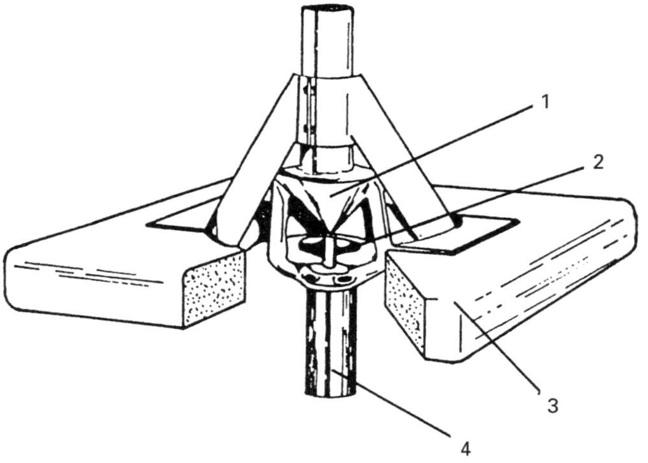

Fig. 7.12. Vertical hydroejector; 1: cone; 2: propellor; 3: float; 4: electric motor.

Aerators also have a role in preventing thermal stratification within the pond by inducing water movement which also contributes to transporting surface water, saturated with O_2 through photosynthetic activity, to the bottom (p. 43). Aerators with inclined axes and paddle wheels as well as submerged pumps also induce water movements. New systems, 'circulators', are beginning to appear on the market: these are specifically designed to prevent or disrupt thermal stratification and are composed of a submerged helix which rotates and sets the water body in motion. All of these aerators require an electricity supply to the site. The electricity is supplied through well-insulated cables that should be checked frequently. Where there is no electricity supply to the site or there is a breakdown, it is possible to use surface aerators powered by generators or tractors.

It is critical that aerators are installed correctly in ponds; the installation should take into account the type of generator, the size and shape of the pond, distance between banks, orientation with respect to current, direction of the prevailing winds and whether the aerator is based on the paddle-wheel principle or the AIRE O_2® type. There should be some flexibility, for example to take account of changes in wind direction. An example of the use of two types of generator in the former East Germany is shown in Figure 7.14; the daily operating regime throughout the year is shown in Table 7.9. The information on the positioning of the aerators, times of operation and the type of aerator should be determined and tested *in situ* within the pond: when they have been installed frequent measurements should be made using portable oxygen meters. In ponds used for intensive rearing it may be beneficial to install continuously operating equipment attached directly to the aerators. These set the aerators going when the level of O_2 drops to a limit previously determined by the fish farmer (Figure 7.15). The electrodes should be cleaned frequently (at least twice each week).

a

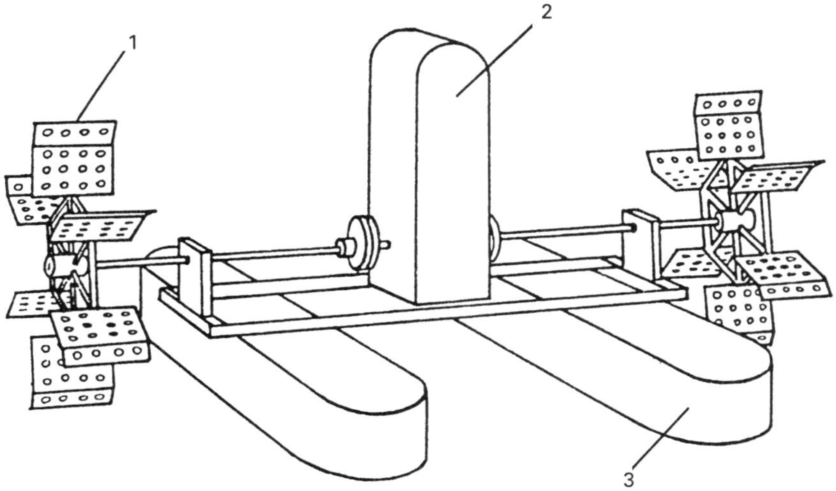

Chinese paddlewheel

b

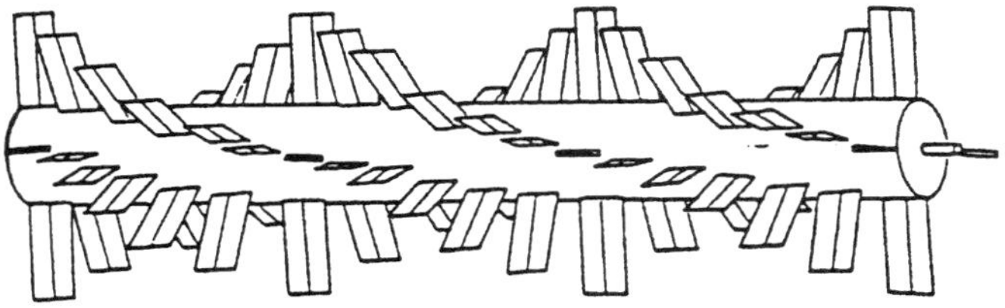

Cylinder with paddles
(House Manufacturing AEMCO)

Fig. 7.13. Rotating paddle aerator which creates a current by disturbing the surface of the water. 1: paddlewheel; 2: motor; 3: float (there are various models such as the Milanese®, Italy and AERPAL®, Germany).

7.2.3 Cutting back the vegetation

The growth of plants can have negative effects on the productivity of fish culture in ponds. Plants can act as obstructions on dykes, interrupting the passage of machinery and sheltering predators. In channels plants impede the flow of water and in the ponds themselves they are usually a trophic dead end and disrupt fishing operations. Various types of equipment can be used to cut them back. Strimmers or rotary mowers have the advantage of cutting and spreading the vegetation and can cover 150–200 m^2/h. Traditional mowers have a greater capacity (5000–9000 m^2/h) but usually collect the vegetation after cutting. Equipment with a fixed mowing blade (Figure 7.16) is used to mow the banks of ponds and channels. In ponds, such blades

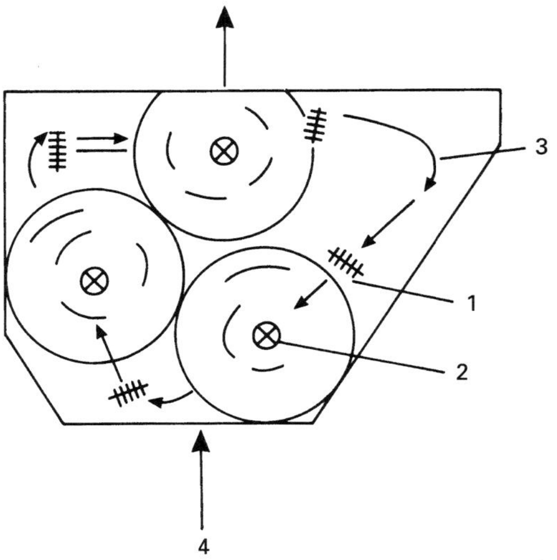

Fig. 7.14. Positioning of aerators in a 10 ha pond in the former East Germany; 1: paddle wheel aerates (Figure 7.13a); 2: vertical hydroejector (Figure 7.12) and aerated areas; 3: water current generated by paddle wheel aerators, 4/5 inflow–outflow of water from the pond (from Knösche 1990).

Table 7.9. Operating regime of aerators in Germany shown in Figure 7.14.

Time of year	Operating hours of aerators
1/6–31/7	23 h 00–5 h 00
1/8–15/8	20 h 00–6 h 00
16/8–31/8	21 h 00–7 h 00
1/9–harvest	19 h 00–7 h 00

can be mounted on a flat-bottomed boat for use in shallow water (up to 60 cm). The 'Esox' model, manufactured in the former Czechoslovakia, is used currently across Central Europe. In France, a company called Veyssière makes a boat for mowing with a cutting width of 2.13 m capable of operating at depths between 15 and 85 cm, controlled remotely; the action can be adjusted as the boat progresses. More sophisticated equipment collects plants after they have been cut (IPV model from Hungary). These mowing boats are generally towed through the pond but some are self-propelled.

7.2.4 Distribution of feed to growing fish[3]

Because of the increasing intensification of production in pond-fish culture increas-

(3) Th. Boujard and J. Marcel.

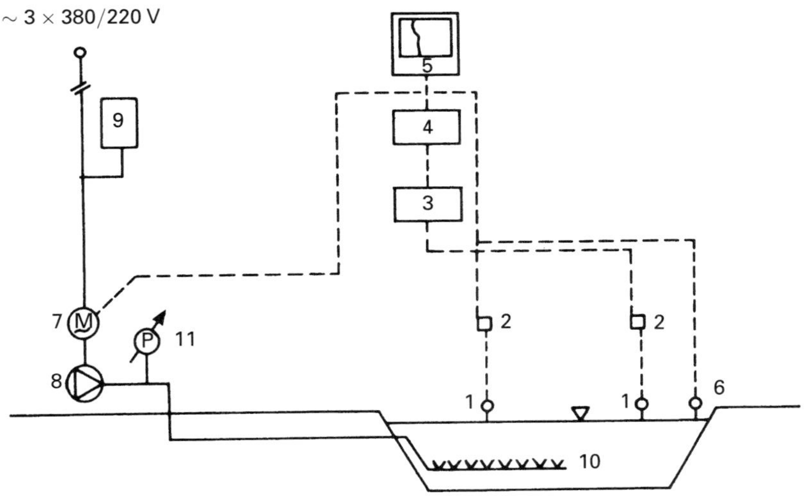

Fig. 7.15. Diagram of a system for recording levels of O_2 using an air compressor; 1: oxygen electrodes; 2: relay; 3: multiplex; 4: signal transformer; 5: recorder; 6: temperature probe; 7: motor; 8: compressor; 9: timer;, 10: air diffusion; 11: manometer (L. Varadi, pers. comm.).

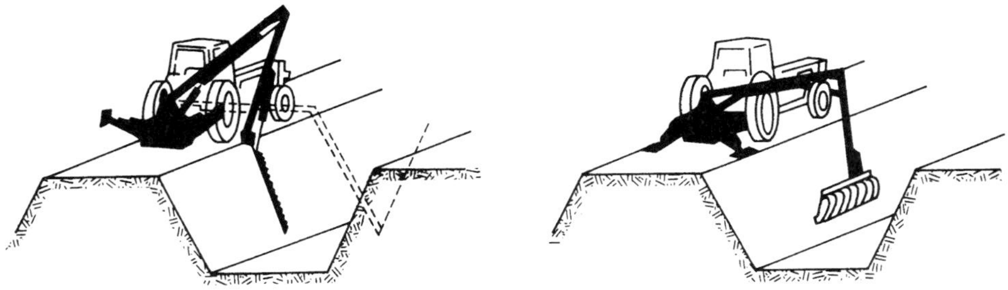

Fig. 7.16. Tractor-mounted equipment for clearing banks.

ing use is made of manufactured feeds. Hand feeding is the simplest method of distribution. This allows visual control and instantaneous response to changes in the voluntary food intake of the fish, and gives an optimum ration, close to that of fish feeding *ad libitum*. However, hand feeding is labour intensive and does not always result in the best conversion efficiency. In practice the timing and frequency of hand feeding by the farmer does not necessarily correspond to the fishes' physiological rhythms. It has been demonstrated that the time of day when the feed is given can affect growth performance, feed utilisation and even the body composition in some species (see also interactions between the method of distribution and digestive physiology, p. 174). For example, goldfish show a faster growth rate when fed during the day rather than at night (see review by Boujard and Leatherland, 1992). Such information does not yet exist for other species of pond fish so it is not possible to plan the distribution of feed based on published information. Automatic feeders are beginning to be used and these allow the quantity given to be adjusted to the biomass of the fish or to the state of the environment (O_2, temperature and plankton).

A wide range of equipment is available for the distribution of feed in ponds or large rearing systems (several hectares in area). Choice will be based on access to the site, whether or not electricity is available, area, stocking density, species, age class and feeding behaviour characteristics.

Belt feeders

These are boxes containing 3 to 5 kg of feed to which is attached a conveyor belt; This is operated mechanically by clockwork and can function for 12 or 24 h, distributing feed continuously. These are ideal for small quantities of the smallest-size feed particles and are used in hatcheries, nurseries and fry ponds. They can be installed on pontoons around the periphery of the ponds (five per hectare).

Projection feeders (Figure 7.17)

These are hoppers (capacity 15 to 250 kg) which have, at their base, a system for dispensing feed using blades that are turned by a small motor. The machinery is attached to an electronic system which regulates the rate at which food flows, possibly programming the interval between feeds and the part of the day over which feed is given. The feed can be projected through an angle of 360° covering a maximum area of 18×30 m. A supply of energy is required: this can be mains electricity or batteries (solar or dry). These feeders are installed on movable brackets or on floating supports. The object is to keep them away from the banks (distance >10 m). Experiments (three rearing cycles) have shown that a feeding station (two feeders installed on a bracket) arranged at an angle in a 4 ha rearing pond satisfies the dietary requirements of 20 tonnes of fish with normal growth rates and feed conversion efficiencies. The feeders operate from 9.00 h to 16.00 h, at a frequency determined by the daily quantity of feed to be distributed. Because fish congregate

Fig. 7.17. Feed blower (Aqualor) (photo Etablissements A. Heyman).

Fig. 7.18. Tube-based feeder in Israel (photo J. Marcel, ITAVI).

around the feeding point this can also become the harvesting zone (using a net). Feeders like these specially designed for fish culture (FFAZ type) are available. Feeders used for game birds in woodland can also be adapted for fish farming.

Tubular feeders (Figure 7.18)

This system is based on the same principle as the one described above, i.e. almost continuous feeding at a single point, with a 'turnover' of the biomass of fish. This is used particularly on Israeli fish farms and consists of a distribution system installed directly underneath a silo. Feed falls continually to the end of a tube, which is pierced with holes along its entire length. It is conveyed by a continuous screw mechanism, pneumatically or by a water current. An electric motor and a programming unit control the operation of the screw. A single silo and single feeding point are used for 10 ha grow-out ponds. This zone is also used for partial harvests.

Distribution using a vehicle with blower unit attached

This system, used in catfish farms in the USA, consists of a hopper mounted on a pickup truck, or pulled by a tractor which carries a turbine and an arm which projects the feed into a pond. An electronic system coupled to a microcomputer allows control of the quantity distributed. The vehicle moves slowly along the length of the bank. In this system feed is given only once each day at the most favourable time (13.00 h–14.00 h). This procedure, coupled with the use of floating feed, allows a daily check on the feeding behaviour of the fish.

Distribution by boat at feeding stations

This is particularly used for the distribution of cereals in semi-intensive systems. It is very labour intensive and cannot be used every day because of the wastage associated with the type of feed given. The feeding points are identified by a small fence around which the boat-mounted feeder passes. Wastes must be controlled (p. 167).

Self-feeders

These work on the principle of releasing diet when the fish want to feed. The most simple is a cylindro-conical tank whose base is closed loosely with a small circular plate, articulated with a rod at its middle point. The fish pushes this rod and causes granules to be released. It is difficult to determine the exact origin of this type of feeder, they can be easily constructed by the fish farmer, for example from a barrel, and are very cheap. Some feeders are simply constructed from a 20 cm diameter tube, 6 m in length, which acts as a reservoir, with the distribution mechanism at the bottom part but these are of limited capacity (150 kg) and have to be frequently refilled. These are extremely simple in their operation and require no form of energy input. However, the use of these feeders presents some difficulties in large ponds and lakes. They are used in salmonid nurseries to feed *ad libitum* or in units with a small area of water. The first difficulty is the continuous operation of the rod by small waves even when fish are not present to consume the feed (feed is therefore wasted) and also the small area over which feed is distributed means that only fish directly under the feeder will obtain feed. This means that a large number of feeders per hectare must be installed. The use of such feeders is largely confined to fry ponds where the wind speed is low and the biomass of fish is high. These feeders replace belt feeders when the fish reach 10–15 g. Because of their simplicity these feeders attract a lot of interest in spite of the difficulties highlighted above. More elaborate models are being developed and research is continuing with the objective of improving their performance.

Self-feeders have two basic component parts: the demand detector and the interface between detector and distributor. The detector is a simple steel rod which, when touched by a fish, opens an electric circuit of 5, 12 or 24 V through contact with a collar. More complex systems, which are also much more reliable, use vibration detectors attached to the rod, photoelectric cells or metal detectors, which are triggered when the rod is moved by a fish. The detectors (and thus the fish) control the feeders directly through an electronic control box or indirectly by means of a microcomputer. The use of a computer as an interface gives greater flexibility to the system. In practice, this can record feeding demands and thus analyse *a posteriori* the quantity of food demanded and the time at which the demands are made (Figure 7.19). In addition, the operator can select and programme a set of chosen criteria so that feed is not distributed under certain conditions. This could, for example, be the minimum time since the last demand, the number of demands needed to trigger a single distribution, the time window outside which demands for food are unrewarded, the minimum level of dissolved oxygen or the maximum temperature above or below which the release of food is blocked.

These demand feeders allow the autoregulation of the timing and size of a meal, resulting in the best possible balance between the true needs of the fish and the distribution of feed. They also allow an instantaneous response to the environment

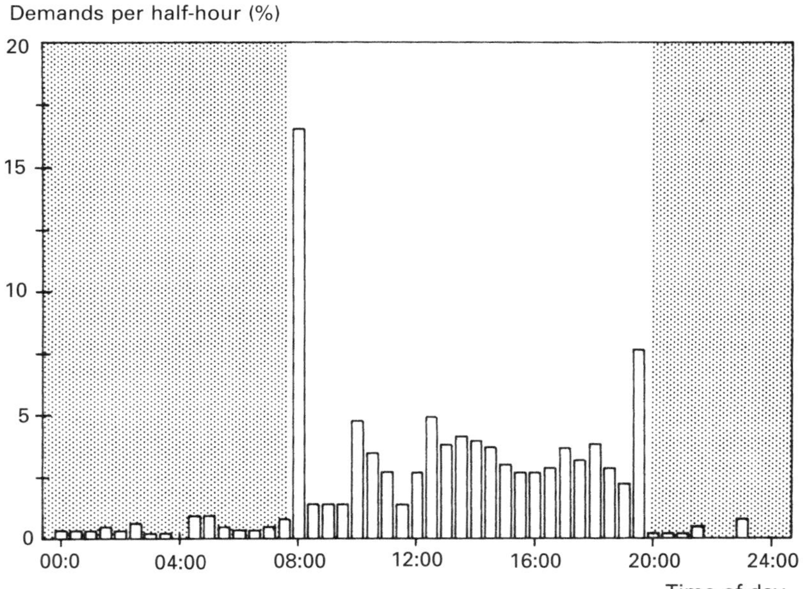

Fig. 7.19. Daily cycle of demand for feed by a batch of 39 carp fry with a mean weight of 20 g, under artificial 12 h light: 12 h dark photoperiod. The two main periods of feeding activity correspond to the artificial dawn and dusk, periods during which the light is filtered. The grey zones correspond to periods of darkness (Mambrini and Boujard, unpublished information).

and its influence on appetite. For trout the use of self-feeders which are adjusted in such a way that only a few pellets are released at each demand by the fish gives a performance as good as that achieved by careful hand feeding (Cho, 1992). This is based on the fact that small variations in water quality and poor health in the fish are accompanied by a decrease in feeding activity. Some authors have suggested recording the daily intake of feed. Analysis of the records in comparison with those from previous days can give an indication of the health of the fish and the state of the environment (Anthouard and Wolf, 1988).

7.2.5 Catching the fish

At various stages in the production cycle it may be necessary to remove fish from the pond: routine sampling throughout the year to check on growth rate and health, reduction of biomass particularly during the 1st year, selection at any time to obtain market-sized individuals for sale and the final harvest (see p. 85). Several techniques are available, but it is generally necessary to capture the fish alive, even in the case of those for consumption. Then, the fish are kept in running water for several days before sale in order to improve their quality. When the fish are to be processed, it is essential that they are killed immediately before filleting which means that holding

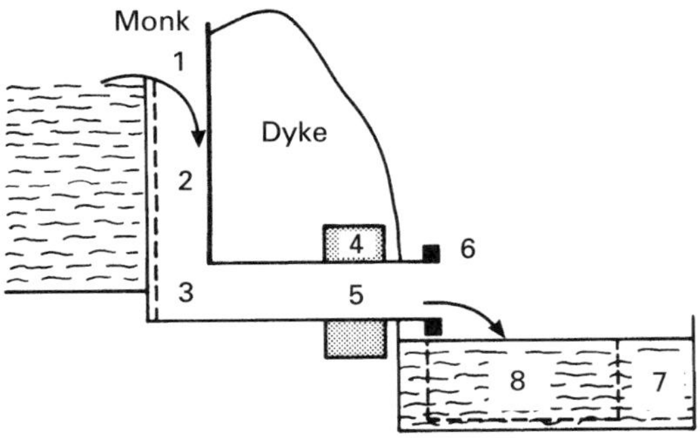

Fig. 7.20. Diagram of a downstream fishery. The fish are trapped by the current coming from the pond and fall into the tank, which may be common to several ponds (see Figure 7.21); (1) lateral grooves into which the planks regulating the level of water in the pond are slotted; (2) passage within the monk; (3) deepening at the base of the passage in relation to the pond which allows total evacuation of water and fish; (4) concrete block used for anchoring the outflow pipe; (5) evacuation pipe passing the external face of the dyke; (6) ring around the pipe allowing the attachment of a harvest net; (7) tank for holding fish; (8) cage for holding the fish which can be raised up to remove the fish.

tanks must be positioned close to the factory. The harvesting method must neither kill nor injure the fish and should also minimise stress. Particular care must be taken in the capture of small fish which are the most fragile.

Equipment used for crowding fish

Traditional methods of fishing are based on the complete emptying of the pond; fish can be harvested within the pond upstream of the monk, in a depression where they are concentrated at the end of the emptying process (internal fisheries). They can also be captured immediately downstream of the pond where the fish are trapped beyond the monk and harvested in various structures such as cages or tanks (external fisheries) (Figure 7.20). The external fishery may be shared between several ponds (Figure 7.21). There are also fisheries based on the counter-current principle which take advantage of the behaviour of fish which tend to swim against a current that has been artificially induced in the pond (Figure 7.22).

Seine nets suitable for use in open water are currently used in ponds that cannot be drained completely and also in ponds during summer (p. 184). Some species are difficult to catch with a net; these include silver carp that form large shoals and swim behind the net to the part of the pond already fished. To prevent escapes, Chinese fish farmers place an aerial net above the water attached to the floats of the capture net. Some rules should be obeyed for successful harvest with a net; it is easier in small ponds of regular shape with a flat bottom, no vegetation, no bank erosion and without excessive silt. The net should be 1.5 times longer than the width of the pond. Over 150 m in length the cost of manufacturing nets increases considerably; the pond should therefore be no more than 100 m wide. Bigger nets are used in old ponds and bodies of water that cannot be completely drained. For example, in Wuxi Province, China, a 110 ha lake is fished with a 2000 m long seine net. Several methods of

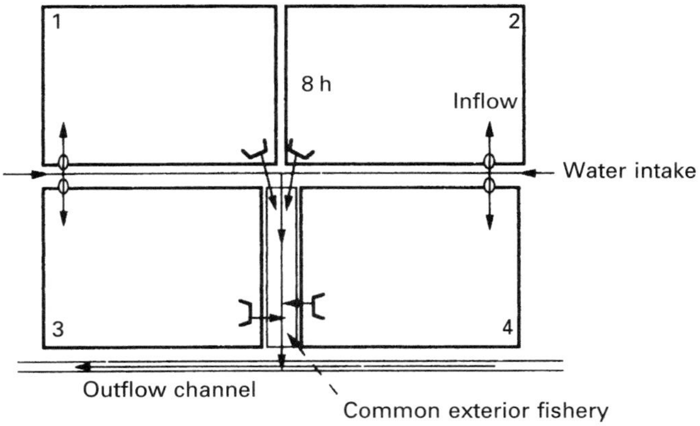

Fig. 7.21. Diagram of a group of four 8 ha ponds with a common harvest point (Szarvas Fish Culture Unit, Hungary). The harvest tank is 40 m^2/ha of pond with a depth of water of 0.6–1 m, and a slope of 20 cm from upstream in downstream.

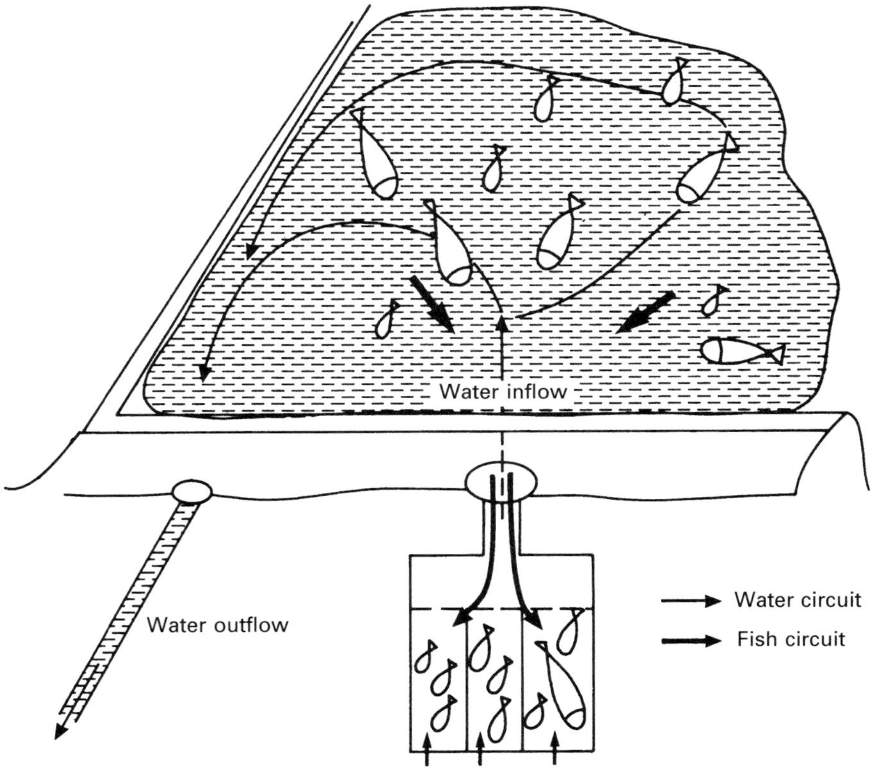

Fig. 7.22. Diagram of a counter-current fishery installed in a Polish pond. When the pond has been emptied (A) by $\frac{3}{4}$, the counter-current water is directed into the pond (B). Fish move spontaneously to the water inflow and are directed into the different compartments of the fishery. Partial grading is possible because of different-sized grills at the entrance of each of the compartments.

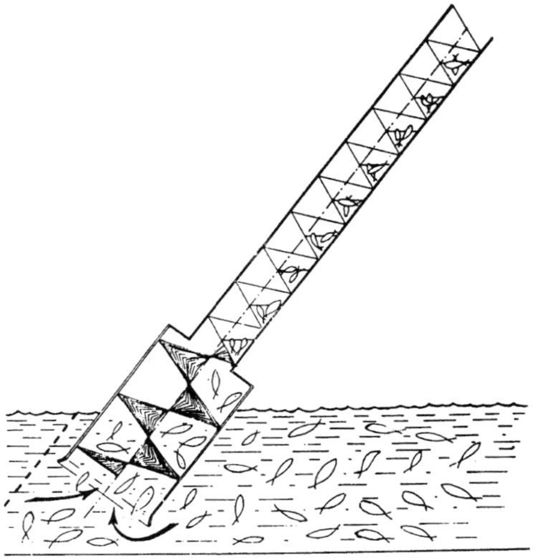

Fig. 7.23. Diagram of a fish-harvesting system based on the principle of Archimedes screw.

pulling the net are used: these generally use tractors operating from the banks of small ponds and motor-powered boats for larger ones.

Equipment used for removing the fish

Once the fish have been concentrated in the fishery or in the net they must be removed and transferred to transport tanks or directly into the grading equipment. There are several types of system used for harvesting fish:

—dry removal in cages (trolley) containing 50–250 kg of fish, moving on 10–14 m long rails (used in Israel);

—large landing nets moved by a hydraulic arm from a transport lorry with the ability to lift over 3–4 m and transfer 100 kg of fish (catfish in the USA and carp in Central Europe). A system similar to this is also used in Israel to capture fish that have previously been concentrated at the feeding station (p. 185).

Such arrangements are very stressful; fish are dried out and piled on top of each other. Preferable methods use an Archimedes screw (Figure 7.23), an airlift pump (Figure 7.24) or a vacuum pump (Figure 7.25).

7.2.6 Grading and transport of the fish

Pond fish are traditionally graded by hand because several species of fish are reared together and there is a great range of sizes. Many different types of equipment are used; some, in Hungary, are shown diagrammatically in Figure 7.26. The graders generally used for salmonids are not usually suitable for large carp (>1 kg).

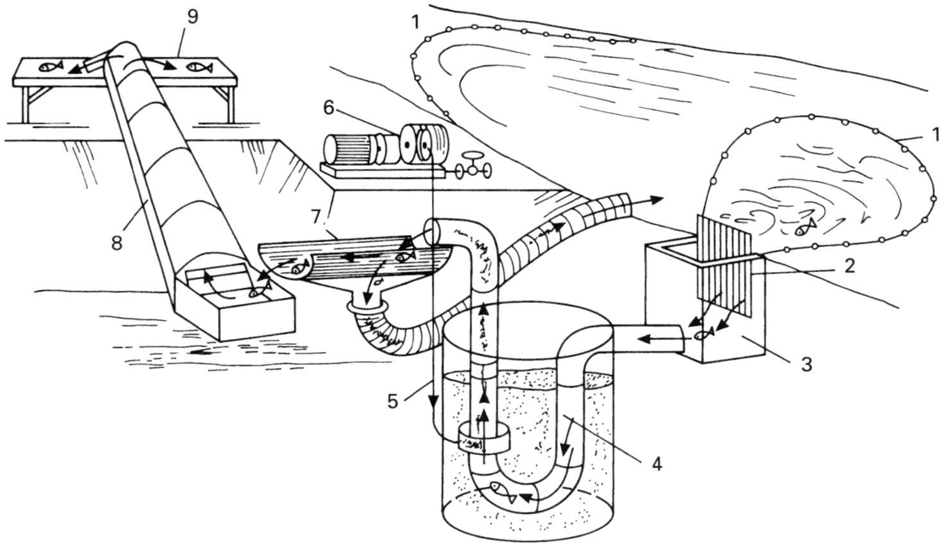

Fig. 7.24. Diagram of a system for harvesting fish using an airlift system used in Germany and Hungary; 1: fish are concentrated using a seine net; 2: grill; 3: concrete chamber for holding fish; 4: air pump buried in the gravel; 5: air pipe; 6: air compressor; 7: perforated slide keeping the fish dry; 8: conveyor belt; 9: grading table.

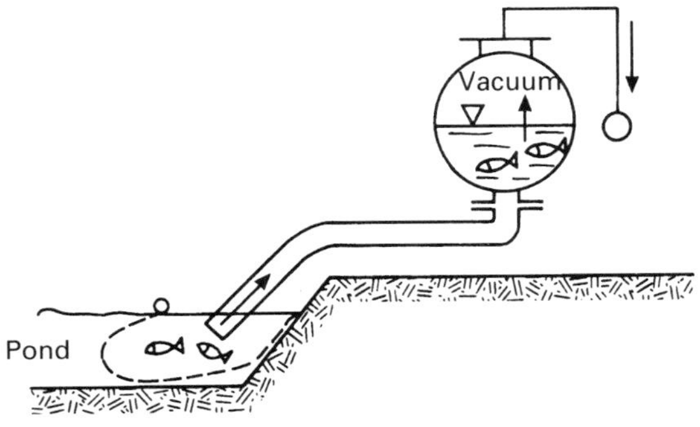

Fig. 7.25. Diagram of a system for harvesting fish by vacuum pump.

However, with the trend towards the production of portion-sized carp, (<1 kg) the use of such equipment is becoming possible. Such graders are already used to grade juvenile carp (1 and 2 years old) reared in monoculture systems in Central Europe.

Transport of live fish[4]

Two different systems are used for the transport of live fish:

[4] Information provided by R. Berka and L. Varadi.

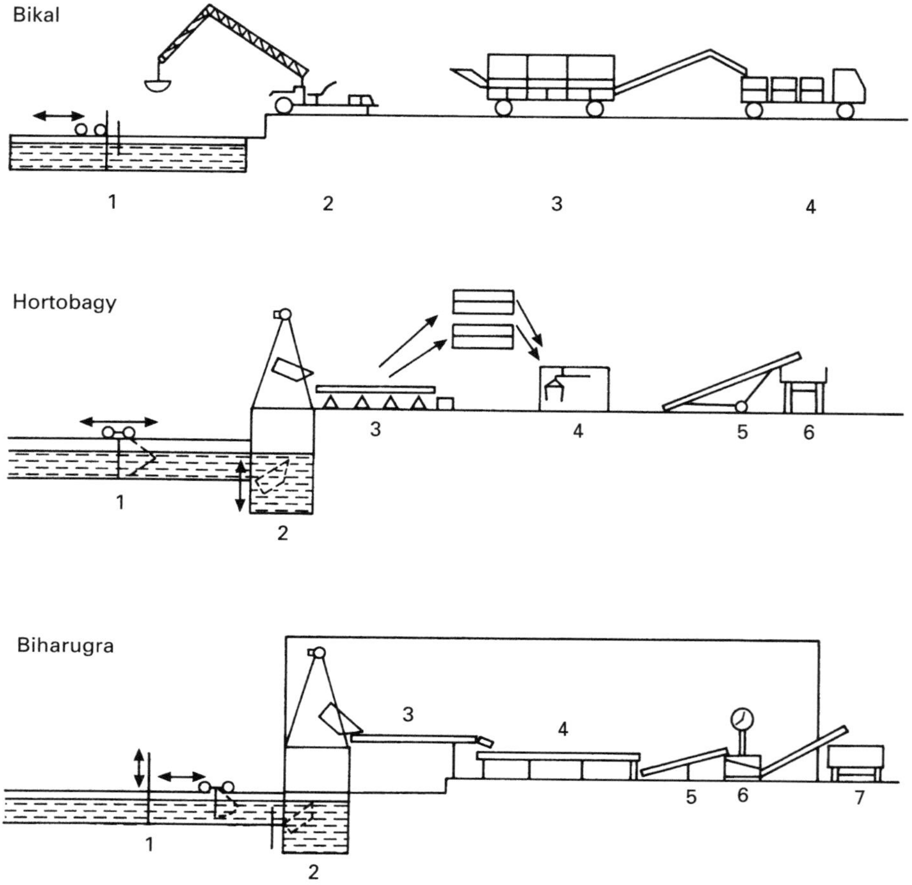

Fig. 7.26. Systems used for harvesting and grading pond fish in several different fish-culture units in Hungary.

Bikal—1: cement capture tank with removable grill to concentrate the fish; 2: landing net operated by a mechanical arm; 3: grading platform; 4: transport vehicle.

Hortobagy—1: similar to Bikal; 2: fish are lifted in a bucket; 3: grading table; 4: weighing (roman balance); 5: conveyor belt; 6: transport vehicle.

Biharugra—1 to 2: similar to Hortobagy; 3: conveyor belt; 4–5 grading table, container and conveyor belt; 6: scales; 7: transport vehicle.

—Closed containers (for example closed plastic sacks containing water, sometimes buffered and/or with added salt and oxygen) are most frequently used for the transport of juveniles.

—Open transport tanks into which oxygen is continuously bubbled and in which the water is sometimes renewed.

The major factors to be considered for fish transport

The health and physiological state of the fish. Fish that are in a poor physiological state, or starved and weakened through dietary deficiency or disease, withstand transport very poorly, even at reduced density. Fish that have been recently fed should not be transported because of their increased metabolic rate and the accumulation of faeces that will act as a substrate for the development of bacteria. If such fish must be transported, the biomass should be reduced by one-half. A fasting period of 48–72 h is recommended; this reduces the consumption of oxygen and the accumulation of ammonia; excretion of ammonia is reduced by one-half in fish that have been starved for 72 h. Temperature shocks must be avoided; for example with fish which have been reared in heated water and are being transported in winter. Any temperature drop should be less than 5°C per hour. It is possible to lower the temperature by adding ice: 25 kg in 1 m^3 drops the temperature by 2°C. Dry ice (solid CO_2) must never be added. Total daily temperature change should never exceed 12–15°C. In winter, part-grown cyprinids or those that have reached commercial size can be readily transported at 1–2°C. In summer it is preferable that temperatures should remain below 10–12°C for long journeys: spring water should be used in transport tanks for such trips.

Oxygen. is the most important factor. Requirements during transport depend on:

—temperature–requirements double between 10 and 20°C;
—weight–a large fish consumes less per unit weight than a small one;
—the degree of stress and agitation–requirements are greatest immediately after packing;
—pH and the concentration of CO_2 and waste metabolic products in the water.

Oxygen requirements also vary between species. Relative consumptions for several species of fish are as follows (carp = 1): trout 2.8; zander 1.8; roach, perch 1.5; bream 1.4; pike 1.1; eels, tench 0.8. In practice it is essential to ensure that the minimum level of oxygen never drops below 5 mg/l during the transport of cyprinids. Supersaturation with oxygen may occur when liquid oxygen is used but this does not appear to have any major adverse effect (trout can survive levels of 35 mg/l) at least for limited periods of time such as experienced during transport. However, in order to avoid excessive expense, the objective should be to maintain the oxygen level at 6 mg/l. The input of oxygen should be doubled when the fish are being loaded and during the 1st hour of transport.

CO_2. During transport the level of CO_2 increases because of the respiration of both the fish and the microorganisms in the water. For each ml of O_2 consumed the fish produces 0.9 ml of CO_2. In an open circuit, especially if air is being bubbled through and the top is not airtight, CO_2 will be eliminated. CO_2 may accumulate in airtight containers and plastic bags: here the concentration of CO_2 should never exceed 150 ml/l for thermophilic species such as cyprinids (CO_2 reduces the oxygen-binding capacity of the blood). The accumulation of CO_2 also results in the acidification of the water; optimum pH is close to neutral. Variations in pH can

be prevented by the addition of buffers such as Tris (1 to 2 g/l), but this increases costs.

Water quality. It is common practice to renew water during transport by pumping from a river at the same time as water is released. This procedure should be strictly avoided as it provides an excellent means of spreading disease into rivers and infecting fish in farms downstream. A better practice is to take on, at the start of the journey, water from the farm where the fish were reared, so long as it is of suitable quality. If not, tap water can be used. This water is likely to contain chlorine or residuals from bleach used in disinfection; levels above 0.01 mg/l should be avoided. Chlorine can be easily eliminated by bubbling air through the water. Certain compounds can be added to water to enhance the conditions for transport. Some have been used in practice and are referred to in the literature; salt is used to reduce osmotic stress (5–6 g NaCl/l for fresh-water fish), buffers (see above), antibacterial compounds to reduce the development of bacteria (acriflavine at 2.5 mg/l), anaesthetics to reduce the movement of the fish and their metabolism (MS 222 at 25 mg/l). The selection of these additives will depend on the duration of the journey, the value of the fish to be transported and the destination of the fish; some products such as anaesthetics cannot be used if the fish are going directly for human consumption. An evaluation of the costs and benefits of such improvements to transport conditions should always be made.

Various methods for the transport of live fish

Transport of eggs and juveniles. Transport of juveniles within the farm (from the hatchery to the first-fry ponds) can be done simply in a plastic bucket at densities of 100 000 larvae/l for journeys which take up to 10 min. Larger juveniles, for example those being removed from the first-fry ponds, should be moved under oxygen in tanks of 1.5 to 2 m^3 such as those shown in Figure 7.27 which can contain 1 to 1.5 million larvae or 100–150 000 4-week-old fry, for journeys lasting a few hours.

For longer distances, eggs, alevins, fry and older juveniles are most frequently transported in plastic bags that are partially filled with water and then inflated with oxygen and sealed. Adding oxygen allows a reduction in the weight of water and extends the duration of transport. There are also rigid transport tanks that are more robust and durable.

Containers are either in the form of bags which are pre-sealed at their base (dimensions 80–110 × 35–45 cm) or a tube of around 40–50 cm width which is cut as needed and sealed at the base. The plastic film used is of varying thickness but must be transparent so that the behaviour of the fish can be observed, and be non-porous to avoid the escape of oxygen. Thickish bags (0.6 mm) are more robust and can be used singly. If thinner bags are to be used they should be doubled. During transport the bags containing fry should be kept in wooden crates or boxes or expanded polystyrene boxes. If warranted by climatic conditions (exceptionally high temperatures), ice may be used for cooling (10–20% of the volume of the transport bag), preferably surrounding the bag containing the fish. The surrounding case must be watertight (Figure 7.28).

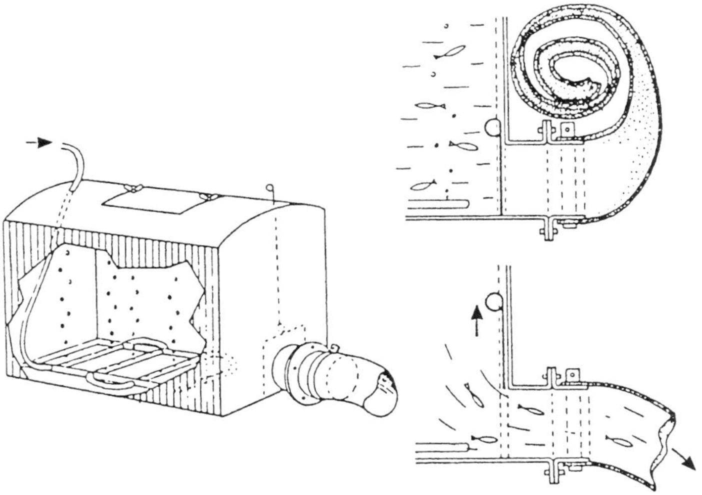

Fig. 7.27. Tank with oxygenation system for local transport of fish using a flexible, folding large diameter pipe, during transport (a), emptying position (b) (after Horvath *et al.*, 1984) (see General bibliography).

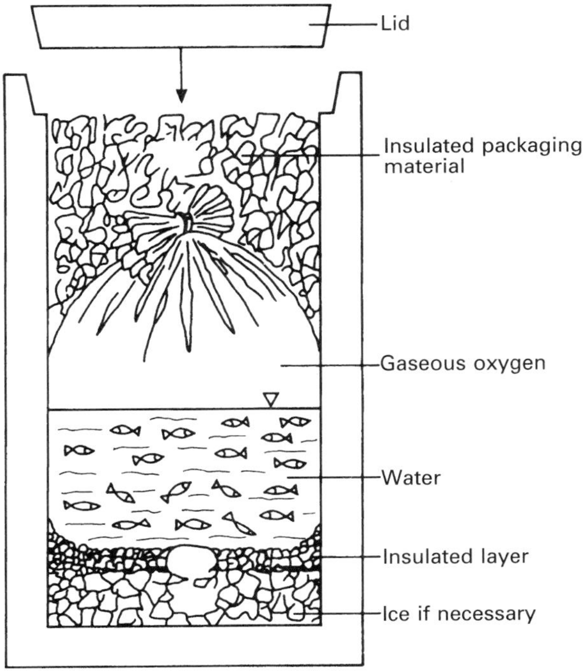

Fig. 7.28. Positioning of a bag with juveniles in an expanded polystyrene box for transport.

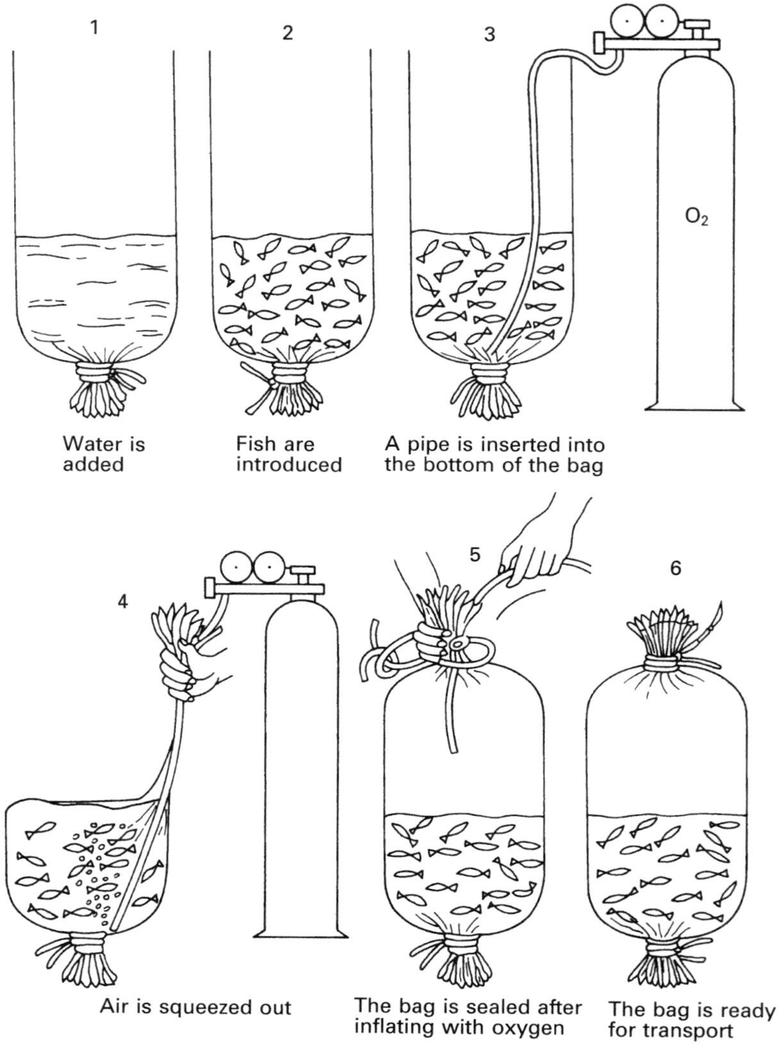

Fig. 7.29. Procedure for filling bags for transport:
(1) introduction of the water, (2) introduction of the fish, (3) insertion of the pipe to the bottom of the bag, (4) squeezing out the air before filling the bag with oxygen, (5) sealing the bag after inflating with oxygen, (6) bag, ready for transport (from Horvath *et al.*, 1986).

Water used during transport should be taken from the same rearing tanks as the juveniles to be transported but should be free from suspended organic matter. The bag should be carefully prepared, placed in the transport box and filled before starting to capture and count fry; this should be carried out as rapidly as possible.

Filling operations are shown in Figure 7.29. Around 20 l of water are put into a 50 l bag. The oxygen (around 30 l) is introduced by means of a tube inserted through

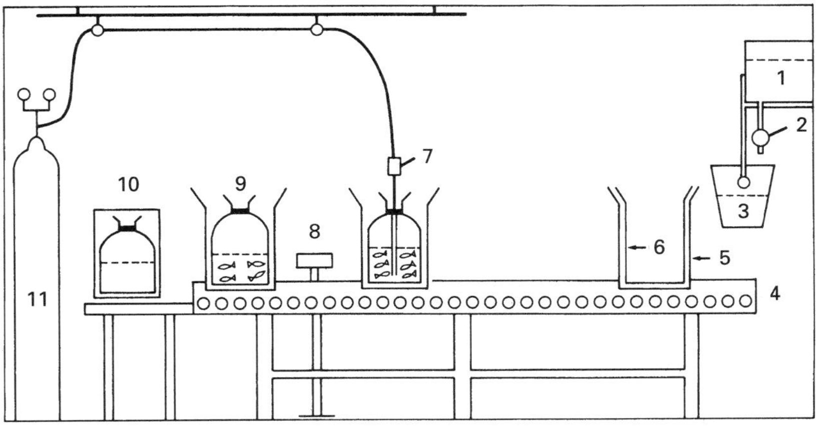

Fig. 7.30. Diagram of a production line for filling and packing bags containing fish for transport (example from the Tabor hatchery in the Czech Republic). 1: header tank which is temperature regulated and aerated; 2: tap for filling the bags; 3: 20 l bucket, balanced; 4: work table with rollers for moving bags with fish; 5: boxes; 6: polythene bag; 7: valve regulating the inflow of oxygen; 8: stock of elastic bands; 9: bag closure; 10: packaging and dispatch; 11: oxygen cylinder.

the upper opening; this is gripped to keep the bag firmly closed. When the bag has been inflated completely and the opening has been closed it should be folded and clamped shut with a strong rubber band or a metallic fastener such as those used for potato sacks. The sides of the bag should be supported. In large hatcheries, the procedure for filling can be organised as a production line; this happens in various countries such as the former Czechoslovakia and Israel (Figure 7.30).

At the end of the journey the sides of the bag should still be rigid and the fry behaving normally. Any dead fish or ones lying on their side on the bottom or corkscrewing indicate that at some stage during the journey conditions have been poor; this should be investigated and problems rectified for future occasions. Any temperature shock when the fish are being released should be avoided; larvae are unable to tolerate temperature changes of more than 1°C (2°C for older fry). The bags should therefore be floated in the receiving pond: when the temperature difference has been reduced to 2–3°C the bag can be opened so that water from the pond enters progressively. When 10 l has been added to a 20 l bag the juveniles can be released. During this whole process the fishes' behaviour should be monitored carefully in order to monitor any weak points in the process and ensure that the quality of water in the receiving pond is adequate.

Table 7.10 shows the densities of fish allowable in relation to temperature and journey time for different species (7-10A, yolk-sac larvae, 7-10B, 2–3 cm fry). The numbers are given in thousands for bags containing 20 l of water and 30 l of oxygen. In A the figures for rainbow trout are presented for comparison. Yolk-sac larvae and fry of carp and other cyprinids should not be transported at temperatures below 15–17°C. It appears that losses increase during low-temperature transport. The journey time should not exceed 24 h for carp larvae and even 20 h for juvenile silver carp and

Table 7.10. The number of larvae and fry (in thousands) which can be carried in a 50 l polythene bag containing 20 l of water and 30 l of oxygen for several species of pond fish and salmonids (from Berka, based on the standards used in the former Czechoslovakia).

(a) Yolk-sac fry

Water temperature (°C)	10				15				20				25			
Duration of transport (h)	4	8	12	24	4	8	12	24	4	8	12	24	4	8	12	24
Rainbow trout	25	20	15	10	20	15	10	5	15	10	5	3				
Char	40	30	25	20	30	25	20	15								
Powan, *Coregonus clupeoides*	80	60	50	40												
Coregonus peled	120	80	70	60	100	60	40	30								
Pike	80	50	40	30	50	30	25	20								
Carp					200	150	100	50	120	80	60	40	100	80	60	30
Tench					100			30	60	40	30	15	60	40	30	15
Grass carp									60	50	40	30	40	30	25	15
European catfish									60	50	40	30	40	30	25	15
Aspius					*100*	*80*	*60*	*40*	*80*	*60*	*40*	*20*				
Dace					100	80	60	40	80	60	40	20				
Barbel					100	80	60	40	80	60	40	20				
Nase					100	80	60	40	80	60	40	20				

(b) Fry of 2–3 cm

Water temperature (°C)	10				15				20				25			
Duration of transport (h)	4	8	12	24	4	8	12	24	4	8	12	24	4	8	12	24
Pike	5	3.5	3	2	3	2.5	2	1								
Zander	4	3	2.5	1	3	2	2	1	2	1.5	1	0.5				
Carp					15	12	10	8	12	10	8	6	10	8	6	4
Grass carp									10	8	6	5	8	6	4	3
Catfish									8	6	5	3	5	4	3	2
Aspius					*10*	*8*	*6*	*4*	*8*	*6*	*5*	*3*				
Dace					10	8	6	4	8	6	5	3				

Note: Every 12 h the oxygen should be replaced or the biomass should be halved.

Table 7.11. Biomass of market-size carp in kg/m^3 which can be carried for long distances in transport tanks where the dissolved oxygen is maintained at saturation for different species over a range of temperatures from 0.5 to 30°C (after Horvath, pers. comm.).

Temperature (°C)	0–5	5–8	8–10	10–15	15–20	20–25	25–28	30
Carp, tench	700	600	450	400	350	280	220	180
Grass carp	750	650	500	450	400	310	250	200
Silver carp	300	250	200	150	100	80	—	—
Big head	700	650	500	450	400	300	220	180

grass carp. For older juveniles, the density should be altered in relation to food intake. It is better for fry to be fasted before transport but if feed is given, the density should be reduced by at least 50%. The density can be increased if the oxygen can be replaced during transport or antibiotics given. Yolk-sac larvae are sensitive to mechanical shocks and to severe shaking: any sudden movements or agitation during transport should be avoided. Older individuals are better able to withstand rough journeys. Some degree of agitation is beneficial as it ensures that oxygen continues to dissolve in the water.

Transport of growing fish and adults. Fish can be transported over long distances in insulated containers of 2–4 m^3 volume. These are carried on lorries equipped with a system for distributing O_2 in either gas or liquid form. Table 7.11 shows the densities of fish that can be carried in such tanks. Large fish such as broodstock can be transported by air in plastic bags inflated with oxygen placed in standard boxes (30 × 30 × 90 cm) under the same conditions as juveniles. Recommended quantities are given in Table 7.12. Broodstock weighing between 4 and 5 kg are transported with a single fish in each sack.

Post-harvest transport and transport of fresh fillets

Fish which have been newly killed and gutted or fillets are transported either fresh or frozen. There are well-established techniques for the transport of the frozen product. This is also true for overland transport of fresh fish and fish products but for transport by air (in the USA this is used for 4% of aquatic products) additional precautions must be taken. Directives have been established in the USA by Air Transport America and the National Fisheries Institute and are summarised here. The objective is to ensure the best possible preservation in watertight packaging to prevent leaks that might damage equipment and any other products being transported. Whole fish should be pre-chilled to 0°C and kept at this temperature during packing. As well as helping preservation this limits the melting of the refrigerant in the containers. Packaging should include a layer of polythene and, at the base, an absorbent layer to retain liquids and prevent the fish becoming soaked

Table 7.12. Recommended biomasses for the transport of cyprinids in containers or plastic bags inflated with oxygen for air transport; the values given are for bags or containers with between 4 and 16 l of water and 45–50 l of oxygen for transport temperatures between 5 and 10°C (from Horvath *et al.* 1982, Publ. 16 HAKI in Hungarian).

Fish size		Volume of water*	Transport duration 8 h		12 h		24 h	
cm	g	l	Individuals	kg	Individuals	kg	Individuals	kg
3–5	2	16–16.5	1250	2.5	1200	2.4	1000	2
5–7	8	13.5–14.5	600	4.8	600	4.8	500	4
7–10	25	8.5–11	400	10	400	10	300	7.5
10–12	30	9.5–11	300	9	300	9	250	7.5
7–12	27	9–11.5	300–350	8.2–9.6	300	8.2	250	6.9
12–16	50	4–11	300	15	250	12.5	150	7.5
16–20	90	9.5–14	100	9	80	7.2	50	4.5
20–25	250	6–12	40–60	10–12.5	40	10	20–25	5–6.3

* The plastic bags are placed in standard boxes (30 × 30 × 90 cm). The final weight (packaging, bags, water, oxygen, fish) is around 20 kg.

in fluid. Blocks of solid carbon dioxide or ice in polythene bags should be positioned all around, or at least above and below to prevent the temperature from rising. The fish should be packed at the last possible moment before the journey and a strict hygiene regime observed. These recommendations also apply to fillets and should be even more strictly observed, as these are thinner than whole fish and therefore liable to warm up more rapidly. Similarly, for the initial refrigeration of fillets, it is important to avoid freezing, which could happen if temperatures drop slightly below 0°C.

The materials used should be capable of withstanding the problems associated with transport. Different companies have different weight limits for individual containers, usually around 75 kg, sometimes more. Ice is only allowed as a refrigerant if it is contained in thick, sealed polythene bags. The use of solid carbon dioxide that gives off gaseous CO_2 is restricted: a certain quantity is allowed in cargo planes and the quantity included should be recorded on the packages and declared. Solid carbon dioxide is not allowed in the holds of planes that carry live animals.

Containers should be insulated. Various materials are used; in decreasing order of effectiveness, these are: polystyrene or polyurethane foam, shredded paper and corrugated cardboard. Packaging should be robust and impermeable (embossed cardboard, expanded polystyrene) and be sealed securely. Clear instructions should be carried on the outside or on an information sheet that is accessible at all times: fresh fish, keep at 0°C, refrigerate in case of delay in delivery, high or low quantity of solid carbon dioxide, as appropriate.

7.2.7 Personal computers

Such equipment is becoming common in fish culture and is very easy to use. Computers are used to maintain business records and to record information such as stocking, feed distribution, growth, production, strains and performance of broodstock. Various software packages are useful, e.g. Excel®, Lotus®. The processing of this information aims to optimise production from the technical but even more from the economic point of view, taking into account the financial characteristics of the business. The computer system can be linked to various databases (which it can also supply), for example on the subjects of disease or weather. By using the weather database the computer can run models such as the prediction of the development of levels of dissolved oxygen during the night (p. 46) or growth (p. 187 and Muller and Zsigri, 1996). There are various programmes for assistance with security (SAS, Italy) or with stock management in intensive fish culture, for example AQUASTOCK®, PISCI® (IGER), SDAP® (SEDIA) and AQUAGEST® (Faivre) and OPPED (GEDIT-GIE by A. Fauré) for salmonid culture. MAP® (AET-IFREMER) (Table 7.13) has been developed for the bass *Dicentravclius labrax*. This software can be adapted for carp reared in intensive or semi-intensive conditions.

The MAP model is based on a strategy of production costs and financial options and integrates technical, biological, biophysical and economic (investments, debts, taxes) information. A Lotus 123 (version 2)® spreadsheet is divided into worksheets.

Table 7.13. Diagram of MAP (Model of Aquaculture Production) developed by IFREMER (Palavas) and Applying Evolving Technology (M. Spratt) for bass in cages and readily adapted for the intensive production of carp in small ponds. MAP is divided into seven sections and can be operated on an IBM PC with 600 kB available memory after the installation of Lotus.

1. Control of the business
   - economic control
     —economics: start of simulation, VAT, taxes, debts ...
     —subsidies: nature, % investment, number of payments ...
     —credit: borrowings, interest rate, duration ...
   - control of production
     —number of batches, stock in different ponds ...
     —quantities and characteristics of the stored feed ...
     —variable costs: medicines, disinfectants ...

2. Biological and technical details
   —annual and daily water temperature profile ...
   —actual and predicted daily growth rate ...
   —feed table ...
   —number and frequency of samples taken ...
   —feeding and cumulative growth performance, conversion rate ...

3. Production details
   —production; results
   —variable costs
   —summary of production

4. Environment
   —nitrogen, ammoniacal nitrogen, CO_2, suspended solids
   —oxygen consumption
   —current speed

5. Fixed costs
   —salaries, number, level, employment costs ...
   —general expenses, energy, transport, veterinary services ...
   —inventory of redeemable investments ...

6. Accounts and outcomes
   —investments
   —depreciation
   —expenses (associated with employees)
   —revenues
   —accounts

7. Macrocommands
   —transfer of information
   —drawing together and simulation

Investments, salaries and operating costs are detailed on these spreadsheets which provide information on profits and losses. This information is summarised and added to the results at the same time as other expenses. The model is designed so that each part of the spreadsheet has a function. Bringing together the parts allows changes to be made to the model in a stepwise fashion so that simulations can be performed and predictions made.

Technical or biophysical elements including the temperature profile of the site, feed tables for daily inputs and daily growth rates are used to operate the model. The user estimates and defines the number and the price of the fish, their initial weight, mortality rate and also the price of feed, etc. The model will establish their growth rate, feed consumption, costs and a final balance over 5 years of production.

In practice there is no operational model for the management of a pond ecosystem. However, it is hoped that the small models applicable to ponds will soon be developed to predict the development of components such as the micro-algae and zooplankton by considering the climatic conditions, the density of fish and the feed given. A model of this type is beginning to be used in the former Czechoslovakia by Prikryl (1990, 1993). An aid to pond fertilization has been developed by Lannan (1993).

Computers also help in the conception and design of ponds where fish-culture units are being extended or created; for example Excel® allows the optimisation of earth-moving work (p. 217). Complementary reading on the use of computers in fish culture can be found in Muench *et al.* (1986), Rosenthal (1990), Van Dam (1990), Fries (1994) and Lorenzen *et al.* (1997).

BIBLIOGRAPHY

Air Transport association/National Fisheries Institute, undated. *Guidelines for the Air Shipment of Seafood.* Washington, USA, 10 pp.

Anthouard M., Wolf V.A., 1988. Computerized surveillance method based on self feeding measures in fish populations. *Aquaculture*, **71**: 151–158.

Aqualog Edition 1994. *Catalogue technique de l'Aquaculture.* La Seyne, France, 260 pp.

Boujard T., Leatherland J.F., 1992 Circadian rhythms and feeding time in fishes. *Env. Biol. Fish.*, **35**: 109–131.

Boyd C., Martinson D.J., 1984. Evaluation of propeller-aspirator pump aerators. *Aquaculture*, **36**: 282–292.

Cho C.Y., 1992. Feeding systems for rainbow trout and other samonids with reference to current estimates of energy and protein requirements. *Aquaculture*, **100**: 107–123.

FAO, 1983. *Pisciculture continentale; l'eau.* Coll. FAO formation 4, 111 pp.

FAO, 1986. *Pisciculture continentale; le sol.* Coll. FAO formation 6, 174 pp.

Forsman L., Virtanen E., Salminen M., 1990. Salt addition decreases transport stress in freshwater pike-perch. *EIFAC 90/Symp:* 53B, 8 pp.

Fries J.N., 1994. SAMCALC. A computer program for fish culturist. *Progr. Fish Cult.*, **56**: 62–64.

Ginot V., 1986. Modéliser l'écosystème étang: In: R. Billard, J. Marcel, (eds), *Aquaculture of Cyprinids*, INRA, Paris. 255–268.

Ginot V., 1990. Modélisation nycthémérale de l'oxygène dissous en étang. Thése Univ. Claude Bernard, Lyon I, 154 pp.

ITAVI, 1989. *Annuaire de la Pisciculture*, 40 pp.

Knösche R., 1990. Aeration of fishponds. *EIFAC 90/Symp.*, 27E: 29 p.

Lannan J.E., 1993. *User's Guide to PONDCLASS® Guidelines for Fertilizing Aquaculture Ponds*. Pond Dynamic Program, Oregon State University, Corvallis, Oregon, 60 pp + disquette.

Lorenzen K., Xu G., Cao F., Hu T., 1997. Analysing extensive fish culture systems by transparent population modelling: big head carp culture in a Chinese reservoir. *Aquacult. Res.*, **28**: 867–880.

Marcel J., 1989. Les systèmes d'aération. *Echo-Système*, **12**: 13–16.

Muench K.A., Thomsen R.D., Croissant R.D., 1986. Computers in aquaculture. *Aquaculture Engineering*, **5**: 199–217.

Muller F., Zsigra A., 1996. Determination of a model optimum for fish production. *Aquaculture Hungarica (Szarvas)*, **5**: 227–234.

Muller-Fuega A., 1993. Computer mon ami. *Aqua-revue*. **46**: 25–28.

Prikryl I., 1990. Computer control of the fish production in the Czechoslovak ponds. *EIFAC 90/Symp.*, 27E, 2 p.

Prikryl I., 1993. Computer model of fish growth in ponds. Production, environment and quality, Bordeaux. In: G. Barnabé, P. Kestemont, (eds), *Aquaculture '92'* European Aquaculture Society, Sp. Publ. 18, Ghent, Belgium, pp. 309–313.

Rosenthal H., 1990. Personal computers in aquaculture. In: H. Rosenthal, E. Grimaldi (eds), Technology improvements in farming systems, *Proceedings of the 4th International Conference on Aquafarming 'Aquacultura 88'*, Verona, Italy, October 14–15: pp. 187–210.

Ruthway A., Kepenyes J., 1986. A new method for calculation of water requirement in fish production. *Aquaculturea Hungarica (Szarvas)*, **5**: 201–218.

Tucker C.S., 1985. *Channel Catfish Culture*. Elsevier, Amsterdam, 657 pp.

Van Dam A.A., 1990. Modeling of aquaculture pond dynamics, *EIFAC/FAO Symp.*, 27E.

Varadi L., 1983. Mechanized feeding in aquaculture. In: *Inland Aquaculture Engineering. ADCP inter-regional training course in inland aquaculture engineering*. Budapest, 6 June–3 September, pp. 445–460.

Varadi L., 1986. Mechanized harvesting in pond fish culture. In: R. Billard, J. Marcel (eds), *Aquaculture of Cyprinids*. INRA, Paris, 305–314.

Varadi L., 1990a. Integrated animal husbandry. *EIFAC/FAO Symp.*, 27E, 17 pp.

Varadi L., 1990b. Harvesting techniques *EIFAC/FAO Symp.*, E27, 20 pp.

Processing, marketing and economics

8

Processing and marketing

8.1 PROCESSING[1]

8.1.1 Body composition and yield after filleting

Information on the relative mass of different organs of carp is available. Figure 8.1 gives the values (excluding gonads) and demonstrates that the proportion represented by muscle increases regularly from hatching to a weight of 800 g to 1 kg and stabilises above this at around 70%. Gonads begin to develop when the broodfish reach weights of 500 g (males) and 1 kg (females). When fully mature these represent 6 to 10% of body weight in males and 20–30% in females. In the former Czechoslovakia, fish processors work to a standard recovery rate of 59% for flesh (muscles) from whole 1 kg carp (including skin and gonads). In France, yields for individuals over 1 kg of around 30–35% have recently been reported. It is likely that there are significant differences between races and populations of carp depending on their morphology, feeding and method of rearing. IFREMER (Nantes) have measured the potential yield from silver carp to be around 45%.

8.1.2 Chemical composition, sensory evaluation and *post-mortem* changes in composition

Different authors have reported variations in chemical composition of whole carp: water (69–81%), protein (12–20%), fats (3–12%), ash (1.1–1.3%) (Table 8.1). The flesh contains the principal amino acids essential to man. These are the same as in most species of fish: valine, isoleucine, lysine, methionine, threonine, tryptophan and phenylalanine. The proportion of unsaturated fatty acids is around 80% (the other 20% are saturated), with the major components being linoleic, linolenic and arachidonic acids. However, the composition is strongly influenced by the dietary regime (Table 6.6). The protein level is high because carp feed naturally on zooplankton and benthos rather than on a cereal-based diet. Artificial diets (cereals

---

(1) Including information supplied by R. Berka and B. Fauconneau.

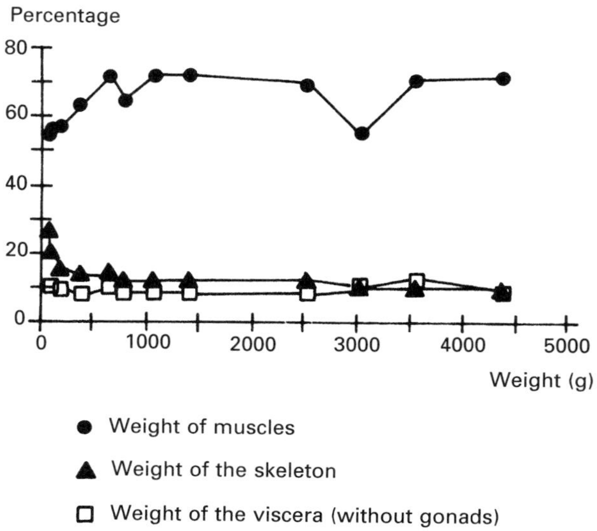

Fig. 8.1. Change in weight of muscles, skeleton and viscera (in %) for carp with increasing weight. Percentage is expressed in relation to total live weight (excluding gonads) (after Yapo, 1990, Thesis, Université de Paris 7).

and pellets) result in high fat levels in the fish. Some organoleptic characteristics are given in Table 8.1.

The physical and biochemical characteristics that prevail after death are highly dependent on the temperature and conditions of rearing, storage and killing of the fish. *Rigor mortis* is one of the most important characteristics to be taken into account and it should be delayed as long as possible after killing in order to delay any undesirable biochemical changes for as long as possible. Any treatment (processing) of the muscle such as mincing, filleting or slicing must take place before *rigor mortis*. The occurrence and duration of *rigor mortis* depends on the general physiological state of the carp, the temperature of the water before harvest, the method used for slaughter and, above all, temperature (Table 8.2). When carp are stressed or subjected to a prolonged fast before slaughter there is very often a depletion in glycogen and Ca^{++} reserves in the flesh which causes an acceleration of the rate of onset of *rigor mortis* and a lower minimum pH. If the fish are kept in water after slaughter the appearance of *rigor mortis* is advanced. At 1°C, bacterial development is weak for 4–5 days while at 10°C it starts within the first hour of storage.

8.1.3 Filleting

Carp should be filleted very soon after slaughter to ensure that the flesh is in *pre-rigor* (i.e. before *rigor mortis* has set in). Filleting should take place in conditions of total hygiene to avoid any bacterial contamination. When large quantities of fish are to be

Table 8.1. Some characteristics of the flesh of common and silver carp (B. Fauconneau, 1991, Rapport MRT); mean values ± SD. A more detailed analysis is given by Vacha *et al.* (1993).

		Common carp	Silver carp	Rainbow trout (for reference)
Chemical composition (%)				
Water		8.1 ± 0.3	76.7 ± 1.5	74.4 ± 0.7
Protein		12.2 ± 0.9	14.3 ± 0.6	15.5 ± 0.3
Lipid		1.9 ± 0.6	3.5 ± 1.8	4.7 ± 0.7
Physical characters (N*)				
Maximum force (raw)		36.4 ± 2.2	31.4 ± 0.4	25.4 ± 5.9
Maximum force (cooked)		7.2 ± 2.0	6.5 ± 3.3	5.1 ± 0.7
Sensory evaluation (note 12)				
Softness		8.3 ± 2.0	5.4 ± 1.9	8.2 ± 3.0
Smell intensity		6.6 ± 2.4	4.5 ± 2.3	6.2 ± 3.3
impression		8.4 ± 1.9	8.0 ± 2.7	6.1 ± 2.8
Colour		7.6 ± 2.7	6.0 ± 3.1	8.7 ± 2.8
Taste intensity		8.8 ± 1.6	6.6 ± 1.9	7.0 ± 2.7
impression		9.1 ± 1.9	9.1 ± 1.6	4.6 ± 3.4
Texture		8.3 ± 2.0	5.4 ± 1.9	6.6 ± 3.2
Definition of terms†	Smell	Flesh	Neutral	Hay, grass, sand
	Taste	Fine	Neutral	pleasant, grass

* N: maximal force at the first level (maximal derivative) of the compression of cooked samples.
† Tests carried out by the employer's federation of the 'haute cuisine française'.

processed, mechanisation is required and large, purpose-built factories are required. This allows for the optimum use of the various by-products of filleting.

Carp should always be removed from the water alive and not allowed to die in the pond water because of the risks of contamination. They should be transported alive

Table 8.2. Delay in the appearance and duration of *rigor mortis* of carp after killing in relation to storage temperature.

Temperature (°C)	Delay of onset after killing	Duration
35	3–10 min	30–40 min
15	2 h	10–24 h
10	4 h	30 h
5	16 h	2–2.5 days
1	35 h	3–4 days

to the processing plant in clean water, preferably spring or drinking water. When production is intensive and especially if organic fertiliser has been used, it is preferable for the fish to be held in tanks of clean water for a few days before slaughter. This gets rid of any microorganisms and various compounds that might give an off-flavour to the fish.

Electric stunning is the most common method in use for slaughtering freshwater fish (see works on trout and catfish). This procedure is used for carp in the former Czechoslovakia and in France. A few tens of kilograms are put into a plastic tank of around 100 l capacity which is perforated to allow surplus water to overflow. An electrode is placed among the fish in the tank and an electric shock applied for 1 to 3 min for carp weighing 0.8 to 1.5 kg. The effect of electrocution on the flesh is poorly understood: some sources suggest that it has adverse effects and recommend other methods such as transferring on ice or to a bath of water at 0–4°C and filleting in a cold room or anaesthesia in a bath saturated with CO_2.

If necessary the fish are then treated with jets of water at a pressure of 20–26 atmospheres for 3 min to remove the scales. Equipment for this purpose used in the former Czechoslovakia can be used to process 100 kg of fish in 3 min.

The various species of cyprinids are generally filleted by hand; there is currently no equipment suitable for automatic filleting available. Bones are removed by passing the fillet under a machine consisting of an assemblage of cutting discs that cut through bony structures without causing significant changes to the muscle structure. The skin is then lifted off mechanically and the fillet is ready either for sale as it is, or for more elaborate processing. Hacker in Cooper (1987) has described the technique used for hand-filleting, with information on intramuscular bones and their location (Figure 8.2). It is useful to know this as they can be removed with the help of the sharp discs without cutting into the fillet. In addition, it is essential that the fillet is placed in the machine in such a way that the line of bones is perpendicular to the cutting discs (Figure 8.2).

Large quantities of waste are produced during filleting and it is necessary to find ways of exploiting such by-products. The gonads, liver and skin can be sold but the rest of the head and the skeleton has little market value and is sometimes converted

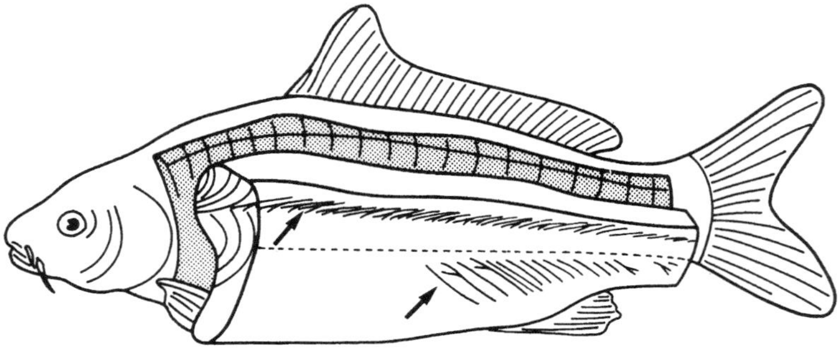

Fig. 8.2. Diagram of a carp showing the position of pinbones (arrows) (from Hacher, in Cooper, 1987, reprinted with permission).

to fishmeal or even fertiliser. The hypophysis can be collected and dried to provide a source of gonadotropin for the induction of ovulation. The dry weight of a hypophysis from a 1 kg carp is around 3 mg and this can be sold for around 150 to 300 €/g. The muscles of a 1 kg carp represent 70% of the body weight without gonads (Figure 8.1): significant quantities of flesh remain attached to the vertebral column and the head and these can be recovered in the form of a paste (see below). In the former Czechoslovakia fish wastes are ensiled by mixing with formic acid (3–3.5%) and liquefied for 2–7 days at 18–20°. The ensiled liquid (55%) is mixed with cereal flours (45%) to make a paste which is fed to livestock (pigs, poultry and cattle) or even to fish. However, it must be remembered that pathogens can be transmitted by silage: heat inactivation is not always complete and the addition of a disinfectant such as Virkon at 1 : 100 (weight : volume) has been advised (Smail *et al.*, 1993).

The flesh of carp and other cyprinids can be presented in a wide range of forms. The fillet can be sold fresh or frozen (2 h at −35 or −40°C and stored for a maximum of 5 months at −18 to −20°C without any deterioration in quality and the protein content. The weight loss is estimated at 2% during this storage time. Rapid freezing over liquid nitrogen (−5°C/minute down to −80°C) preserves the structure of the muscle fibres better.

8.1.4 Smoking[2]

Whole carp or fillets are smoked using standard techniques. Hot smoking as practised in Central Europe consists of immersing the carp in brine (maximum 120 g salt per litre) for 5 to 10 h, then transferring to a kiln at 70–130°C for 0.5 to 3 days depending on the size of the piece of fish and the temperature. Cooling should be rapid in order to restrict the development of microorganisms. Weight loss is estimated at 10–12%.

Cold smoking is preferred, especially for fillets. It is essential that bones are removed from the fillets using cutting discs (see above). Carp bones cannot be removed by hand when the fish is fresh.

The fillets are placed in a brine bath: dry salting is not possible as the salt becomes trapped in the cuts made by the discs when removing the bones. Brining takes around 4 h. The fillets are then placed on a grid and dried overnight at 4°C; they are then reshaped by hand pulling together the edges of the cuts left by the discs. The grids are then placed in the smoker where they are left for around 4 h at a temperature that fluctuates between 12 and 45°C (maximum 50°C) and a low relative humidity (75–80%) so that the fillet continues to dry during smoking.

Coarse sawdust (0.5–1 cm) from beech or oak should be used. Conifers are not suitable although it is possible to add fragments of various woods such as rosemary and blackcurrant to flavour the smoke. After smoking is complete the fillets are left to dry for a further 7–10 h at 4°C and then vacuum packed whole, sliced or in a

---

(2) Information provided by E. Rolland.

block. Slicing or cutting is done by hand using a flexible blade; the flesh of carp is not oily but sticky and cannot be sliced mechanically. One experienced operator can process 300 kg of fillets per day. It is essential that this processing should be in clean, aseptic conditions. Smoked fillets can be consumed as they are but are often used as a base for salads, for example seasoned with hazel-nut oil, or as part of more elaborate dishes.

8.1.5 Other forms of processing

Carp fillets can be prepared in a wide variety of ways: marinated, filleted and made into terrines or kebabs. They can be cut into strips (goujons), cubes for fish fondue, soaked in various marinades, sometimes before smoking, or served in cold dishes. One method used in Central Europe is to cook pieces of fish together with various seasonings and vegetables (40–60%) in a range of sauces and to sterilise them at 100–118°C for 35–50 min. Such products can be kept for up to a year. Similar products (medallions) of excellent quality are made in France but not yet widely available commercially.

Many dishes are based on minced carp flesh or pulp. This is made from fillets supplemented by a significant amount of muscle which remains attached to the axial skeleton, recovered using a flesh/bone separator (Figure 8.3). Once minced, the flesh must be rapidly refrigerated and treated. It should be washed in order to eliminate soluble proteins and thus aid processing. The addition of stabilisers (sugar, salt and sodium citrate) allows short-term storage. Chopped flesh can be stored at −30°C after cryoprotectors (sucrose, sorbitol and certain amino acids) have been added. It appears that fatty acids are best preserved under these conditions, showing no signs of rancidity. Fish flesh is incorporated in a wide range of charcuterie produce. Some products are based on fish alone; pâtés, sausages, quenelles and fish meatballs while others, such as terrines, incorporate various vegetables. Additional treatments may follow such as the smoking of sausages or fish meatballs; these must incorporate binders to stabilise the shape of the final product.

Fish flesh can be added to other animal meats in classic charcuterie products or conserves. One of these, produced in Central Europe, consists of a mixture of carp (22–37%), beef (22%) and pork (24%) with added eggs and spices which is canned and sterilised. Minced carp flesh goes well with minced beef from the point of view of texture and taste.

In general, carp flesh can be viewed as a primary raw material from which a wide range of products can be created. These can be packaged attractively to respond to consumer demand, which should be kept under continuous review.

There are many recipes available for the preparation of carp; these can be found in general cookery books and in leaflets distributed by fishmongers. Some have been brought together in specialist works: around 30 are given by Hecker in Cooper (1987) and the Chambre des Métiers de l'Ain (1983) edited a booklet of recipes.

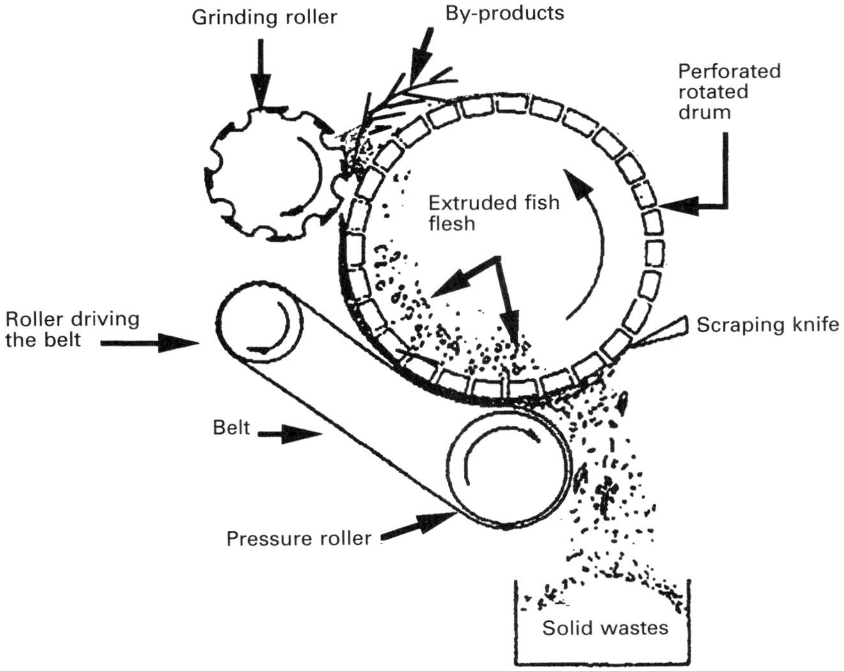

Fig. 8.8. Principle of the operation of a fish-flesh extractor with a belt and moving drum. The grinder is compressed between a belt and a perforated drum: the flesh passes through the perforations in the drum. This is generally used to obtain minced flesh from fish which have been filleted and eviscerated. This equipment also extracts the flesh from around the vertebral column after the fillets have been removed. (Frentz *et al.*, Bull. liaison CTCCV, 4: 267).

8.2 THE MARKET FOR CARP—A FRENCH EXAMPLE[3]

A recent survey of a random sample of French consumers based on semi-direct interviews has shown that carp is viewed as a prestigious but static product (GEM and Faculté de Psychologie de Nanterre):

—the people questioned mostly associated carp with medieval-type festivities (castles, feasts, knights, banquets, etc.). This apparently surprising finding can be easily explained; the "culture" of carp for non-fishermen is limited to academic literature (*Roman de Renart*, *Fables de la Fontaine*, *Lettres de Mon Moulin*, etc.), referring to a past time when carp was held in high esteem. In the absence of more modern references it is this image of carp that finds its way to the collective consciousness. In food marketing this type of product

[3] M. Jacquinot.

which only has associations with the past is called a "dead" product because of its bleak commercial prospects.

—the people questioned also expressed views on the organoleptic and other properties of the fish including; "very poor," " it smells of mud," "it's bland, insipid," "it is tasteless," "it's fatty," "it's full of bones," "it's huge and I don't know how to cook it." It appears to us that the large majority of the people surveyed who expressed their views so prolifically and so negatively with regard to carp have probably never eaten it. Their opinions are thus based on preconceptions and stereotypes rather than personal experience.

What should be concluded from these observations? Carp have a poor image: all people questioned knew of the fish and other studies have also revealed the poor opinions. Carp are seen as an old-fashioned product with innumerable faults. Under these circumstances is it reasonable to continue the interest in the species? This question is posed frequently and the answer must be affirmative, for two reasons:

- The product is unfairly denigrated; some of the criticisms made of carp seem to be based on prejudice. Blind taste tests carried out using cooked carp-based dishes and simple court bouillons attracted flattering comments from consumers, many of whom had difficulty in believing that they were eating carp.
- This species has some genuine qualities that can be promoted beneficially as part of a marketing campaign:
  —carp are an "ecological" fish: the rural charm of fishponds is a positive feature; after "free-range chickens" why not a "pond fish" or similar label? A section of French agro-food production takes advantage of similar labels.
  —guarantees of origin: among the requirements of the concerned modern consumer is the need to be reassured as to the origin of food products purchased. In France all carp can claim prestigious origins: Sologne, Limousin, la Dombes and Camargue for example.
  —fine taste: at a time when smelly cheeses are vanishing for want of buyers, the blandness attributed to carp can be used in its favour.

8.2.1 Towards a new range of carp-based products

Carp have some undeniable commercial trump cards but suffer from a series of handicaps which all of those involved in aquaculture would like to remove:

—"it smells muddy": even professionals recognise that this is true for some fish. However, the taste disappears if individuals are allowed to empty their guts before slaughter. Because of this it is essential to define a protocol for producers which will eliminate the risk of unwanted flavours. If a change of image for a product is needed, the slightest deviation from the desired standard cannot be tolerated.

—"big fish, difficult to prepare": modern consumers increasingly shrink from preparing their own fish. Among the arguments advanced are not only the unpleasant smell, mess and dislike of the preparation procedure but also the loss of the home-based preparation and cooking, which the growth in the sales of books on food and food-related topics has failed to halt. Because of the size, anatomical configuration and the disappearance of oral tradition on the subject of preparation, carp are vulnerable to this sort of criticism. It therefore seems to be essential that carp should be sold as semi-prepared or fully prepared products; the supply of live carp on the banks of ponds after the big autumn harvests is all that remains of a traditional market. For most consumers, it is necessary to come up with new products, adapted to new-style cooking practices (p. 266).

—"it's full of bones": the presence of numerous bones is, unhappily, part of the true characteristics of this species. However, the absence of bones is one of the fundamental requirements; their elimination from products put on the market is an absolute necessity. The specification for novel carp-based products should cover the possible techniques for the treatment of bones (removal, crushing, cutting, softening or partial dissolving) (p. 268).

—"this is an old-fashioned product": records show the presence of carp at banquets during the middle ages and Lacepède, at the start of the 19th century, referred to carp and herring as "the two fish which are transported to all of the markets, and are seen on all tables, that everyone recognises, seeks out, distinguishes and appreciates the tiniest nuances of their taste," but this is not sufficient to persuade modern consumers to buy. Historic dishes are only eaten under certain circumstances. The image of the product must therefore be modernised without misrepresentation: this is a task for the advertising industry.

8.2.2 The new "marketing-mix" for carp

The challenge for those associated with the production of carp appears clear: either to keep the traditional pattern of production and marketing while limiting commercial ambitions to peripheral markets (ethnic minorities, farm shops) or to give carp a radically new position and therefore considerably expand the market.

Which option should be chosen? It is important to keep the true character of the product, its roots (rural production, gastronomic nobility and history) but also to modernise the image; carp is not and should never become a cod substitute. It is not odourless, tasteless fish flesh but a fish with a distinctive personality, a speciality. Given the current status of production in the West (small quantities, poor fillet yield, production costs still relatively high), it does not seem sensible to go further in promoting this fish. This specialist product should be adapted to the requirements of modern consumers (and restaurateurs); after all, the best, highest-priced poultry is not sold in ready-to-cook form.

Carp producers in Western Europe should also take into account the growth in supplies coming from Eastern Europe resulting in two separate consequences:

—*intensification of competition*: as with a number of food products, the disappearance of the Soviet market has caused producers and exporters to turn to new markets in the European Union;

—*diversification of sales*: the new economic framework will cause the Eastern European businesses to be more creative to adapt to the market.

The marketing of carp in Western Europe should therefore take account of the following different parameters:

—the need to modernise product image;
—exploitation of the characteristics of the product;
—presence of competing supplies produced at lower cost.

This will result in the production of a marketing strategy. Traditionally a marketing strategy has five elements: the product, the price, the distribution channel, sales effort and communication. What new marketing-mix can be developed for carp?

Product strategy. Forget live carp kept in aquaria at fishmongers and supply consumers, caterers and restaurants with semi-prepared and prepared products (boneless fillets, pâtés, terrines, prepared dishes). In addition it seems to be a good idea not to confine the marketing exercise to a single species, but to exploit a whole range of freshwater fish (carp, pike, tench, perch, catfish, black bass, etc.).

Price strategy. Carp cannot and should not be positioned alongside the highest-priced fish fillets. Bearing in mind the special features (costs of production too high, organoleptic qualities not conforming to usual specifications), carp fillets must be sold and priced as speciality products.

Distribution strategy. Several niches appear to offer themselves for this product:

- restaurants: not only traditional restaurants in lake and pond regions but also theme restaurants looking for new primary food sources;
- caterers: at small-scale and industrial level require a supply of semi-prepared products such as fillets and pâtés;
- major distributors: in the first instance concentrating on hypermarkets in the traditional regions of production.

Sales strategy. This is a crucial point which will determine the success of the operation; in practice it is necessary to find an equilibrium between vital sales effort (carp cannot be viewed as a self-selling product) and the requirements of management. Thus the producer of pond fish who organises supplies to local and regional restaurants risks disproportionately high distribution costs in relation to the quantities sold; partnerships with wholesalers may allow an increase in the range of products sold and thus an increase in turnover. In general, it is necessary to be careful about direct sales or distribution; such methods of selling are frequently accompanied by high transport costs.

Communication strategy. The modernisation of the image of carp is essential for increased production and sale of the product. This presupposes effective communication. Without doubt this is an area fraught with problems; because the volume of sales is so small there is no big budget for classic, large-scale promotional campaigns. A public relations-type approach should be investigated; this would involve supplying editorial material, attending media events and stimulating debate with chefs and others.

8.2.3 Convincing experiments

Around 8000 tonnes of carp are produced each year in France; around 5000 tonnes of this production is destined for the restocking market. The rest is mainly sold alive, mainly in the regions where it has been produced or exported to Germany. These traditional markets are declining consistently; the industry owes its survival to a few pioneers who are preparing the way for new markets for the sale of carp fillets. A co-operative in the Dombes region has established a filleting plant where around 400 tonnes of carp are processed each year. This has enabled the establishment of a regular supply to local caterers and some further away and an export trade representing 60% of sales (Table 6.7). Other projects and initiatives have unfortunately met with less success: one plant has abandoned the production of cooked carp dishes after several years' activity. This setback has been attributed not to the quality of the product, which was universally liked, but to the poor image of carp.

These experiences, both good and bad, give an indication of the future for carp production and marketing. The development of this industry requires the initiation of an intensive and voluntary marketing policy, which includes producers, and regional, national and community-based public bodies.

8.2.4 A plan for European carp production

The new directions within the Common Agricultural Policy clearly define among their priorities: research on the exploitation of land with increased regard for the well-being of the environment and countryside, lower productivity and conformation with quality criteria. French pond-based aquaculture fits these objectives well and many of the ponds are located in environmentally and socially sensitive areas. It thus appears that there is an opportunity deliberately to encourage this activity and favour its development through the modernisation of production techniques and through the communication that is essential to commercial success. The rearing of carp may be the start of a new relationship between man and the environment: carp may yet show itself to be a modern fish.

BIBLIOGRAPHY

Chambre des Métiers de l'Ain, 1983. *La carpe de Dombes; une recette par semain.* 86 pp.

Cooper E.L., 1987. *Carp in North America.* Amer. Fish. Soc., Bethesda, Maryland, 84 pp.

Fauconneau B., Alami-Durante M., Laroche M., Marcel J., Vallod D., 1993. Growth and meat quality relations in carp. Carp Conference, Budapest. *Aquaculture* **129**: 265–297.

Smail D.A., Huntly P.J., Munro A.L.S., 1993. Fate of four fish pathogens after exposure to fish silage containing fish farm mortalities and conditions for the inactivation of infectious pancreatic necrosis virus. *Aquaculture*, **113**: 173–181.

Vacha F., Vavreinova S., Holasova M., Tvrzicka E.T., 1993. Analyses of fish of different freshwater fish species. Representative preliminary results. Production environment and quality. In: G. Barnabé, P. Kestemont (eds), *Bordeaux Aquaculture '92*. European Aquaculture Society Sup. Publ. 18, Ghent, Belgium, pp. 315–322.

9

The economics of pond-based fish culture[1]

The economic approach to the culture of fish in ponds was almost unheard of until recently but has now become of major interest. This can be explained by:

—the reawakening of interest in fish culture in ponds during the 1980s. This is a traditional activity, carried out extensively, which provides a possible means for diversification within agricultural enterprises, provided some technical improvements are made;
—very recently (in France) the culture of single species of fish on their own has begun in small ponds.

A classic approach to the economics of an activity such as fish culture has several stages that can be analysed:

—within the context of development of the activity and its constraints: technical know-how, availability of suitable sites and good-quality water, availability of investment, product markets and marketing methods;
—establishment of a production unit: calculation of investment required, search for finance, development of provisional cash-flow projections, risk assessment and necessary insurance;
—operating routines and management of the business, based on technical and economic analysis which allows profitability to be evaluated using different indicators, with the object of acquiring tools to aid decision-making within the business.

This chapter is based around this last point, i.e. the routine operation of the business and its management; the other headings are either covered in previous chapters (marketing and sales) or should be dealt with on a case-by-case basis which is beyond the scope of this work.

Unfortunately there is very little available information on the management of this type of fish farm because:

---

[1] C. Mariojouls.

—traditional fish culture has been considered as a marginal activity and subjected to little economic analysis;
—more modern forms of fish culture in ponds or tanks are only in their infancy in Europe.

The paucity of information on this topic makes it difficult to give advice that can be used directly by the fish farmer, because of the different technical and economic characteristics of each operation. The objective is to highlight information and gaps in the management of fish farms and the common characteristics of this activity. Finally, with the aid of available information, the chapter will demonstrate methods to be used and results predicted.

9.1 CHARACTERISTICS OF POND-FISH CULTURE SUITABLE FOR TECHNICAL OR ECONOMIC ANALYSIS

9.1.1 Traditional fish culture

The traditional culture of fish has been practised in France for a very long time. The characteristics of this form of aquaculture do not lend themselves readily to straightforward economic analysis but must be analysed in any consideration of the profitability of the activity. The characteristics can be summarised under three headings:

As with most agricultural activities, the culture of fish in ponds is strongly dependent on the natural environment:

—climate has a major influence, particularly through rainfall and temperature conditions;
—rearing takes place within a complex ecosystem that is difficult to manage;
—the choice of production site has a strong influence, as the basic profitability of the pond is related to the nature of the substrate and the quality of the water.

Production techniques are not standardised and performances are variable:

—the size of the ponds, the degree of management and the methods of operation are highly diverse, particularly in ponds which cannot be fully controlled by complete emptying;
—in most ponds, different species of fish are stocked in varying proportions;
—the fishes' life cycles are still only managed to a low degree; thus spawning (when it takes place in the ponds), survival and yields are very unpredictable. In addition to this, harvests are concentrated in winter;
—although some intensification techniques based on the improvement of the productivity of the ponds through the application of mineral and organic fertilisers (p. 160) or on supplementary feeding (p. 167) are practised they are still relatively unusual and the improvements in yields are not properly monitored.

There is a lack of economic information on pond fish culture:

—ponds can be operated with various objectives: simply for ornamental purposes, for commercial hunting and fishing (which may be rented out), for fish culture or for one or more of these activities. This diversity makes the economic evaluation of fish production very difficult;
—the economic aspects of fish production itself are often poorly understood, either because the activity is considered to be marginal and not worthy of study, or because it is part of an agricultural operation and assessed with the rest of the enterprise.

9.1.2 Intensive fish culture

Intensive fish culture, based on the control of rearing conditions, can be distinguished from traditional fish culture in that it is less dependent on the environment and, theoretically, has a more reliable biological and technical performance. However, because this activity is in its early stages, there is a general shortage of technical and economic data.

9.2 METHODS OF MANAGEMENT AND AGRICULTURAL ECONOMICS

The conceptual models developed for the analysis of agricultural businesses can be adapted for use in the analysis of fish culture. Specialist manuals should be consulted (references are given at the end of this chapter). This chapter outlines the most important analytical tools.

The first approach is the analysis of the *financial accounts* which enables the balance sheet and profit-and-loss accounts to be established, from which the intermediate balance can also be obtained (Table 9.1).

The second approach is to produce *management accounts*. In these, costs are classified in two ways:

—economic costs. These include operating costs or direct costs, which are proportional to the level of activity, and overhead or fixed costs which remain the same whatever the level of activity;
—functional costs, which distinguish costs of mechanisation, maintenance, etc. (Table 9.2).

In a business with several different activities, it should be possible to establish cost accounts, that is for each activity to distinguish the revenue and costs associated with that activity. This will allow the derivation of margins. For any given activity it should be possible to distinguish:

- Enterprise gross margin = revenue − operating costs
- Enterprise direct margin = revenue − (operating costs + fixed costs)
- Net margin = revenue − (operating costs + fixed costs + common fixed costs)

Table 9.1. Analysis of overall accounts (source: IGER, "*le mot juste*", 1989).

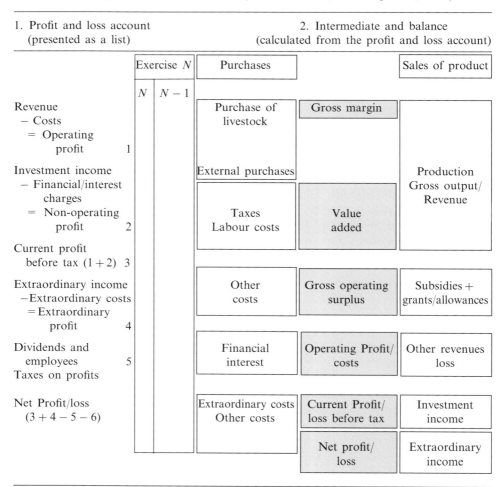

1. Profit and loss account (presented as a list)

		Exercise N	
		N	$N-1$
Revenue – Costs = Operating profit	1		
Investment income – Financial/interest charges = Non-operating profit	2		
Current profit before tax (1 + 2)	3		
Extraordinary income –Extraordinary costs = Extraordinary profit	4		
Dividends and employees	5		
Taxes on profits			
Net Profit/loss (3 + 4 – 5 – 6)			

2. Intermediate and balance (calculated from the profit and loss account)

Purchases		Sales of product
Purchase of livestock	Gross margin	Production Gross output/ Revenue
External purchases		
Taxes Labour costs	Value added	
Other costs	Gross operating surplus	Subsidies + grants/allowances
Financial interest	Operating Profit/ costs	Other revenues loss
Extraordinary costs Other costs	Current Profit/ loss before tax	Investment income
	Net profit/ loss	Extraordinary income

The gross margin is the most reliable indicator as it is linked only to the activity being studied and is generally well defined. The calculation of enterprise direct margin can also be useful. The net margin requires an allocation of general overhead costs associated with all enterprises to each specific enterprise, and the arbitrariness of this process can diminish its significance. For preference, the gross margin should be chosen as the method of comparing different examples.

The choice of accounting method is likely to be a function of the objectives pursued; it must be understood that financial accounts are a tool for evaluating the economic state of the business, while the management accounts and their analysis give a better picture of its technical and economic performance.

Table 9.2. Calculation of margins; application to fish culture, management accounts and cost accounts (source: IGR, "*le mot juste*" 1989).

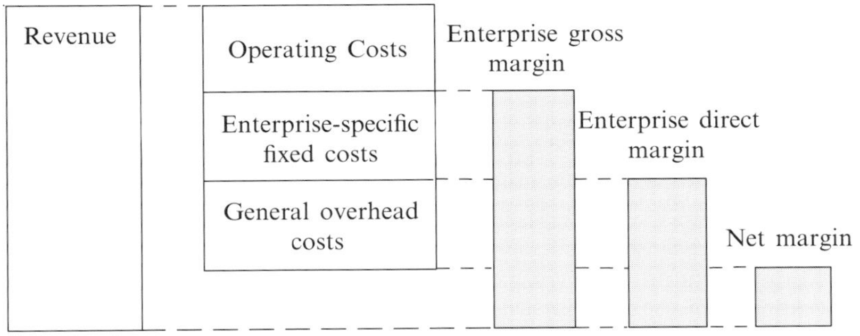

Operating or variable costs:
Fertilisers and other additions, fish for stocking, feed, treatments, work carried out by third parties, various supplies, insurance, taxes, remuneration and social-security costs (temporary personnel) and various other operating costs.

Enterprise–specific structural costs:
Fuel, leasing costs, maintenance and depreciation of specific equipment (e.g. mowers), depreciation of charges for digging ponds, social costs, salaries of permanent staff on the fish farm.

General fixed costs:
Fuel, leasing costs, non-specific maintenance and depreciation costs, remuneration and social costs not linked to the fish farm. The precise list of costs must be established for each production unit, taking into account real rather than theoretical conditions.

9.3 AN APPROACH TO THE STUDY OF THE ECONOMICS OF POND-BASED FISH CULTURE

9.3.1 Essential information

Characteristics of the business

The first stage is a description of the business and its operation. The parallel which has been established between agriculture and fish culture can also be used for the technical description of a fish farm, remembering that this will take account of the degree of intensification which is somewhere between two extreme poles. Extensive fish culture can be likened to plant production, for which the concepts of agronomy are appropriate (technical specifications, yields). Intensive fish culture requires a more technical approach (rearing methods, performance indicators and biotechnical information such as conversion indices, weight of individuals, etc.).

Having recorded the characteristics of the production system (land, materials, manpower, capital), technical options for the management of ponds (e.g. addition of fertilisers) and rearing practices (feeding, treatments, management of spawning) the objectives of production are defined (species and stage of development). Inputs and outputs need to be identified and quantified.

Accountancy for aquaculture

The establishment of accounts for a fish-culture unit requires the recording of costs and revenues of the business. There is a huge disparity in the literature in the calculation and classification of costs, as well as in the choice of indicators of economic or techno-economic performance. This lack of standardisation of approaches makes the interpretation and use of previously published results difficult for comparative purposes as well as for the establishment of enterprise accounts by an operator or a researcher into fish culture. We recommend that practitioners should draw on agricultural business accounting methods (see above) and emphasise below the main methodological difficulties in constructing accounts for a fish-culture business that are apparent from some recent studies. In the absence of standardised methods, particular attention should be paid to the accounting conventions chosen, whether for the generation of a business account or for the analysis of results already obtained.

9.3.2 Principal difficulties with methodology

Scale of the study: pond or entire holding

Working at the level of the pond allows an evaluation of performance (in the biological and technical sense) and also operating costs and revenues. The whole business can be viewed as the sum of data from each pond.

Duration of production cycles and fallow periods

In pond-fish culture there are many examples of production cycles extending over several years, with harvest and sale of fish taking place only at the end of the final year. It is possible to choose to work on a calendar-year basis or over the whole production cycle by considering a "mean fish-culture year", obtained by dividing all the costs of production over the duration of the cycle. For example, taking a single pond over a 3-year cycle, it is necessary to wait for the end of the cycle to obtain the profit from which an average year can be calculated. If there are three comparable ponds, each 1 year out of phase, then a result for the cycle can be obtained by calculating revenue costs for a single year. The first choice of an actual year will take account of inter-year variations in economic performance and cashflow for the fish-culture activities, while that of the average year allows an overall view of the situation. One aspect of this problem is how to take account of fallow periods when ponds are left dry. As an example, in the culture cycle as used in the Dombes region (fish rearing/plant production) plant production can be considered as part of the overall cycle on the fish farm. One year the pond is used for plant production (study (a) in Table 9.3). In a farm where fallowing is entirely for the benefit of fish culture, with or without seeding, the accounts for the year will show the loss due to the period when the pond was dry. The use of an "average fish-farm year" is less distorting.

Distribution of costs

When fish farming is carried out as a single activity, it is possible to evaluate costs directly. When it is part of a range of activities, those linked directly to the culture of fish or fixed costs are common to a range of activities. In a monoculture, the costs are evaluated directly (operating costs such as supplies, fixed costs, materials used for fish culture, costs of excavating ponds). In a combined activity, non-specific fixed costs (overheads) will be allocated between the activities according to conventions that must be defined.

The methodological difficulties in the calculation of costs classically encountered in agricultural economics are as follows:

- —in monoculture or pluriactivity (multi-enterprise business) the valuation of internal transfers (stocking, recapture of fish, use of cereals produced on the farm as feed) which can be obtained from the costs of production, where they are known, or from the sale price of the products concerned;
- —in pluriactivity (combined farming) the distribution of overhead costs, the classic (traditional) conventions are as follows:
  - pro rata according to the area occupied for each crop or grazing for livestock enterprises for costs linked to finances (farmers mutual insurance, rent or taxes),
  - pro rata according to the total number of hours of use of equipment for costs associated with mechanisation,
  - pro rata for turnover for general costs.

As a matter of practice, external purchases are included with the supplies (improvements, artificial fertilisers). This is not the practice where the purchase is included within the costs of mechanisation, or in structural costs as for example with organic fertiliser based on animal wastes that come from another part of the business.

Investments, depreciation and financial costs

In existing traditional extensive pond systems, concepts of investment and depreciation are practically never taken into account. All the same, there is currently a resurgence of interest in pond culture in France and a large number of under-exploited ponds have been identified: this has stopped creation of new ponds and the costs entailed.

However, for a modern, intensive fish farm, it is essential to be able to manage water and stock in order to determine the likely profitability of any investments. The size of initial investment needed will vary according to the site and the objectives of the business and may be extremely high (from 4545 to 9090 €/ha for a pond of

Table 9.3. Gross margins in fish culture calculated from study data. Values 1988/89 converted into €

Type of enterprise	Agro-fish culture (a)						Agro-fish culture (b)						Trial (d)
Location	Dombes			Sologne			Brenne				Theoretical (c)		Northern France
No. of the enterprise	1	2	3	4	5	6	A	C	E	F	Case 1	Case 2	Mean of 4
Area of water (ha)	32	25	163	4.5	22	13							1 to 5
Type of production*	CF pl sp	CF pl sp	CF pl sp	roach	roach dom	CF carp dom	CF pl sp	CF pl sp	leaves	leaves	CF carp dom	CF roach dom	small pike MNR
Culture methods†	OF, MF AF	OF, CA AF	OF, CA AF	OF, CA CA	MF	OF	OF	OF	OF	OF, OF			
Net yield kg/ha/year	310	135	110	250	87	190					167	167	Final weight 1.3 to 3 g Yield: 1.4 to 2.3/m²
Product (€/ha/year)	430	268	167	740	196	435	381	83	322	907	304	365	6007
Marketing method	whole-sale	whole-sale	whole-sale	whole-sale	whole-sale	whole-sale	whole-sale	whole-sale	whole-sale	wholesale and angling club	wholesale	wholesale	
Operating costs													
(€/ha/year)	26.4	20.4	36.8	39.1	7.7	27	79.5	20	97	239	82	82	212
• Stocking	35.4	12.1	6	30.3	12.1	30.5		5.6		30.4			
• Additions					20.4					90			
• Fertilisers		36.8								138.8			
• Feed	5												
• Treatments			87.1		33.6		37.7						
• Extra labour													
• Total	66.7	69.4	59.1	69.4	57.1	87.9	117	24.8	97.1	422			212
Gross margin (€/ha/year)	363	194.7	163	670	139.4	347.4	264	58.8	225	485	222.7	283.3	580

*CF: Controlled fishing; MNR: managed natural reproduction.
† OF: Organic fertiliser; MF: mineral fertiliser; CA: calcium added; AF: artificial feed.
(a) Boiron, 1989; (b) Etasse, 1988; (c) Marcel, 1989; (d) Manelphe and Bry, 1988.
pl sp = many species; dom = dominant.

1–5 ha) (p. 227) It should be emphasised that the investment required is always high for fish culture in comparison with other agricultural or industrial activities.

While producing business accounts, the necessary investments for the management of land will be depreciated over a period of 15 or 20 years. The financial costs, made up of the interest on borrowing, will be accounted as financial charges.

Risk and insurance

In reality the risks of loss of stock are not generally taken into account in pond culture but should certainly be considered in intensive, controlled production with expensive, selected stock. Risk cover is currently being studied in some of the first French monoculture projects. Insurance in aquaculture is based on a periodic evaluation of the stock; this is done monthly in trout farming. The annual premium represents 3–5% of the value of the stock and the rate of excess varies according to the situation, from 10–40% of the value of the stock at the time of the accident.

9.4 SOME EXAMPLES AND RESULTS FROM POND-CULTURE FARMS

9.4.1 Facts relating to agro-fish-culture businesses based on extensive production

There is too little factual information available on pond-based fish culture to provide infallible economic reference points. Some results from different types of production are presented here as an example (Table 9.3). These are:

—studies from actual agro-fish-farming businesses in Dombes, Sologne and Brenne: studies (a) and (b);
—studies of monoculture: simplified data from traditional fish farms, study (c), study of the production of pike juveniles produced by the management of natural spawning, study (d).

In these studies, methodologies used for the establishment of accounts for fish culture are too disparate to attempt a comparison of the end results. From the available information we have therefore chosen gross margins for the fish-culture activity (Table 9.3) in order for comparisons to be made. It should be noted that the results are extremely variable in relation to predicted production and its value (carp, roach, pike), the techniques used (whether or not fertilisation and feeding are used) and the method used to sell the fish (wholesale or retail). Variations in return for similar techniques can be explained by the actual productivity of the pond or the technical competence of the farmer or even varying factors which cannot be completely controlled.

9.4.2 Indicative data on production costs established from a pilot farm using intensive production methods(2)

In a 4-ha pilot unit where carp are produced in an intensive monoculture system under conditions described above (p. 181), a summary evaluation of production costs was made after 2 years of operation. An extrapolation was made for a 40-ha farm (gross production, 160 t per annum) which has a filleting plant attached. The technical characteristics of the operation are shown in Table 9.4.

The costs of production of carp for market are slightly reduced if the production cycle is shortened to 2 years; 1.2 € at 2 years and 1.24 € at the end of the 3rd year. If

Table 9.4. System for the production of carp in ponds with water management (based on the pilot plant operated by R. de Courson (4 ha), Champagne, Ardennes, France), information from J. Marcel.

Equipment and structures (for a 40 ha production unit producing around 160 t per year)
- Hatchery: 1.5 million feeding fry
- 3 ha fry ponds (each with a maximum surface area of 1 ha)
- 7 ha of ponds for early rearing (1–4 ha)
- 30 ha of ponds or tanks where water can be controlled (1–4 ha)
- 1 filleting plant with a production capacity of 40 t of fillets

Rearing

3-year rearing cycle

1st autumn, mean weight of individuals:	40–50 g	Production: 3 t/ha
2nd summer	400–500 g	5 t/ha
3rd summer	700–1200 g	4 t/ha

Diet: 30.8% protein; 5.45% lipid; 4.25% crude cellulose; 29.75% cooked starch, distributed at a rate of:

	1.5% body weight at	18°C
	2.0%	20°C
	2.5%	22°C

Conversion coefficient: around 1.5

Water quality:
- —dissolved oxygen supplied by photosynthesis when the water temperature is less than 20°C and the biomass of fish is below 2.5–3 t/ha; above these values oxygen is supplied by mechanical aerators
- —ammonium (NH_4^+) is eliminated by phytoplankton at levels $<$1 mg/l (where there is a deficit of phytoplankton, superphosphates are spread at a rate of 50–80 kg/ha of 14/48).

Harvesting: selective net fishing in open water for commercial-sized fish

Filleting: skin and bones removed automatically, filleting by hand; yield from filleting: 32–35% for carp

Staff: one production worker, one filleter, one manager and occasional workers to help with harvesting.

---

(2) J. Marcel.

Table 9.5. Production costs for whole carp or fillets. The fish were reared intensively according to the 4 ha pilot farm operated by R. de Courson, Champagne, Ardenne, extrapolated to a 40 ha unit with technical features as shown in Table 9.3 (information from J. Marcel).

	Costs of production (€/kg)	Main sources of expenditure (%)			
		Feed	Stock	Man-power	Depreciation
CO^+ Carp fry					
Final weight 20 g	1.86 (0.25)*	42.7	34.8 (214.285)†	8.4	12
Final weight 40–50 g	1.49 (0.44)*	53	19.0 (95.238)†	10.4	15.1
Final weight 100 g	1.34 (0.89)*	59	9.6 (42.850)†	11.6	16.9
Part-grown carp					
150 g	1.34	44	22.7	11.3	14.6
350 g	1.27	46.2	18.8	11.8	15.4
Market-size carp					
Sold in autumn					
and winter	1.20 (33.81)‡	49.1	13.9	12.5	16.3
—age: 2 years	1.24 (34.60)‡	47.4	14.7	12.1	15.7
—age: 3 years					
Sold in summer	1.41 (37.75)‡	42	31.7	9	11.5
—age: 3 years					

*Unit price
† Number of individuals/ha stocked
‡ Final cost/kg fillet

the fish are sold during the course of the summer of the 3rd year this results in higher costs (1.41 €/kg) because of the shorter cycle and the need to stock a greater number of fish initially: this leads to a contribution of 32% towards production costs compared with 14–15% in the other examples (Table 9.5).

The cost of filleting works out at 1.51 € per kg fillet in a processing unit dealing with 120 t marketable product per annum from the associated fish culture unit. Yield from filleting is around 32–35%. 3 kg of whole fish are required to obtain 1 kg of fillet; adding the cost of filleting gives a final mean cost of fillets of around 5.3 €/kg (Table 9.5).

9.5 CONCLUSION

The culture of fish in ponds and the search for new methods of development is beginning, at least in some places, to take on the form of a planned, modern operation that is facing up to the challenges associated with providing a return on capital investment. As happened in agricultural businesses in the 1960s, the fish farmer has had to acquire, and understand how to analyse, the information needed for decision-making in order to adhere to a production plan.

In summary, we have shown the difficulties in methodology that exist for the development of a management plan for fish culture and the need to make progress in the standardisation of methods and the acquisition of information in this sector. This is necessary not only as an accompaniment to the current technical progress but also to aid comparison with agricultural businesses. When agriculturists are considering diversification into extensive fish culture, it is probably essential for them to analyse its operation with appropriate analytical management tools; traditional extensive fish culture is equally capable of supplying produce to meet market demands as those from intensive systems.

Parameters taken into account in the calculation of production costs

(Values are in € at 1993 prices)

—Construction of fry ponds (1 ha) 15 151 €/ha
—Construction of growout ponds (>4 ha) 6060 €/ha
—Centralised feeding station (for a site of a minimum of 4 ha):
- 6–8 t silo with fittings: 2272 €
- battery operated projectile feeder: 1136 €

or self-feeders with 30 kg hoppers: 95 €
—1-hp aerators (hydroinjectors): 1029 to 1181 €
(depreciation of equipment over 5 years)
—Feed for the second-fry stage, 35% protein, 0.53 €/kg
—Growers feed, 35% protein, 0.39 €/kg
—Labour (*): 11.8 €/h, all-costs included (agricultural wages)
—Insurance costs: 182 €/ha

Production of 4–5-week-old fry:

—Construction costs of fry ponds, spread over 3 years (**), ponds or tanks: 1 ha spread over 10 years
—5 belt feeders/ha
—1 million first-feeding fry per hectare (1.51 €/1000)
—Feeding of 465 kg diet at 1.06 €/kg
—Survival rate: 50%
—Treatment with dipterex: 68.2 €/ha
—1/3 of the structural, maintenance and fertiliser costs (60.6 €/ha)

Costs of production of 4–5-week-old carp fry are around 9€ per thousand or 4.39 €/kg.

Production of 1-summer-old fingerling carp (from 4–5-week-old fry)

—1/3 of the annual interest on the costs of construction of fry ponds
—1 hp/ha aeration (turbo-jet, paddle wheel or air hydroinjector)
—6 self-feeders per hectare
—Final gross production: 3 t/ha
—Survival rate: 70%
—Feed conversion coefficient (based on gross production): 1.5
—1/3 of the annual financial costs
—Insurance, variable: 90.9 €/ha
—Manpower: 0.19 €/kg of fish produced

The calculations show that the costs depend on the final weight of the carp fingerlings 1.86 €/kg and 0.03 €/individual for 20 g fish and 1.34 €/kg or 0.13 €/individual for 100 g fish (Table 9.5).

Production of carp in early on-growing stages (150 or 350 g) or growout stage (900–1200 g):

—Annual depreciation (over 10 years) of pond construction costs (area >4 ha)
—0.5 hp/ha aeration
—1 centralised feed distribution station/4 ha
—1 pump/10 ha (2272 €)
—Fishing nets, 3030 €/40 ha
—Gross production: 5–6 t/ha
—Survival rate 70%
—Feed conversion coefficient (based on gross production): 1.5
—Variable costs:
- electricity for the aerators: 75 days operation for 5 h each day
- fuel, telephone and other costs: 75.7 €/ha

—Material for maintenance: 7% of the value
—Annual financial and insurance costs
—Insurance:1.15% of turnover (around 75.7 €/ha)
—Manpower: 16 h/t fish produced: 96 tonnes per unit annual production

* Marcel J., 1993. *Echo-système*, **23**: 5–15.

† The production of juvenile carp of 4–5 weeks and of 1 summer in the same pond is preceded by the production of pike juveniles: annual costs can thus be spread over three productions.

BIBLIOGRAPHY

Anonyme, 1987. *Guide comptable des exploitations agricoles*. IGER, Paris.

Anonyme, 1989. "*Le mot juste*". IGER Ed. (Institut de Gestion et d'Economie Rurale), Paris.

Boiron B., 1989. *L'intégration agro-piscicole; Etude technico-économique en Dombes et en Sologne, analyse des perspectives et limites*. Mémoire de D.A.A. de l'INRA, Paris-Grignon, 120 pp + annexes.

Boulon P., Denis P., 1989. *Intensification de la pisciculture en Dombes*. Mémoire de fin d'études ISARA, Lyon, 62 pp.

Cemagref, 1988. *La diversification de l'exploitation agricole par la pisciculture d'étang*. Ministère de l'Agriculture, Paris, 70 pp.

Etasse E., 1989. Approche de la mise en valeur de la pisciculture par fertilisation organique en Brenne (Indre); example des intégrations agro-aquacoles. Mémoire de M.S.T. Aménagement et mise en valeur des régions, Université de Rennes I, 59 p + tableaux, figures, annexes.

Leopold M., 1986. Economic assessment and decision-making of carp farming. In: R. Billard, J. Marcel (eds), *Aquaculture of Cyprinids*. INRA, Paris, pp. 457–466.

Lhostis D., 1980. Analyse de l'efficacité économique de la cypriniculture. In: R. Billard (ed.), *La pisciculture en étang*. INRA, Paris, pp. 397–406.

Manelphe J., Bry C., 1988. Reproduction Naturelle Aménagée du brochet en petits étangs: aspects économiques de la production de juvénile. *Bull. Fr. Péche Pisc.*, **310**: 45–58.

Marcel J., 1989. La pisciculture extensive. *Dossiers Economiques*, Tome II. ITAVI, Paris, 40 pp.

The author would like to thank Professor B. Revell, Harper Adams College of Agriculture for assistance with technical terminology in this chapter.

Appendices

Rearing species other than carp in ponds

A.1 REARING JUVENILES[1]

It is now possible to control the spawning of many species of cyprinids (p. 63) so as to advance the time of spawning. This is done mainly by the manipulation of photoperiod and temperature, thus benefiting from the whole of the warm season for the growth of fry. Achieving control of larval production means that the species can be produced according to market requirements. In addition to this, the demand for ornamental fish such as goldfish, koi carp, golden orfe (*Idus idus*) is at its greatest from March to June/July for stocking garden ponds. Demand for other fish such as the gudgeon (*Gobio gobio*) and the European minnow (*Phoxinus phoxinus*) is directly linked to the start of the sport-fishing season.

In addition to this, the stocking of feeding fish into ponds strongly improves survival rates of fry and reduces variations between years observed when unfed fry are stocked. For some species such as gudgeon and minnow, early stocking of fed fry (reared in tanks for 1–2 months) reduces the rearing time to a single year. For species with a high commercial value such as ornamentals or bait-fish, intensive rearing of larvae is clearly justifiable.

A.1.1 Equipment used for rearing and feeding

Overall, the same considerations as for carp concerning diet (preparation, nutritional quality, method and frequency of distribution) and rearing equipment (tanks, filtration system) apply for most cyprinids. Problems associated with the small size of the larvae, particularly their digestive system and their nutritional needs, are generally similar to those encountered with common carp. At present little success has been achieved rearing such larvae on an entirely artificial diet; the survival and growth rates have been generally below those achieved with dietary regimes including living or frozen organisms such as rotifers (*Brachionus plicatilis*) or

---

[1] P. Kestemont.

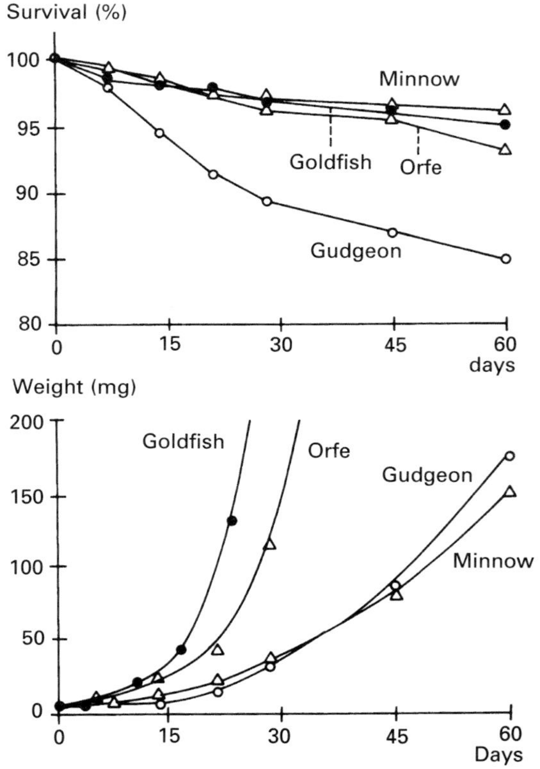

Fig. A.1. Percentage survival and weight gain in juveniles of several species of cyprinid receiving a mixture of live prey and artificial diet (cf. Table A1).

Artemia nauplii. In view of the costs of production of natural feeds, their use is only justified for cyprinids whose commercial value is high, and then only for short periods. A mixed diet in which a fraction of the artificial feed is replaced by natural feed leads to an overall reduction in the mortalities that result from nutritional deficiencies. Highly satisfactory growth and survival rates have been achieved by distributing a mixed diet to different species of cyprinids. The proportion of natural feed in the diet (rotifers and artemia) is progressively reduced as the larvae grow (Figure A.1, Table A.1). Commercially available microencapsulated diets are available; these have a high nutritional value, are very stable in water but are generally very expensive. Giving a low proportion of natural feed allows the efficient use of medium-quality manufactured diet which costs substantially less.

A.1.2 The rearing cycle

The maintenance of optimum environmental conditions similar to those that have been described for carp (p. 125) is essential for all species of cyprinids. While most of

Table A.1. Weight gain, conversion efficiency (CE) and recommended ration (in % increase in weight/day and in g/week) over the first 4 weeks of larvae rearing for goldfish, orfe, gudgeon and minnow.

Period	Weight (mg)	Proportion (%)		Ration		Conversion efficiency
		A. salina	Artificial diet	(% W/day)	(g/week/10 000 larvae)	
Goldfish						
1st week	1.2–9.0	50	50	50*	80	1
2nd week	9.0–40	25	75	40*	300	0.97
3rd week	40–160	10	90	30*	850	0.71
4th week	160–250	0	10	20*	3000	0.85
Orfe						
1st week	1.9–9.0	50	50	50*	95	1.3
2nd week	9.0–2.5	25	75	40*	200	1.25
3rd week	25–60	10	90	30*	450	1.3
4th week	60–150	0	100	20*	1080	1.2
Gudgeon						
1st week	0.5–2.1	50	50	50	18	1.2
2nd week	2.1–5.5	50	50	50	40	1.4
3rd week	5.5–1.3	25	75	25	86	1.1
4th week	13–30	25	75	20	160	0.95
Minnow						
1st week	1.9–5.4	50	50	30	40	1.1
2nd week	5.4–13	50	50	25	90	1.2
3rd week	13–26	25	75	20	240	1.9
4th week	26–55	10	90	15	580	2

* The rapid growth of juvenile goldfish and orfe requires a readjustment of the rations every 3–4 days.

Table A.2. Length of incubation and yolk-sac resorption and recommended temperatures for rearing larvae of various cyprinids.

Species (°C)	Incubation		Resorption of the yolk (degree-days)	Rearing temperature
	T (°C)	degree-days		
Goldfish	19–21	90	80–90	24–26
Orfe	15–18	100	80–90	24–26
Gudgeon	17–20	120	60–80	20
Minnow	16–17	100	130	20

the larvae are thermophilic, each species has an optimum temperature which represents the best compromise between on the one hand, growth and on the other the development of pathogenic organisms and resistance to disease. Table A.2 shows, for several species, the temperatures and duration of larval development and resorption of the yolk until the start of first feeding.

One of the characteristics of larval rearing is the speed at which the fish are growing, doubling their initial weight within a few days. This requires frequent adjustments to be made to daily ration size so that the larvae are always fed to satiation. The daily ration increases rapidly as rearing progresses but the fish should not be overfed, in order to avoid deposition of uneaten food and the likely deterioration in water quality. Figure A.2 shows the influence of daily ration and

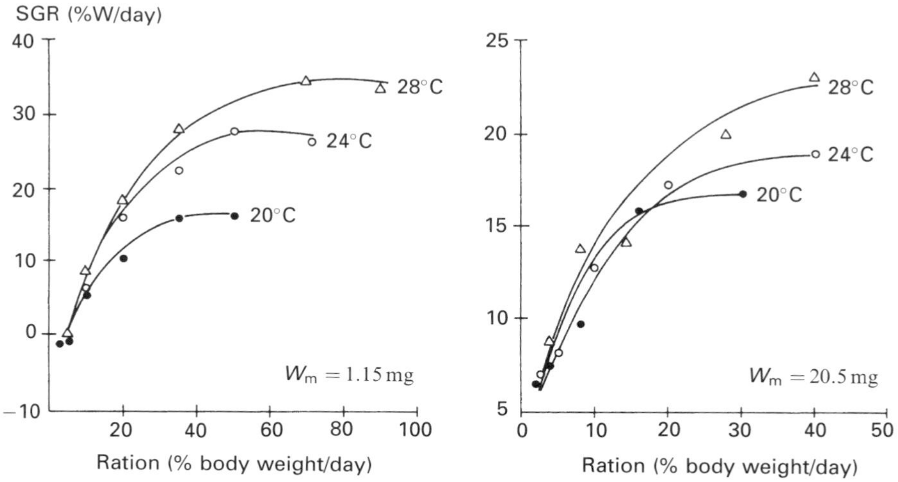

Fig. A.2. Specific growth rate (SGR) in % weight gain per day for young goldfish in relation to mean weight (W_m, as a function of the feed distributed per day (% of body weight) (cf. Table A1), at three rearing temperatures.

temperature on the specific growth rate of goldfish fry from the start of exogenous feeding (weight 1.15 mg) to 10 days after first feeding (weight 20.5 mg). Another factor which must be taken into consideration is the size of the feed particles (live or frozen prey or granules). This should be compatible with the width of the buccal opening in the larvae. In gudgeon, the mouth is around 250 μm at the start of exogenous feeding. This allows the fry to capture and ingest *Artemia* nauplii easily as well as larger zooplankton such as *Daphnia magna*. At this stage, *Brachionus plicatilis*, which can be mass produced, can be used during the first few days of rearing. *Artemia* nauplii are also consumed without any difficulty when given in a frozen state. Live *Artemia* nauplii can be used successfully for the fry of orfe or goldfish which have larger mouths.

Certain species, such as goldfish, demonstrate huge differences in growth rate within the same batch; this rapidly results in problems with cannibalism as the smallest individuals become prey to the larvae with more rapid growth rates. A pattern of feeding to satiation and regular regrading will generally limit such predation. In addition to this, the allometric growth of the mouth reduces the capacity for predation and ingestion of smaller fry after a few weeks. Cannibalism has also been observed, although to a lesser extent, in the fry of minnows and orfe when feed is limited.

By way of example, Table A.1 shows the quantity of feed (expressed in dry weight based on a mixed diet of *Artemia* and artificial feed) to be given according to the age and weight of the fry. The proportion of each type of feed is shown. After the 3rd or 4th week (depending on species), the diet can be exclusively artificial using commercially available trout feed. This type of feed is not always recommended for slower growing species such as gudgeon or minnow until the end of the 2nd month.

A.2 SOME EXAMPLES OF SYSTEMS USED FOR THE PRODUCTION OF CYPRINIDS[(2)]

The systems used for the production of cyprinids are extremely varied. In Europe, the common carp is the species most frequently reared in ponds and its production cycle generally extends over 3 years (p. 157). In Europe, various other species are associated with carp culture in ponds and the culture techniques are described below as they are of interest from the point of view of farm diversification. The most complex systems have been operated for many years, particularly in China and, to a lesser degree, Asia generally. These are based on associations of complementary species used to exploit different levels of the trophic web. This optimises the use of organic manure coming from farm operation (concept of integrated culture). The rearing of cyprinids is practised with different degrees of intensity in lakes, reservoirs or rivers; the fish may be free-swimming, in cages or enclosures. Monoculture units for culturing carp in earth ponds are currently being developed (p. 179).

[(2)] R. Billard.

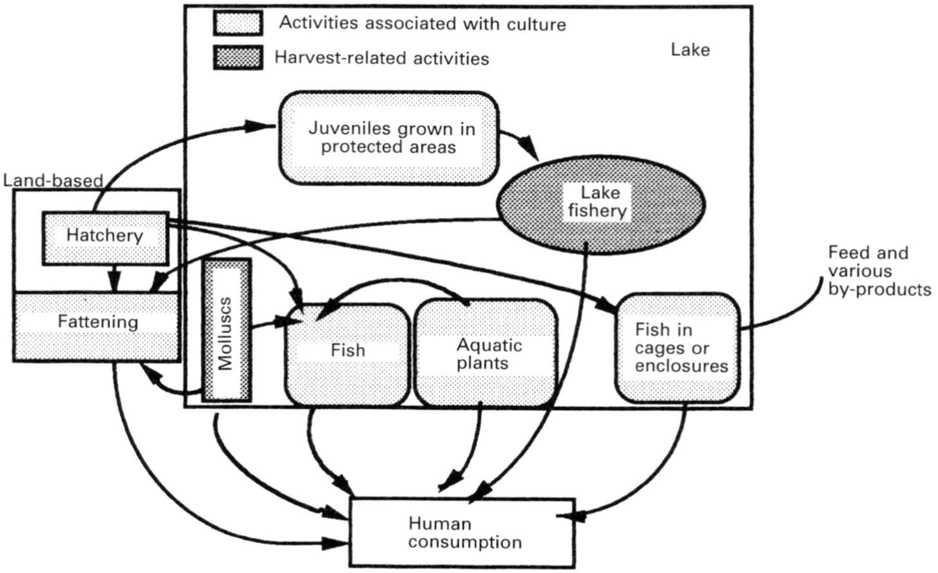

Fig. A.3. Diagrammatic representation of the integration of different production systems in Chinese lakes and reservoirs.

Cyprinids are reared in open environments such as lakes, reservoirs and rivers. The fish may be at liberty and captured by a variety of means or contained within a cage. In the first type of system there is "pasture", the food is derived within the system (plankton, benthos or macrophytes) and polyculture is frequently used. In the second type of system part of the food is delivered to the cages or enclosures and can come either from the waterbody itself or from outside. These systems have been described by several authors (Billard, 1985; Li, 1986).

The use of lakes and reservoirs as pasture (ranching) is common practice in China where the major problem is the presence of piscivorous species that prey on free-living juveniles in open environments. The solution is to continue growing juveniles either in a hatchery or in a cage or enclosure *in situ* in a lake or reservoir (Figure A.3) and to release them when they are large enough to escape the attention of predators. This size will vary depending on the species of piscivorous fish present; this should be investigated. In many reservoirs piscivorous fish are no more than 50 cm long and only eat prey of 10–12 cm or less in length. In such situations fish should only be released into the reservoir when they have reached a length of 13 cm (Li, 1986). An association of species is generally based on bighead carp (60%) and silver carp (40%). Gross production will be from a few hundred kilograms to 1 tonne per hectare depending on the productivity of the natural environment (there is no fertilisation in open waters) and management conditions including stocking. An increase in the number of fish stocked per hectare will be followed by an increase in yield (Figure A.4) but should take into account the growth rate and the required

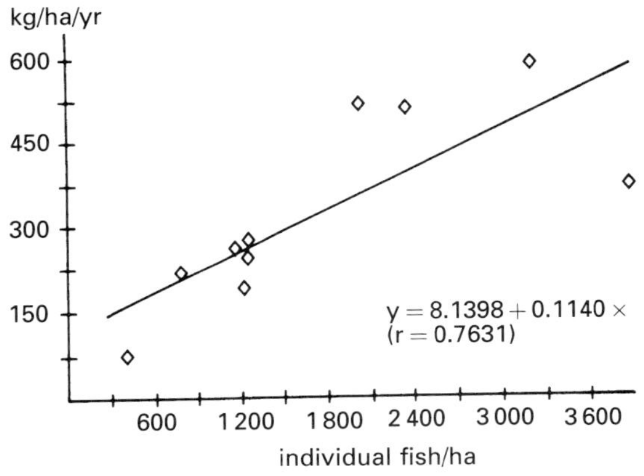

Fig. A.4. Relationship between net production and stocking density (individual fish/ha) in fish stocked in the Nanshake reservoir in China (La Sifa, pers. comm.).

commercial size. Ranching is also practised in water-storage reservoirs (used for irrigation) or in rivers of various sizes (China) and oxbow lakes, which are often eutrophic, in countries such as Hungary, where production of 600–800 kg/ha/year have been reported for common and silver carp together.

In several countries fish are reared in cages or enclosures in lakes. In China, communal on-growing takes place in enclosures but the feed coming from the plankton or benthos is insufficient and more must be supplied from outside sources. This may be in the form of granules, cereals or byproducts of various processes associated with agriculture or from the lake itself, such as molluscs and aquatic plants, either gathered from the wild or cultivated (Figure A.3). There is thus an integration between the various gathering/culturing activities on land and in water.

There are, however, instances where no food is given. A farm rearing carp in cages in Lake Cao Hai in China (300 km^2) has been described by Hai and Zweig (1987); 2400 cages of 28 m^3 (4 m × 7 m × 1 m) produce a total of 900 t/year or 13.5 kg/m^2 (or m^3) of cage without any additional food supply. This lake is not very deep (1.5 m in the part where the cages have been installed) and receives effluent directly from the town of Kunming (100 000 m^3 per day) and inputs from upstream where there is intensive agriculture. In the rainy season a transparency of 21 cm has been recorded (annual rainfall is 1040 mm of which 10–20% falls in the dry season). The lake is in a sub-tropical region (latitude 25°N) at an altitude of 1900 m; temperature fluctuates between 8.8 and 22.2°C with an annual mean of 16.7°C. The cages are installed throughout an area of 7 km^2 at around three cages per hectare. The total production should therefore be related to the whole area and is around 1.3 t/ha. The fish are stocked at a rate of 22 to 28 individuals/m^2 with a mixture of silver carp (65%) and bighead (35%) of between 13 and 16.5 cm length. These fish reach commercial size (800 g) after 18 months of rearing; the growth rate is equivalent to 1.5 g per day. This may be an unusual scenario because of the recycling of domestic effluents, but it demonstrates the possibility of exploiting highly eutrophic waters through fish farming.

There are also systems which combine cage rearing with ranching. In Japan, intensive carp rearing is practised in reservoirs of 1–30 ha area and starts with the intensive rearing of juveniles in cages attached to pontoons. After the fish have been released from the cages, feeding continues at the same site in a trap where the fish will take food out of habit and can thus be captured for sampling and sale when they reach commercial size. Any fish escaping from this system are captured with a net; this assumes that the topography of the bed of the reservoir is flat enough to allow complete fishing out. Production reaches 5–10 t/year but can be as high as 20–30 t if the number of aerators is increased. Production of 1 tonne of carp requires the distribution of 1.3–1.7 tonnes of granules with 30–40% protein content (Yamaha, 1991).

The final stocking densities are variable and depend on water quality and the level of dissolved oxygen. In Japan a large production unit of 57 cages (5 m × 5 m × 2.5 m) produces 100 t per year in a reservoir where temperatures vary between 3 and 30°C. Aerators are used. At the start of rearing each cage is stocked with 3000 fish. This farming operation has led to the eutrophication of the reservoir (Yamaha, 1991). Cage rearing in lakes and reservoirs is practised in many Asian and European countries and even in rivers in Central Europe. A wide variety of cages are in use (see Beveridge, 1987). A higher degree of intensification is found where fish are reared in running water as in salmonid culture. An example from Japan (Yamaha 1991) shows that 100 m^2 ponds, 1.2 to 1.5 m deep produce 10 t/ha/year when water flows through at a rate of 100 l/min and 20 tonnes per year with flows of 200 l/min. This production system is profitable as long as the temperature of the water remains above 15°C for at least 4 months of the year.

A.3 REARING ORNAMENTAL CYPRINIDS IN PONDS[(3)]

A.3.1 Characteristics of ornamental species

Several types of ornamental cyprinids are reared in ponds, particularly goldfish, koi carp, orfe and golden tench (*Tinca tinca*). They are reared in all continents and their farming has a long history; goldfish were reared in the Sung Dynasty in China (420–479 BC). Goldfish have been exported from China to several countries, first to Japan and then Indonesia, Mauritius, Reunion, arriving in Portugal and then England during the 18th century. About this time the Portuguese began culture and marketing, followed by France and Prussia. After this, because of huge demand, particularly in Britain, production grew considerably in France and Germany and, towards the end of the last century, in Italy. Since then the production in Italy has had a major impact on the markets throughout Europe and has now reached 19 million individuals per year (14 million in 1960 and 1965, 22 million in 1970, 13 million in 1970 and 15 million in 1985).

Goldfish are very hardy; their biology is similar to that of carp. They prefer still,

[(3)] P. Melotti and M. Natali.

warm, shallow waters that are rich in plant life. They can withstand cold waters and have a remarkable resistance to falls in temperature, down as low as is needed for ice to form on the surface of the water in winter, and temperatures over 30°C in summer. The most favourable temperatures are between 18 and 25°C. They are also able to withstand levels of dissolved oxygen as low as 0.5 mg/l. Goldfish are omnivores but with a tendency towards a carnivorous diet. Their favoured diet comprises several species of protozoans, particularly infusoria, worms (*Chaetogaster*, *Tubifex*), molluscs and their eggs (*Bithnia*, *Lymnea*, *Planorbis*). In addition crustaceans (*Daphnia*, *Sida*, *Latona*, *Bosmina*, *Cyclops*) and various insects, particularly in their larval stages (*Culex*, *Anopheles*, *Stegomyia*, *Chironomus*) form a significant part of the diet. If there is a shortage of animal prey goldfish can feed on microscopic plants such as diatoms. When there is sufficient food available goldfish can spawn as 1-year-olds. Generally spawning takes place in May or the first fortnight in June. The number of eggs released varies between 50 000 and 100 000 per kilogram live weight. Incubation time varies between 3 and 7 days depending on water temperature. At hatching, larvae are 3 to 5 mm long and, after 2 to 4 days, resorb their yolk and begin to search for zooplankton. Under optimum rearing conditions, at a moderate density (<100 000 individuals/ha), growth is very fast and individuals can reach 12 cm in length at 1 year.

A.3.2 Rearing goldfish and koi carp

There are several industrial scale goldfish farms, mainly in the USA, Israel and Japan that ensure regular production of a standard product in response to market demand, particularly for the aquarium and laboratory markets (goldfish are widely used in experiments). The producers pay particular attention to the management of clearly identified genetic stocks: there are several defined varieties, some very elaborate such as the shubunkin and big eye and others such as the comet variety which are most commonly mass produced. Much attention is paid to rearing conditions, the quality of the environment and to feeding, which contributes towards the quality of coloration. Farmers generally prefer to use natural prey because of the quantitative and qualitative supply of pigments in plankton.

Goldfish are reared under a variety of conditions. In general, varieties with a high commercial value are reared indoors under controlled conditions and the temperature of the water is regulated. More common varieties are often reared in the same way as carp, entirely outdoors from spawning, which is easily induced in plant-rich ponds (see below for an example of traditional production in Italy). However, to ensure a standardised and regular production, it is better to used hatchery-based spawning techniques as used for carp, with the first-fry stage either in the hatchery–nursery (p. 293) or in small tanks for first fry, using the same techniques as for carp (p. 126).

The success of a farm producing goldfish is largely linked not only to the quantity of fish produced and to their being placed on the market at the time when demand is greatest (demand is lowest in summer) but also to the quality of the fish; colour, morphology and hardiness, which is their capacity to survive transport and to adapt to their new environment without significant mortality. It is essential to organise a packing station in preparation for transport (p. 255). Various additives are often put in the water; these include anaesthetics, tranquilisers, antibiotics and buffers (Table 3.6 and p. 249).

Koi carp are reared under the same conditions as goldfish. There are several small farms producing high-quality individuals and also industrial-type rearing units. The objective is to produce standard individuals sold at a moderate price with low production costs and to seek the most effective organisation of the rearing process. In general, the same production techniques are used as for carp. Hypophysation is performed in the same way but the percentage of females that ovulate is lower, often below 50% and fecundity is significantly lower (<100 000 eggs/kg).

The first- and second-fry stage can be reared indoors with artificial feeding or in small outdoor ponds, managed so as to stimulate plankton production. Operators consider that koi fry are more fragile than those of common carp. They are particularly sensitive to temperature variations and to pathogens, requiring a strict management regime.

As with goldfish, the most important characteristic of koi carp is their colouration and also their capacity to withstand the sudden changes in the environment which happen when the fish are transferred to aquaria or ornamental ponds. Colour is determined by feed and by the genetic makeup of the population. Feeding based on plankton is considered to give the best, most vivid, colours. Genetic determination is poorly understood (p. 110) and the selection method used is empirical. Each farmer has his own "recipe" of parental crosses producing relatively high proportions of particular colours or associations of colours. The publicity given to some producers suggests a well-defined genetic basis for some colours. Recently studies on the genetic basis of colours in koi carp have been carried out by Wohlfarth and Rothbard (1991). The results of crosses between parents of a single colour show that white and orange colours are controlled by two alleles on the same locus, orange being dominant to white. It is likely that other colours are alleles at a single locus. More detailed investigations are needed to rationalise production of single-colour koi. Recessive colours such as white can be obtained directly *en masse*. In contrast, for dominant colours such as orange, progeny testing is required in order to select parents; the process is thus more complex. However, once a suitable parent has been found, it can be used for several years as carp live for a long time. A technique for the production of gynogenetic diploids and triploids has been developed by Cherfas *et al.* (1990) (p. 114).

A.3.3 Traditional techniques used for rearing ornamental fish in Northern Italy

Several techniques for rearing goldfish are used in Italy. The most intensive of these, developed recently, uses methods similar to those for carp where the young are reared in ponds for several weeks and fed on a balanced complementary diet. Juveniles are produced either in a traditional hatchery and first-fry ponds or in spawning ponds (where the larvae remain) which then serve as first-fry ponds.

In the traditional method, the whole of the life cycle is completed in small ponds that are either the "maceri", otherwise used for hemp or "piane", in which rice is cultivated and which are now used for extensive farming. Complementary feeding may be used.

'Intensive' rearing

In this technique, spawning is allowed to take place naturally in ponds of 50 to 500 m^2 in areas that have a maximum depth of 60–80 cm; the bottom is uniformly level, bounded by peripheral troughs and seeded with grass several times each year. The pond is filled with water 15 to 20 days before the projected spawning day. When the temperature of the water reaches 18–20°C broodfish are introduced. These have been selected on the basis of their morphological characteristics, colour and health. They are stocked at a rate of 0.5 to 2 females per m^2. A sex ratio of 1–2 males per female is used. Broodfish are bathed in 4% formalin for 15 min before stocking in the pond. After spawning behaviour has been observed the level of the water is lowered so that the central part of the pond and the vegetation to which the eggs are attached emerges. Ideally, this operation should be in the evening so that direct sunlight does not reach the eggs. This operation causes the broodfish to gather in the peripheral troughs where they can be captured. After this the level of the water is allowed to return to normal. This method has the advantage of separating parents and juveniles and avoiding the transmission of parasitic and bacterial diseases as well as predation on juveniles by adults.

An alternative method is to spawn the fish artificially in a hatchery using, as for carp (p. 72), hypophyseal extracts of carp or LHRH-A and domperidone or pimozide for females that have reached the appropriate stage of maturation. Ovulation occurs 8–10 h later and ova and sperm are obtained by massaging the abdomen and fertilising artificially. The techniques of dissolving the adhesive layer and incubation described for carp apply also to goldfish. After 10 to 15 days, when hatching and resorption of the yolk is complete, the fry are transferred to the first-fry ponds, which measure between 200 and 1000 m^2 and have a maximum depth of 60 cm. These ponds are fertilised by the addition of phosphates (100 kg/ha) and nitrogenous fertilisers (120–150 kg/ha) split into 4 weekly applications. Larvae are stocked into the first-fry ponds at a rate of 200–500 individuals per m^2 of pond; this is maintained up to the end of the first part of summer.

During the last few days of June the fry are captured and released into second-stage fry ponds, the area of which can be between 500 and 4000 m^2 and the depth 80 to 130 cm. The flow of water is between 2 and 10 l/s/ha. The density of fry is between 200 000 and 500 000 per hectare. These fish are kept for a further 2 to 15 months depending on market demand.

The technique of intensive rearing requires manufactured diet (carp type) to be given manually or by means of automatic feeders. During the first few months of life complementary feeding is generally offered in the form of finely ground particles (p. 139), after this the fry progress to a diet of crumbs. When they have reached a length of 10–12 cm they are given pelleted diet. Ground and crumb diet have protein levels of between 38 and 42% (animal and plant proteins in equal quantities). In the larger pellets, protein values drop to 32–36%; the protein is mainly of plant origin. The daily ration varies between 4 and 8% of live weight for the first 2 months, dropping to 3–4% during the following period.

Traditional rearing in "macero" and in "piane"

Although widespread in the past, this type of rearing system is seldom used today. The ponds are filled up in early spring so as to create the most favourable conditions for stocking around 600–800 spawners per hectare, from mid-April to mid-May. Spawning occurs spontaneously and, from the second half of July, most of the juveniles are captured in nets, together with their parents (95%). The others are harvested later while hemp is being soaked, when they come to the surface because of shortage of oxygen. The unsold fish are returned to the "macero" towards the end of September after the soaking of hemp is over and conditions are again suitable for rearing. The fish remain here until spring. Production in the "macero" is highly variable, from between a few thousand to over 400 000 fish per hectare.

Rearing in "piane" (old rice fields) is carried out in bodies of water ranging in size from between a few hundred m^2 to several hectares with a depth of around 1 m. Production is based entirely on natural plankton. Sometimes there is no water turnover: at best the flow rate is 3–4 l/s/ha. Spawning occurs naturally in spring after 2–3-year-old broodstock have been introduced at a density of 40–100 kg/ha. Harvest, using nets, takes place in September. Production levels are very variable, from 50 000 to 150 000 per hectare.

In Italy, in addition to goldfish (the ornamental fish with the highest production and sales) modest numbers of other species are reared. koi carp have been bred from broodstock imported from Japan in 1980–81. The annual production is only a few hundred thousand individuals and the quality is below that of the original broodfish. The production techniques are almost identical to those used for carp but the ponds are smaller in area. Generally, spawning takes place spontaneously but sometimes hormonal stimulation and artificial fertilisation are required. The fish are usually sold at no more than 15–20 cm in length and are almost all destined for central and northern Europe.

Another species aimed at the ornamental market is the albino catfish, *Ictalurus melas*: this is a mutation that appeared on farms resulting in individuals that are almost completely albino, yellow or blotched. These fish are produced in the same way as catfish. Annual production is estimated at 100/150 000 individuals sold to both internal and external markets.

There is also a small production of albino and golden tench in Italy: these are almost all sold on the home market. In central and northern European countries the golden orfe, a small cyprinid that stays close to the surface, is produced. Techniques of production are based on natural spawning that takes place in spring, followed by pond rearing using manufactured diets. The orfe needs cool, well oxygenated water and cannot withstand the high temperatures typical of southern Europe in summer.

A.4 FARMING OTHER SPECIES TOGETHER WITH CARP IN PONDS[4]

Rearing together, in the same pond, several species with different dietary requirements is a promising way of utilising the resources of the pond in such a way as to maximise the production of fish. A thorough understanding of the biology of all of the species is needed in order to optimise synergy and to reduce antagonistic or competitive interactions that occur between species or species and the environment. This leads to the establishment of combinations of fish according to stocking density and appropriate time periods: these will vary depending on local conditions. The composition of associations of species should also take into account the sale price of each species and the potential revenue from each combination of species.

The objective of this section is not to show the interactions that occur between different species of fish but to give some information on the rearing of the main species currently used as "companion" fish in cyprinid culture. The main species is generally the common carp and the companion species have the function of ensuring that the quantitative, qualitative and economic performance of the pond are met. Companion species may be non-predatory (other cyprinid species), or predatory (mainly pike, zander and wels). Of the predatory fish, only the rearing of zander is described here; pike culture has been described elsewhere (Billard, 1983) see General Bibliography.

A.4.1 Tench rearing

The tench is distributed throughout Europe, apart from northern regions. It is the main companion fish of traditional fish culture and is farmed either for the table or for stocking purposes. Over the last few decades, its use as a companion fish in ponds has declined, mainly because of the increase in production of carp under intensive conditions in monoculture. However, there appears to be a resurgence of interest in tench as several farmers are discovering the value of polyculture.

Biology

Tench prefer muddy, flat-bottomed ponds, that become much warmer in the summer, with plenty of vegetation. They are eurythermal and withstand colder water than carp, but can also tolerate temperatures as high as 35°C. They can tolerate water of very variable quality and low levels of dissolved oxygen. They tolerate pH fluctuations well: lethal levels are 4.5 at the lower end of the scale and 10.8 at the upper. Tench diet is similar to that of carp although more strongly based on macroinvertebrates such as molluscs, worms, crustaceans and insect larvae. The diet also includes algae and a range of detritus.

Female tench release ova sequentially in a series of batches. Spawning occurs when water temperature reaches 20°C or higher. It takes place between May and

---

[4] P. Kestemont.

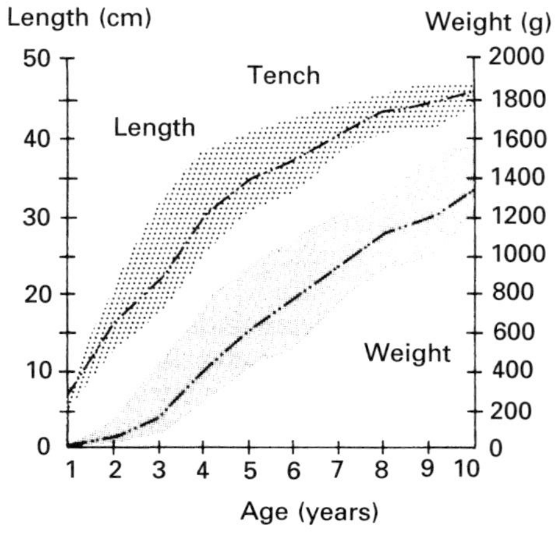

Fig. A.5. Growth in weight and length (mean and extreme values) of tench reared in fish-culture ponds (after Timmermans, 1989).

August, depending on location. The temperature profile during the time when the fish are completing sexual maturation and spawning can be a factor determining the yield from spawning. In heated ponds the spawning season begins about 1 month earlier and each fish spawns between five and eight times during the season, in comparison to the three or four times observed in unheated ponds. In addition to this, each female releases 40–90% more eggs in heated ponds. Female tench mature at 3–4 years (sometimes 2), males at 2–3 years. Fecundity is very high and varies, according to different authors, from 80 000 to over 300 000 eggs/kg body weight. The eggs are very small, the diameter at spawning is 0.4–0.5 mm, 0.6–1 mm after hydration. Incubation time is between 60 and 120 degree-days, depending on the temperature. The larvae, on hatching, are tiny (3.5–4.3 mm long) and become pelagic after 4 to 6 days when the yolk sac is resorbed (see Table 3.1).

Under natural conditions, tench larvae feed mainly on rotifers and then planktonic crustaceans. Growth is generally slow, especially in young fish (Figure A.5). Attempts at mass selection have shown that it is possible to achieve a significant improvement in growth rate and to obtain tench of 250 and 800 g after 2 and 3 years respectively. From the second year, the overall growth of females is faster than that of males: adult females are 100% larger than males of the same age. tench are not generally reared beyond 3 years of age as there is a likelihood of excessive spawning in the ponds.

Techniques used in traditional spawning

Tench spawn in ponds where the water temperature is relatively high and there is an abundance of submerged plants or artificial spawning mats made from branches of fir trees. After spawning has been completed, the artificial substrate can be removed and transferred to ponds which serve for hatching and for first-fry stage. The use of

Dubisch type spawning ponds is almost impossible because of the small size of the tench at hatching and the consequent difficulties of harvesting. Spawning should be in the largest ponds where the fry can remain throughout the summer until collected in the autumn or the following spring. This single-stage fry production can be carried out in ponds specially constructed for this purpose (0.5–1 ha) or in ponds used for carp in their 2nd or 3rd year.

A two-stage fry-production system has also been recommended. In this the ponds are emptied at the end of August and the fry, about 3 cm long, are stocked into the second-fry ponds where they remain until the following spring. This procedure has the advantage of providing control of the biomass of fish in the fry pond. The number of spawners placed in each pond depends on its productivity. In spawning ponds a density of 20 to 40 females (3 or 4 years old, 300–400 g) per hectare is recommended, with similar figures for males. If spawning occurs in a pond, with 2 or 3-year-old carp, the number of pairs of spawners is reduced to between one and six per hectare.

Techniques for controlled spawning (p. 72)

The technique used for artificial spawning of tench is the same as that described earlier in this book for carp and uses dried extracts of carp hypophysis but at higher dose rates (9–10 mg/kg body weight) followed by artificial spawning and fertilisation of ova after a period of 12 to 14 h at 24–25°C. Two successive intraperitoneal injections (2 and 7–8 mg respectively) 8 h apart, produce the best results (p. 80).

Ova should be extracted several times at hourly intervals. Fertilisation and incubation of eggs require the same procedures as have been described for carp.

It has recently been demonstrated that tench will spawn spontaneously in tanks if left undisturbed and that the induction of spawning is therefore unnecessary. Spawning lasts from 4 to 6 h and the eggs at first become attached to the walls of the tanks. Gradually, they lose their adhesiveness and are progressively removed from the tank with the outflow current. A fine-mesh sieve (150–200 μm) placed over the outflow of the tank allows the eggs to be collected and incubated in a Zoug jar, at a concentration of 1 million eggs in a 7 to 9 l bottle and a fertilisation rate of 70–80%.

Intensive production of fry

Intensive production of fry in their first summer is carried out in two stages; a pre-growing stage where the larvae are in ponds of a few tens of m^2, followed by a second-fry stage in ponds of 0.5 to 1 ha. The first-fry pond is fertilised and then treated with chemicals with compounds containing trichlorphon (e.g. Flibol E at 0.5–1 mg/l or Masoten at 0.3–0.5 mg/l) in order to eliminate unwanted species of zooplankton including cladocerans and copepods and promote the multiplication of rotifers (p. 132). After 4 to 5 days, under satisfactory temperature conditions, (23–24°C) the fry can be introduced at a rate of 1–5 million tench juveniles per hectare. These fry will eventually be fed 1 or 2 days after stocking with hard-boiled egg yolk or fine-grained artificial diet (p. 139). After 2 to 3 weeks, the tench (1.2 cm in length) are able to feed on small cladocerans such as *Moina rectirostris*. A single injection of a few thousand *Moina* per 100 m^2 or the spreading of cut vegetation can have a huge effect on natural food availability. Manufactured diet can also be distributed daily. Fry of 3–4 cm length are harvested after 5–6 weeks of culture in these small ponds.

The second-fry pond should be shallower, partially vegetated (submerged plants) and rich in plankton. The advice is that pre-grown tench fry should not be reared together with common carp. Good results have been obtained from polyculture systems with larger tench, silver carp and grass carp. The stocking rate is 40 000 to 60 000 pre-grown fry per hectare. By autumn these will reach 2–3 g with a high survival rate.

Production of 2- and 3-year-old tench

Although there is some demand, mainly in Central and Eastern Europe, tench is seldom reared in monoculture but is generally cultured together with common carp. As the two species have a very similar diet, production of tench will be to the detriment of carp. For this reason the recommendation is for tench production never to exceed 10% of the total production of the pond. Trials have all shown that an increase in the stocking rate of different species (common carp and complementary species such as silver carp, grass carp and tench) results in a reduction in the growth rate of the complementary species (Table A.3). The survival rate of the different species is not always affected by an increase in the stocking rate. The stocking of carp and tench in combination has several advantages if the correct procedure for introduction is followed. In searching for food tench burrow more actively than carp into the bed of the pond, resulting in better oxygenation of the substrate and a recycling of nutrients. This is particularly beneficial in plant-rich ponds where high yields will not be achieved from carp alone.

In ponds for carp in their 2nd year, the density of tench is generally 20 to 30 individuals per 100 carp of the same age, allowing the tench to reach a weight of around 50 g with a survival rate of 60–80%. Production of table-sized tench (3 years, 200–300 g) is carried out in ponds containing carp stocked at a rate of 20 tench per 100 2-year-old carp. At this stage it may be a good idea to separate the sexes in order to reduce the risks of excessive spawning which are characteristic of tench of this age. The association between tench and small pike (6 weeks) is also used; the pike feed on the abundant fry which are produced by the tench broodstock.

A.4.2 Roach and rudd culture

Roach (*Rutilus rutilus*) and rudd (*Scardinius erythrophthalmus*) are farmed in several western European countries, principally with the aim of producing fish for stocking rivers, canals or ponds for fishing; these fish are much appreciated by anglers. They are occasionally sold for live bait in places where the rearing of traditional bait-fish such as gudgeon and minnow is not practised. Roach are generally reared in association with common carp and, because of their slightly higher price, can improve the profitability of a pond.

Table A.3. Combined production of common carp, chinese carp (silver and grass) and tench in ponds at different stocking densities and different individual weights (modified, after Janecek and Prikryl, 1990). Weights—carp: 70 to 180 g—chinese carp: 7 to 8 g—tench: 2 g.

Treatment	Stocking density		Stock composition (%)				Harvest	Net annual production			Tench mean
	(ind./ha)	(kg/ha)	Carp	Chinese carp	Tench	Diet	(kg/ha)	kg/ha)	(% carp)	(% tench)	weight (g)
1	1 882	98	52.4	34.6	12.2	–	662	564	78	22	99
2	6 916	449	57.8	31.6	10.6	Cereal*	1 933	1 484	83.4	6.6	47
3	13 626	823	58.8	3.1	10.2	Granules†	3 493	2 670	82	8	41
4	27 069	1 677	59.2	30.7	10.1	Granules†	5 180	3 508	80.4	9.6	27

* Wheat flour.
† Crude protein.

Biology of roach and rudd

Roach breed prolifically in any body of water that is warm enough in summer and has shallow zones suitable for spawning. Males mature at 3 or 4 years, females at 4 or 5 years. The spawning season for roach extends through April, May and June so long as the water temperature exceeds 15°C and reaches 20°C at the surface during the warmest parts of the day. Female roach release all of their ova at once over a wide range of substrates such as aquatic plants, pebbles and rocks and over artificial substrates which have a similar texture to natural ones. These include bunches of conifer branches and fine-mesh netting. A female can release 20 000 to 100 000 eggs depending on her size. The eggs are between 1 and 1.5 mm in diameter. The rate of fertilisation is generally very high in natural conditions (above 90%). Roach eggs can withstand temperatures of between 12 and 24°C during incubation, which lasts between 5 and 10 days, depending on the temperature of the water.

In contrast to roach, it appears that rudd have several successive spawnings, separated by intervals of a few days to a few weeks. The spawning season extends from May to July when temperatures are between 18 and 24°C. As with tench, rudd breed more abundantly if the pond has been heated, but spawning is inhibited at temperatures of 28°C or above. The female deposits the eggs on various species of aquatic plant such as water crowfoot, reeds and *fontinalis*.

Of all the species of companion fish, roach, and particularly rudd, have the slowest growth rates (Figure A.6). It is often possible to distinguish two size classes among rudd of 1 year spawned in the same pond; each size corresponds to a different spawning time.

Under natural conditions young roach are omnivorous. Adults become increasingly herbivorous, specialising in eating algae, although insect larvae (stoneflies, dragonflies and mayflies), molluscs (gastropods) and crustaceans are all found in the diet together with plants such as duckweed, myriophyllum and diatoms.

Spawning techniques

Methods of natural spawning used for tench can also be applied to roach and rudd. Techniques of artificial spawning developed for carp can also be used for roach and for rudd, with a few adaptations. A solution of dry carp hypophyseal extract can be injected in one or two stages at a rate of 0.3 mg/kg of female for the first dose followed by a second injection of 3 mg/kg 18 h later (p. 72). Ova are removed by manual stripping 13 h later at 22°C. Table A.4 shows the results for two spawning periods in rudd. The gametes are mixed for 15 min in a solution containing urea (1.5‰) and NaCl (2‰). Two successive baths of tannin (0.35‰) for 8 and 3 s eliminate any problems of eggs clumping together. At a temperature of 20–22°C, the duration of egg incubation is around 5 days. A Zoug jar of 15 to 20 l can contain around 300 000 eggs.

First feeding of larvae

Once the yolk has been resorbed, roach and rudd fry (mean weight 1.5 to 2 mg) can

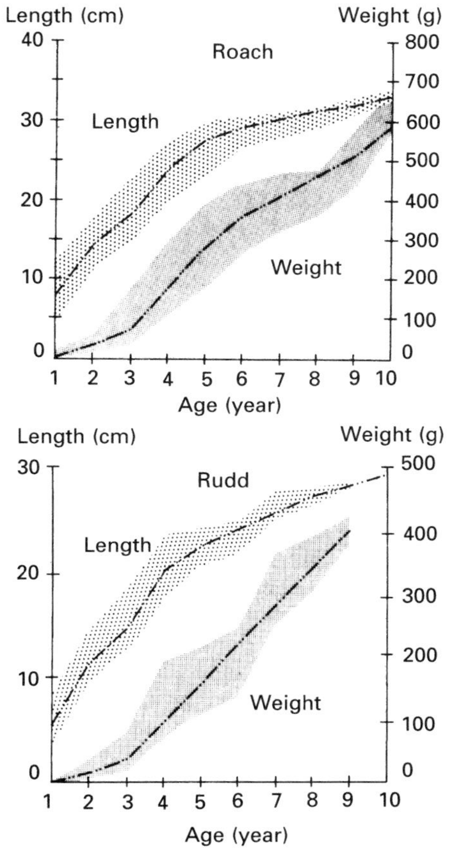

Fig. A.6. Growth in weight and length (mean and extreme values) of roach, *Rutilus rutilus* and rudd, *Scardinius erythrophthalmus* reared in a fish-culture pond (after Timmermans, 1989).

be stocked directly into fertilised ponds using the techniques described for carp or grown on in the pond (p. 125). In the former method it is recommended that the larvae should be fed for a few days using hard-boiled-egg yolk and zooplankton of appropriate size (80 to 125 μm), distributed several times each day. *Artemia* and/or artificial diet specially formulated for fish larvae (p. 139) or shrimps can also be used for 3 or 4 days. When the larvae reach a weight of 5 mg they can be released into the fry pond, Mortality from the egg stage up to the point of stocking into the pond can be as high as 50%.

When rearing takes place in controlled conditions from the stage of yolk-sac resorption until the fish weighs a few grams, consideration of nutrition, quality of the rearing environment, health control and frequency of feed distribution are similar to those described previously for the intensive rearing of carp. This technique for rearing juvenile roach and rudd is rarely used because of the slow growth of the two species and their relatively low economic value.

Table A.4. Egg production in rudd which have undergone hypophysation (after Klein Breteler, 1979).

Females (n)	Weight (g)	First hypophysation		Second hypophysation		Spawning		Number of eggs obtained†	Number of eggs per female
		Date	mg/kg*	Date	mg/kg*	(n)	%		
7	600	15-Apr	0.3	16-Apr	3	1	14	13 000	13 000
63	150	15-Apr	0.3	16-Apr	3	11	17	62 000	5 636
15	650	30-May	3	–	–	13	87	900 000	69 231
10	650	30-May	–	–	–	2	20	1 500	7 500

* Dry carp hypophyseal extract
† 1 g eggs = 1300 eggs

Table A.5. Proportion of individuals (%) of different species stocked in a pond, destined for the table or for restocking (after Marcel, 1987).

	Proportion (%)	
Species	Table market	Restocking
Roach	15–30	60–80
Carp	60–80	5–17
Tench	5–20	4–16
Pike	1–5	1–2

Production of roach and rudd for stocking (see Table A.4)

Production of roach and rudd is mainly directed towards the market for fish for stocking angling waters. Both species are usually sold during their second or third summer. Their growth is slow, lower than that of tench. Early spawning of roach compared with rudd or tench allows the fry to reach a bigger size than either of the other two species by the end of their first-growth season. During the first 3 years of growth in ponds, roach weigh approximately twice as much as rudd of the same age.

There are two choices of production system:

—the production of fish for consumption (carp are dominant) in association with other cyprinids (tench, roach, rudd, gudgeon) which add value because of their higher sale price;

—production of fish for stocking (roach are dominant) in association with other cyprinids (carp, tench, gudgeon) and possibly some predators (pike).

At the time of stocking the proportion of each species will depend on the type of exploitation (Table A.5). The production of roach and rudd of two or three summers, growth (respectively 1 and 2 years) from the first type of system is summarised in Table A.6. Table A.7 gives an example of the stocking rate and production calculated for an association of species favouring the production of cyprinids for restocking.

A.4.3 Rearing of bait-fish

The culture of fish for use as live bait began in the USA, stimulated by high prices associated with sport fishing in that country. In the past it was not difficult to obtain the required live bait from natural waters, but the demand is now such that, in some American states, restrictions have had to be applied to the collection of fish as the practice was threatening the development of some fish stocks in natural waters. This

Table A.6. Example of annual production (mean of 4 years of production) for roach and rudd in polyculture with other complementary cyprinid species (carp, tench, gudgeon) following either the stocking of individuals in their first summer (age 0^+) and 2 summers (age 1^+) (from Gerard, pers. comm. with alterations).

	Stocking density				Harvest (per ha)					Net annual production	
Species	*n*	Bm (kg)	Mw (g)	% Bm tot.	Age	*n*	Bm (kg)	Mw (g)	% Bm tot.	kg/ha/yr	% P. tot.
Age 0^+											
Roach	1400	6.3	4.5	16	1^+	1162	54	47	12	48	12
					0^+	126	1	4		1	
Rudd	1400	3.8	2.7	10	1^+	1 127	39	35	9	36	9
Total	—	38.3	—	100	—	—	449	—	100	411	100
Age 1^+											
Roach	500	23	45.5	17	2^+	410	56	137	15	33	14
					1^+	3225	15	5	4	15	6
Rudd	500	16	31.5	12	2^+	300	27	90	8	11	5
Complementary species	—	94	—	71	—	—	259	—	73	165	70
Total	—	133	—	100	—	—	357	—	100	224	100

Bm = Biomass, Mw = Mean weight, % Bm tot = % total biomass, % P. tot = % total production.

Table A.7. Example of stocking density and production expected in a roach–pike association (after Marcel, 1987, modified).

	Stocking density (per ha)			Growth			Net production	
Species	Age	n	Bm (kg)	Mw (g)	Coef.	Harvest Bm (kg)	kg/ha/yr	% P. tot.
Roach	0^+	4400	22	5	× 7	154	136	66.8
Roach	1^+	200	11	55		17*	6	
Tench	1^+	600	12	20	× 5	60	48	22.6
Pike	4 weeks	250	—	—	× 5	12.5	125	10.6
Pike	0^+	25	2.5	100	× 5	12.5	10	—
Total	—	—	47.5	—	—	256	212.5	100

* 60 kg of roach was consumed by pike. % P. tot. = % of the total production, Bm = biomass, coef. = growth coefficient calculated in relation to survival rate and the growth of each species, Mw = Mean weight.

encouraged the development of the specialised culture of about 20 species used as live bait in sport fishing. Three of these species are of major commercial importance: these are the golden shiner *Notemigonus crysoleucas*, the fathead minnow *Pimephales promelas*, and the goldfish, which is also produced as an ornamental fish. In Europe the culture of bait fish is not really a commercial proposition on its own but is generally a complementary activity, representing only a very small proportion of the overall production from the farm. Most frequently, bait fish are taken from the wild, although minnows, gudgeon and hybrids between common and Crucian carp are reared. Faced with an often-insatiable demand for bait fish, there appears to be a market opening for large producers able to transport fish over long distances and ensure regular supplies to sales outlets. At the same time there is a niche which could be exploited by small producers in the development of a local market and the supply of high-quality bait fish to the public.

In this section, discussion is confined to gudgeon and the European minnow, the species most frequently used as live bait in sports fisheries. Both of these species are cultured for stocking watercourses where overfishing and destruction of spawning sites have reduced their numbers.

Gudgeon

These are mainly sold at a length of 10–12 cm as bait for large predators such as pike, zander and perch and at a size of 4–6 cm for trout fishing. They can also be sold for stocking and for direct consumption (fried). In spite of its high sale price it is only farmed in small quantities and generally in large cyprinid ponds (carp, tench, gudgeon), as a companion species.

Biology of the gudgeon

Gudgeon adapt well to a variety of habitats including streams, rivers, lakes, canals and ponds where the bed is gravelly, stony or sandy and where the current is moderate. It is generally agreed that a temperature of 14 to 17°C is required for females to spawn naturally. Depending on region, the spawning season extends from mid-April to the end of August with a peak of spawning activity in May and June. The gudgeon is a fractional spawner with an asynchronous development of oocytes. Because of this, it is difficult to estimate the fecundity of females. Depending on author, estimates vary between 500 to 5000 ova in 2-year-old females and 4000 to 20 000 ova in 5-year-old females. Gudgeon have a short lifespan and the majority die by their 3rd year.

The diet mainly consists of planktonic invertebrates (rotifers, crustaceans, principally cladocerans, then copepods and ostracods) at the fry stage, and benthic invertebrates (preferably insects, crustaceans, molluscs and worms) at the adult stage. It appears to be an opportunistic feeder with its diet fixed mainly on the abundance of the fauna *in situ*.

Monoculture and polyculture in ponds

Intensive rearing of gudgeon is best carried out in small ponds (10 to 20 m^2) which can be emptied completely and managed from the paths along the banks. These are used for spawning of broodfish and the development of eggs. Different trials have shown that, depending on the techniques used, in monoculture ponds (stocked with broodfish and with the fry kept until autumn) a density of 0.5–1 broodfish/m^2 is sufficient to ensure production of the order of 20 fry/m^2/year (Table A.8).

While most gudgeon spawn at 2 years a small fraction of the population is likely to reproduce during the second summer (1 year old). Thus, stocking the pond with fry produced during the previous year will produce fish of commercial size after two

Table A.8. Effects of the stocking density of broodfish on annual production of gudgeon fry in holding ponds.

Stocking density			Harvest				
Adults		Number of trials	Adults		Fry		Net production (1) (kg/ha/year
n/m^2	Mw (g)		n/m^2	Mw (g)	n/m^2	Mw (g)	
0.5	25.8	3	0.25	36.2	18.0	0.70	210
1	31.0	5	0.5	38.0	20.8	0.88	230
2	30.4	3	0.8	37.5	36.4	0.75	230
3	26.7	2	1.1	39.0	24.5	0.65	220

(1) Production only takes into account the 8 months between stocking and harvest, Mw = mean weight in g; n = number of individuals.

Table A.9. Annual production of gudgeon (mean of six trials) in ponds used for both holding and on-growing (area = 0.2–0.3 ha).

Stocking density			Harvest			Net production (kg/ha/yr)
Age	n/m^2	Mw (g)	Age	n/m^2	Mw (g)	
0^+	4–10	0.7–1.1	0^+	0–9	0.6–1.0	
			1^+	3–9	7–12	350–960
0^+	4–10	0.7–1.1	0^+	5–15	0.6–1.0	
1^+	0.2–1.5	10–15	2^+	3.5–9	15–25	420–750

Mw = mean weight in g; n = number of individuals.

summers (annual production 450–750 kg/ha when supplementary feeding is given) and also ensure an annual recruitment of 5 to 15 fry/m^2 (Table A.9).

As shown in Table A.9, the survival of broodstock (2 years old) of three or more summers is low (between 15 and 70%) but generally high (75–80%) for gudgeon up to the end of their first or second summer.

There are two possible courses of action for the fish farmer:

—either using older gudgeon (three summers or older) which favours a higher production of juveniles because of the higher fecundity of older, larger adults, at the expense of an increased parental mortality rate;

—or, using young spawners (1 or 2 years) which leads to a lower recruitment of fry but increases the survival of the adults which can be sold.

The mean weight of fry at the end of their first growth season (1 year) varies from one year to the next but is relatively constant within the same year, whatever the population density. The weight of the fry is determined more by the earliness of spawning and the abiotic and biotic conditions during the spawning season than by the density of fry from natural spawning. In polyculture with other cyprinids, the annual production of gudgeon reaches 90 to 95 kg/ha, which generally represents 15 to 50% of the total production of the pond (Table A.10).

Intensive rearing in tanks

Intensive rearing of gudgeon is only practised on an experimental scale. Control of the breeding cycle through the modulation of environmental factors such as temperature and daylength results in the availability of mature broodstock throughout the year and the almost continuous production of eggs. Figure A.7 shows a programme for broodstock management based on control of temperature and photoperiod regimes.

Ovulation is induced through intraperitoneal injection of carp dry hypophyseal extracts at a dose rate of 10 mg/g female. The same dose is given to males in order to increase the volume of sperm produced. At a temperature of 20°C, ovulation occurs in 50 to 60% of the treated females. Artificial fertilisation, as described for carp

Table A.10. Production of gudgeon in ponds in polyculture with other cyprinids (carp, tench, roach) = (complementary). Mean results from four trials (after Gérard, pers. comm.).

	Stocking density				Harvest (per ha)			Net production		
Species	Age	n (× 1000)	Bm (kg)	% Bm tot.	Age	n (× 1000)	Bm (kg)	% Bm tot.	kg/ha/yr	% P. tot.
Gudgeon	0^+	11	10	11	0^+	17	21	25	90	29
					1^+	7	79			
Complementary	—	—	81	89	—	—	301	75	220	71
Total	—	—	91	100	—	—	401	100	310	100
Gudgeon	1^+	1.2	11	11	0^+	55	88	28	95	33
Complementary	—	—	88	88	2^+	1	269	72	180	67
Total	—	—	99	100	—	—	375	100	275	100

Bm = biomass, n = number of individuals, % P. tot. = % of the total production.

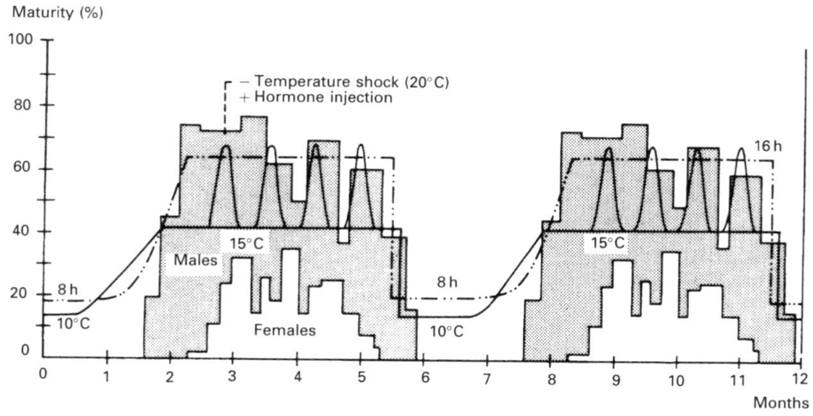

Fig. A.7. Programme for the management of broadstock in a pond with controlled photoperiod (16 and 8 hours of light/day and temperature (10 and 15°C) allowing two spawning cycles per year. The percentage maturity corresponds to the proportion of males and females that are mature.

(pp. 83–85) can be used for gudgeon but because of the difficulty of precise determination of the maturity of females, the fertilisation and hatching rates are highly variable, but generally low. The duration of incubation in Zoug jars, after a short treatment with tannic acid (0.35 g/l) is linked to temperature according to the following relationship:

$$t = D/(T - 10.2)$$

where t = incubation time in days
D = a constant equal to 52 degree-days
T = incubation temperature
10.2 = lower temperature limit for embryonic development.

Resorption of the yolk-sac lasts 60–80 degree-days.

A semi-natural spawning technique using a gravel substrate is being used in trials that are producing promising results. This reduces handling associated with artificial fertilisation and results in a relatively constant hatching rate (50–70%). An injection of a hormone preparation administered to gravid females, combined with a temperature shock (15–20°C), followed by transfer to spawning tanks equipped with gravel substrates, induces spontaneous spawning in most of the treated females within 24 h of the injection. The quantity of ova released by each female appears to be similar to that by manual stripping followed by artificial fertilisation.

Stocking alevins (with yolk sacs) into fry ponds is not recommended as the survival rate is generally low (5 to 20%) because the larvae are so small (0.5 mg). It is therefore necessary to establish first feeding in tanks. This procedure has been described in the section dealing with the production of juveniles using artificial or mixed diet (p. 139). After a month, the pre-fed fry (30 mg) can be stocked into fry ponds (p. 126). After this stage, survival in ponds will be high, around 85–90%.

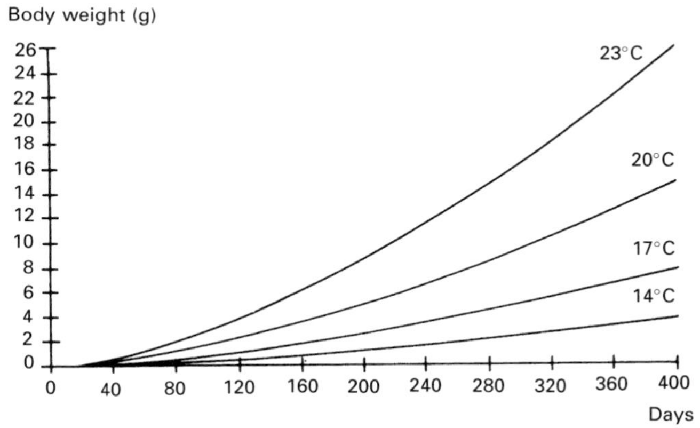

Fig. A.8. Theoretical growth (weight) of gudgeon raised in intensive systems at different temperatures (from Kestemont *et al.*, 1990).

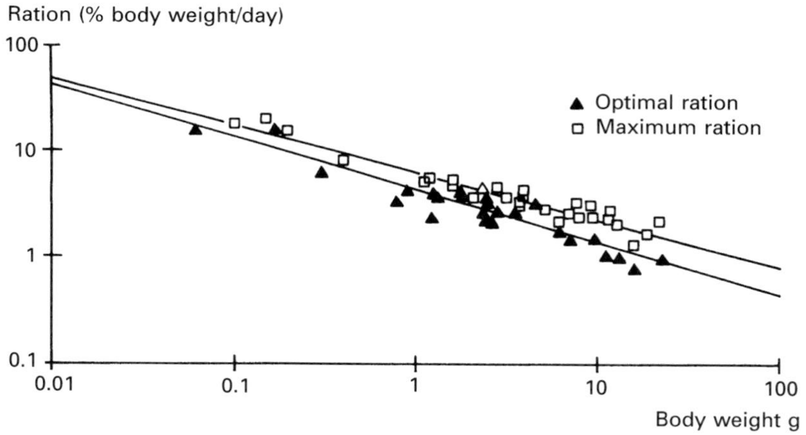

Fig. A.9. Relationship between feed ration and body weight for gudgeon (temperature 20.9 to 24.7°C—granular trout diet) (from Kestemont *et al.*, 1990).

The first results from growing gudgeon in tanks have allowed the determination of rates of weight increase at different temperatures as well as their quantitative dietary requirements and optimum daily ration (Figures A.8 and A.9). Commercial pelleted diet is readily accepted but leads to a high proportion of skeletal malformations in growing fish (40%).

Minnows

Minnows reach a maximum length of 8–10 cm and are frequently found in streams and cool rivers (trout, grayling and barbel zones) and also in lakes and ponds which have a tendency towards being oligotrophic. Minnows are generally 2 or 3 years old when they mature. Females spawn 500 to 2 500 eggs in batches, between April and July.

As a bait-fish minnows are mainly used by trout anglers. They may also be produced for restocking. Small producers are growing minnows in semi-intensive systems, ensuring local needs are met in areas where fishing with live bait is practised.

Controlled spawning

Trials with artificial spawning of broodfish quickly showed that adult minnows required hormonal stimulation in order to trigger the ovulation process and breed successfully. The ovulation rate is between 50 and 65%, similar to that of gudgeon, and fertilisation rate varies between 22 and 94%.

The applicability and yield from techniques of artificial spawning are dubious; alternative methods of semi-natural controlled spawning have been proposed and tested experimentally. At present, it appears that the technique of spawning over a substrate is the best way of producing large numbers of feeding fry suitable for transfer to fry ponds. During the spawning season, minnows become mature in tanks and ponds as soon as the temperature exceeds 15–16°C. Putting down gravel substrates results in mass spawning in the following days without the need for hormone injection. The substrate should be regularly removed to avoid predation on the eggs by male fish. A male : female sex ratio of 1 : 3 can also significantly reduce the level of predation. At 15–16°C incubation takes 90 degree-days (or around 6 days) and resorption of the yolk, up to the first-feeding stage requires 13 further degree-days. The control of the spawning cycle through modulation of external factors (temperature and daylength) allows spawning to be advanced and fry to be obtained earlier in the season.

Production of fry age one summer and saleable fish

Rearing fry directly in the ponds where they have been stocked at first feeding gives a satisfactory survival rate of 45–70%, even at high density (Table A.11). The relatively large size of these larvae (weight 1.9 mg) means they can be stocked in

Table A.11. Production of minnow fry (1 summer) from stocking at the yolk-sac resorption stage at different densities (from Stalmans and Kestemont, 1991).

Density (n/m^2)	Number of trials (n)	Duration (d)	Survival (%)	Mean weight (g)	Production (1) (kg/ha/yr)
15	3	180	48.1	0.52	76
40	4	140	58.9	0.83	511
100	5	130	65.6	0.23	416
150	7	115	45.5	0.49	1047
200	1	130	71.6	0.34	1358

(1) Between June and November.

ponds early in the season, while conditions outside are still not really favourable. In ponds, the commercial size is reached after around 15 months of rearing or during their second summer. Recent trials have shown the possibility of mass production of minnows, under entirely artificial conditions, from the yolk resorption stage up to feeding fry of 30–500 mg. If they are stocked into ponds early in the season, these fry can reach commercial size after a single year of production.

A.4.4 Zander

Sometimes in cyprinid culture it is beneficial to rear the young stages of a predatory species together with the main species (generally carp). The main predators are generally zander, pike and European catfish. These fish destroy the fry of carp and other cyprinids that may have been spawned in the on-growing ponds, or have been introduced into them accidentally. For example, zander are used in Hungary to eliminate unwanted fish such as *Pseudorasbora parva* from ponds. They also provide a significant extra income as they generally command a higher sale price than carp.

The zander, *Stizostedion lucioperca*, belongs to the perch family. Originating in central and eastern Europe, it has been introduced into many parts of Europe and North Africa. It is a piscivorous fish which can be reared as a predator in large carp ponds. It feeds exclusively on small fish except during its 1st year of life, although at a very early stage (1 to 2 months) it can feed on fish larvae. The quality of its flesh and absence of intramuscular bones makes this one of the most appreciated of freshwater fish with a significant commercial demand.

Biology

Zander spawn in spring (March–April) when water temperatures reach 8 to 12°C. Eggs are generally deposited on the roots of aquatic plants, gathered together in the form of a nest by the male, in sites where the depth of water is between 0.4 and 0.5 m. Sexual maturity generally occurs at 2–3 years in males and 3–4 years in females when the fish have reached a length of 50 cm and a weight of around 1 kg. The total fecundity is very high, varying from 100 000 to 1 000 000 eggs depending on the size of the female. After hydration, the eggs are 1.5 mm in diameter. The male guards the nest during incubation and resorption of yolk. The optimum incubation temperature is 12–16°C (extremes 8 and 25°C). At 12°C incubation lasts 13 to 14 days.

Zander larvae are very small and transparent at hatching. They feed on zooplankton (rotifers and then cladocerans) for the first few weeks, and then zoobenthos. When they have reached a length of 35 to 40 mm predation on fry begins. Cannibalism frequently occurs. The growth of zander in ponds is represented in Figure A.10.

Controlled spawning

Zander broodfish should preferably be kept in deep ponds of a medium or large surface area with a partially stony or sandy bottom. The dissolved oxygen content should be high and the temperature should remain below 25°C during summer.

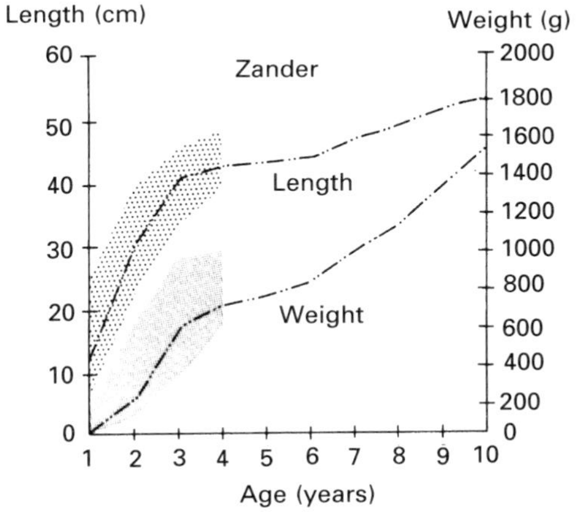

Fig. A10. Growth in weight and length of zander *Stizostedion lucioperca* reared in fish-culture ponds (after Timmermans, 1989).

Zander feed on small cyprinids (roach, rudd, bleak, gudgeon and tench) which should be abundant in the pond.

Spawning can be achieved in two ways:

—natural spawning in ponds using nests;
—artificial spawning: collection of ova, fertilisation and incubation in Zoug jars.

Spawning in ponds requires the capture of broodstock during the year which precedes the planned spawning season. These fish are held in wintering ponds at a 1 : 1 sex ratio. A biomass of fodder fish equivalent to 20–25% of the weight of the broodfish should also be stocked to provide food during winter. In spring the male and female broodfish are separated when the temperature reaches 6 to 8°C. Ponds are then furnished with natural nests (juniper branches or willow roots) or artificial nests (synthetic fibres) 0.5 × 0.5 m in area, dropped onto the bed of the pond where the water is 1.5–2.5 m deep. Two days after introducing the broodfish (a few pairs per hectare), the nests can be checked and the eggs removed and transported over long distances, if required, 24 h after spawning. During incubation, levels of oxygen must be kept high. Three to 4 h before hatching, the nests are transferred to fry ponds. This technique of semi-controlled natural spawning can also be used in small ponds with a concrete bottom, tanks or cages in which spawning substrates have been placed.

Cages (1–2 m^2, 2 m deep) are equipped with natural or artificial nests and stocked with one to two females and two to four males per cage. Around 80% of females spawn, 2 to 30 days after being placed in the cage. The area of the nest is correlated with the size of the female: 65 × 65 cm for large individuals (>2 kg) and 45 × 45 cm for smaller ones. One nest per cage is enough as it is very rare for more than one female to spawn during the same night. The nests are checked each day. After spawning, the nest containing the eggs and the male which has participated in spawning are transferred to an empty cage. The fish that remain in the cage are given a new nest. The male guards the nest and keeps the eggs clean by moving the water with his pectoral fins. On the day before hatching the eggs are transferred to a hatchery where the water temperature has been raised to 18–22°C so as to confine the hatching period to 1 to 2 days. A hormonal treatment (carp hypophyseal extract) can be injected into the females at a dose rate of 2 mg/kg body weight in order to facilitate spawning.

Artificial spawning of zander is not used routinely, because of the high efficiency of natural spawning over a substrate. It has been used successfully after administering carp hypophyseal extract at a rate of 2 mg/kg female. The method of artificial spawning described previously for carp can be used for zander (p. 83).

The fry stage and on-growing

Four to 5 days after hatching, the larval zander can be stocked into a pond or a tank. The traditional rearing method for zander, either in monoculture or in polyculture with carp (of one summer's growth) gives satisfactory results so long as certain rules are observed: preparation of the ponds before fry are stocked by the application of organic or mineral fertilisers, stocking fry at a rate of 2000 per hectare in polyculture, 5000 to 8000 in monoculture and harvesting in the autumn by which time the young zander will have reached 10 to 30 g (5 to 30% survival, depending on the ponds). Sometimes ponds are also used for the first-fry stage. These are stocked at a rate of 1 million eyed eggs or 500 000 pre-fed fry (rotifers and *Artemia* nauplii) per hectare. These fry are captured after 30 days by which time they will have reached a length of 30 to 40 mm with a survival rate of 3 to 5% when eggs have been used for stocking and 20 to 30% when fry have been used. There is a significant amount of cannibalism; this means that young zander should be stocked into cyprinid ponds where there is an abundance of fodder fish.

From the 2nd year onwards, zander are raised in polyculture as accompanying carnivores in ponds where carp are the dominant fish. They consume around 3 to 6 kg of fish in order to grow by 1 kg.

Over several years, research efforts have increased in the field of management of rearing of juvenile zander with natural food in captivity leading progressively to the use of manufactured diets, which would allow intensive rearing in tanks.

BIBLIOGRAPHY

Beveridge M., 1987. *Cage Aquaculture.* Fishing News Books, Farnham, Surrey, UK.

Billard R., 1985. Gestion piscicole intégrée des plans d'eau en Chine. In: D. Gerdeaux, R. Billard (eds), *Gestion Piscicole des lacs et retenues artificielles.* INRA, Paris, pp. 237–246.

Billard R., 1995. The major carps and other cyprinids. In: C.E. Nash and A.J. Novotny (eds), *Production of Aquatic Animals. Fishes.* World Animal Science C8, Elsevier, pp. 21–55.

Cherfas N.B., Kozinsky O., Rothbard S., Hulata G. 1990. Induced diploid gynogenesis and triploid in ornamental (koi) carp *Cyprinus carpio* L. 1. Experiments on the timing of temperature shock. *Israeli J. Aquacult. Bamidgeh*, **42**: 3–9.

Hai L.Y., Zweig D., 1987. Cage culture in Kumming: an effective means of resource recovery. *Aquaculture Magazine*, Sept./Oct.: 28–31.

Janecek V., Prikryl I., 1990. Mixed culture of carp, phytophageous fishes and tench in ponds at different culture intensities. *EIFAC/FAO Symp. on Production Enhancement in Still-Water Pond Culture*, Prague, Czechoslovakia, 15–18 May, E. 24.

Kestemont P., Meillard Ch., Micha J.C., Philippart, J.C., 1990. Développement à l'échelle pré-industrielle de la reproduction artificielle et de l'alevinage intensif du goujon et de quelques autres espèces de poissons d'eau douce. Convention de recherche, Min. Reg. Wal., *Rapp. annual 2*, 89 pp.

Klein Breteler J.G.P., 1979. First experiments concerning controlled reproduction and rearing of fry and fingerlings of rudd, *Scardinius erythrophthalmus.* L. EIFAC Tech. Pap. 35, suppl. 1, pp. 184–188.

Li S., 1986. Reservoir fish aquaculture in China: 347–356. In: R. Billard, J. Marcel (eds), *Aquaculture of Cyprinid*s. INRA, Paris.

Marcel J., 1987. Les empoissonnages en poissons de repeuplement. *Echo-Système, ITAVI, Paris*, **2**: 1–14.

Melotti, P. 1986. goldfish *Carassius auratus* farming in Italy. In: R. Billard, J. Marcel (eds), *Aquaculture of Cyprinids.* INRA, Paris, pp. 369–376.

Stalmans J.M., Kestemont P., 1991. Production de juvéniles de vairon *Phoxinus phoxinus* L. à partir de larves obtenues en conditions contrôlées. *Bull. Fr. Pêche Pisc.*, **320**: 29–37.

Timmermans J.A., 1989. *Données sur la croissance de quelques espèces de poissons dans des étangs de Campine.* Station de Recherches Forestières et Hydrobiologiques, Travaux-Série D, **56**: 3–34.

General bibliography

MAIN WORKS ON AQUACULTURE, AND MORE SPECIALLY CYPRINID AQUACULTURE[1]

Arrignon J., 1970. *Aménagement piscicole des eaux intérieures.* S.E.D.E.T.E.C., Paris, 643 pp.
Arrignon J., 1976. *Aménagement écologique et piscicole des eaux douces.* Gauthier-Villars, 320 pp.
Association Internationale des Entretiens Ecologiques., 1982. Colloque sur la production et la commercialisation du poisson d'eau douce (Dijon, France, 30 mars–1er avril 1982). *Cahiers, Association Internationale des Entretiens Ecologiques*, **11–12**: 1–152; **13–14:** 1–195.
Bachasson B., 1991. Mise en valeur des étangs. *Agriculture d'aujourd'hui: sciences, techniques, applications.* Lavoisier, Technique et Documentation, Paris, 166 pp.
Bagenal T. (ed.), 1978. *Methods for Assessment of Fish Production in Freshwater*, 3rd ed. Blackwell Scientific Publications, Oxford, 365 pp.
Bardach J.E., Ryther J.H., McLarney W.O., 1972. *Aquaculture.* Wiley-Interscience, New York, 868 pp.
Barnabé G. (coord.), 1989. *Aquaculture*, 2nd ed. Lavoisier, Technique et Documentation, Paris, 2 vol., 1308 pp.
Beveridge M., Ross L., 1990. *An Introduction to Aquaculture.* Blackwell Scientific, Boston, Massachusetts, 224 pp.
Billard R. (ed.), 1980. *La Pisciculture en étang:* actes du Congrès sur la pisciculture en étang, Arbonne-la-Forêt, France, 11–13 March 1980. INRA, Paris, 395 pp.
Billard R., Gall G.A.E., 1995. The carp. *Aquaculture*, **129:** 485 pp.
Billard R., Marcel J. (eds), 1986a. *Aquaculture of Cyprinids.* INRA, Paris, 502 pp.
Billard R., Marcel J. (eds), 1986b. Aquaculture of carp and related species: selected papers presented at a symposium held at Evry, France, 2–5 September 1985. *Aquaculture*, **54**: 164 pp.

[1] F. Nadot and M. Margout.

Billard R., Benoit G. (eds), 1987. L'Aquaculture en agriculture. Association pour le développement de l'aquaculture, No. 18, 126 pp.

Billard R., De Pauw N. (eds), 1989. *Aquaculture Europe '89 Bordeaux*, Short Communications, Abstracts, International Conference. European Aquaculture Society, Bredene, Belgium, 342 pp.

Boyd C.E., 1979. *Water Quality in Warmwater Fish Ponds*. Auburn University Agriculture Experiment Station, Auburn, Alabama, 359 pp.

Boyd C.E., 1982. *Water Quality Management for Pond Fish Culture*. Elsevier, Amsterdam, 318 pp.

Boyd C.E., 1995. *Bottom Soils, Sediment, and Pond Aquaculture*. Chapman & Hall, 348 pp.

Chauderon L., 1970. *La pisciculture en étang des poissons de repeuplement pour les cours d'eau de la deuxième catégorie*. Club halieutique, Mas de Carles, Octon, 34 Clermond L'Hérault et Conseil supérieur de la pêche, 143 pp.

Chondar S.L., 1980. *Hypophysation of Indian Major Carps*. Motikara, Agra-3, India, 146 pp.

De Pauw N., Billard R. (eds), 1990. *Aquaculture Europe '89, Business joins science: reviews and panel reports, Bordeaux*. European Aquaculture Society, Bredene, Belgium, 462 pp.

Ghittino P., 1983. *Tecnologia e patologia in acquacoltura*, Vol. 1: *tecnologia*. Tipografia E. Bono, Torino, 532 pp.

Goubier V., 1989. Influence de la fertilisation sur certains compartiments de l'étang à pisciculture. Thèse, Institut régional de recherches appliquées en aquaculture, Université Lyon, 240 pp.

Hepher B., Pruginin Y., 1981. *Commercial Fish Farming, with Special Reference to Fish Culture in Israel*. J. Wiley, New York, 261 pp.

Hepher B., 1988. *Nutrition of Pond Fishes*. Cambridge University Press, New York, 388 pp.

Horvath L., Tamas G., Tolg I., 1984. *Special Methods in Pond Fish Husbandry*. Akademia Kiado, Budapest; Halver Corporation, Seattle, 147 pp.

Horvath L., Tamas G., Coche A.G., 1986. *La Carpe commune*. 1ere partie: *production massive d'oeufs et de post-larves*. 2e partie: *production massive de carpillons en étangs*. Organisation des Nations Unies pour l'alimentation et l'agriculture, Rome, 2 vol., p. 85 and p. 87. Collection FAO Formation: 8 and 9.

Horvath L., Tamas G., Seag C., 1993. *Carp and Pond Fish Culture*. Fishing News Books, Oxford, 158 pp.

Huet M., 1986. *Textbook of Fish Culture*, 2nd ed. Fishing News Books, Farnham, Surrey, UK, 438 pp.

Jensen G.L., Bankston J.D., 1989. *Guide to Oxygen Management and Aeration in Commercial Fish Ponds*. Louisiana State University Agricultural Center, 26 pp.

Jensen G.L., 1989. *Handbook for Calculations in Finfish Aquaculture*. Louisiana State University Agricultural Center, 59 pp.

Jhingran V., Pullin R., 1985. *A Hatchery Mannual for the Common, Chinese and Indian Major Carps*. Metro Manila, Southeast Asian Fisheries Development Center, 191 pp.

Kinkelin P. De., 1985. *Précis de pathologie des poissons*. Institut National de la Recherche Agronomique, Paris, 348 pp.

Kirpichnikov V.S., 1981. *Genetic Bases of Fish Selection*. Springer Verlag, Berlin, 410 pp.

Lannan J.E., Smitherman R.O., Tchobanoglous (eds), 1986. *Principles and practices of Pond Aquaculture*. Oregon State University Press, Corvallis, Oregon, 252 pp.

Li S., Mathias J. (eds), 1994. *Freshwater Fish Culture in China: Principles and Practice*. Developments in Aquaculture and Fisheries Science, **28**. Elsevier, 445 pp.

Marx C.E., 1991. *Elsevier's Dictionary of Aquaculture in Six Languages: English, French, Spanish, German, Italian and Latin*. Elsevier, Amsterdam, 454 pp.

McLarney W., 1984. *The Freshwater Aquaculture Book: A Handbook for Small Scale Fish Culture*. Hatley and Marks, Point Roberts, Washington, 583 pp.

Melloti P., Resta C., 1989. *La carpa erbivora in Emilia-Romagna: aspetti biologici e gestionali*. Amministrazione provinciale di Ferrara, Ferrara, Italy, 87 pp.

Michael R.G. (ed)., 1987. Managed aquatic ecosystems. Amsterdam, Elsevier, *Ecosystems of the World*, **29**: 166 pp.

Michaels V.K., 1988. *Carp Farming*. Fishing News Books, Farnham, Surrey, 207 pp.

Moriarty D.J., Pullin R.S.V., 1987. Detritus and microbial ecology in aquaculture. *Proceedings of the Conference on Detrital Systems for Aquaculture*, 26–31 August 1985, Bellagio, Como, Italy. International Center for Living Aquatic Resources Management, Manila, The Philippines, ICLARM Conference Proceedings: No. 14: 420 pp.

Muir J.F., Roberts R.J. (eds), 1985. *Recent Advances in Aquaculture*, Vol. 2. Croom Helm, Beckenham, UK, 282 pp.

Pillay T., 1990. *Aquaculture: Principles and Practices*. Fishing News Books, Oxford, 576 pp.

Piper R.G., Mcelwain I.B., Orme L.E., 1982. *Fish Hatchery Management*. Department of the Interior, Fish and Wildlife Service, Washington, DC, 517 pp.

Pullin R.S.V., Shehadeh Z.H., 1980. *Integrated Agriculture-Aquaculture Farming Systems*. Manila, The Philippines, ICLARM, 258 pp.

Roberts R.J., 1979. *Pathologie du poisson*. Maloine, Paris, 317 pp.

Ruddle K., Zhong G., 1988. *Integrated Agriculture-Aquaculture in South China: the Dike-pond System of the Zhujiang Delta*. Cambridge University Press, Cambridge, 143 pp.

Ruwet J.C. (Dir.). 1987. *Colloque Aquaculture et développement*. Cahiers d'éthologie appliquée, Liège, Belgium, 146 pp.

Schaperclaus W., 1962. *Traité de pisciculture en étang*, 2nd edn, Vigot, Paris, 620 pp.

Schlumberger O., 1986. *Memento de pisciculture d'étang*. Association toulousaine d'ichtyologie appliquée, Toulouse, 104 pp.
Schlumberger O., 1993. *Memento de pisciculture d'étang*. CEMAGREF, 166 pp.
Sevrin-Reyssac J., 1985. *Bien connaitre son étang... pour mieux le gérer*. Muséum national d'Histoire naturelle, Laboratoire d'Ichtyologie générale et appliquée, Paris, 77 pp.
Shepherd J., Bromage N., 1988. *Intensive Fish Farming*. BSP Professional Books, Oxford, 404 pp.
Shilo M., Sarig S., 1989. *Fish Culture in Warm Water Systems: Problems and Trends*. CRC Press, Boca Raton, Florida, 304 pp.
Steffens W. (ed.). 1979. *Industriemässige Fischproduktion*. Deutscher Landwirtshaftsverlag, Berlin, 376 pp.
Stickney R.R. (ed.). 1986. *Culture of Non-Salmonid Fishes*, CRC Press, Boca Raton, Florida, 201 pp.
TABLE RONDE INTERNATIONALE "BARBUS". 1989. *Compte-rendu de la Table ronde international "Barbus", 25–27 juillet 1989*. Faculté des Sciences, Montpellier, 345 pp.
Winfield I.J., Nelson J.S. (eds.), 1991. *Cyprinid Fishes: Systematics, Biology and Exploitation*. Chapman and Hall, London, 667 pp.

MAIN AQUACULTURE JOURNALS

Belgium

Aquaculture Europe. Ostende: European Aquaculture Society
Travaux—Station de recherches forestières et hydrobiologiques. Serie D. Groenendaal-Hoeilaart: Station de recherches forestières et hydrobiologiques.[1]

Canada

Canadian Journal of Fisheries and Aquatic Sciences. Ottawa.

Denmark

Ecology of Freshwater Fish. Copenhagen: Muiksgaard.

France

Aqua Revue. Tours: M.C.M.[1,3]
Aquatic Living Resources. Paris: Elsevier.[1,4]
Bulletin Français de la Pêche et de la Pisciculture. Boves: Conseil Supérieur de la Pêche.[1,2,5]
Equinoxe. Nantes: IFREMER.[1,2,4,6]
Echo-système. Paris: ITAVI.[1]

Les Etangs: Bulletin de liaison piscicole. Lycée agricole de Cibeins, Mizerieux, 01600 Trevoux.
La Pisciculture Française. Paris: Fédération Française d'Aquaculture.[1,2,3,4,5]

Germany

Journal of Applied Ichthyology. Berlin: P. Parey.[1,2]

Hungary

Aquacultura hungarica. Szarvas: Fisheries Research Institute.[8]

Israel

The Israeli Journal of Aquaculture Bamidgeh. Jerusalem.[1,2,3,4,8,9]

Italy

FAO Aquaculture Newsletter. Rome.
FAO Fisheries Circular. Rome: Organisation des Nations Unies pour l'alimentation et l'agriculture.[1,2,4,6]
FAO Fisheries Report. Rome: Organisation des Nations Unies pour l'alimentation et l'agriculture.[1,2,4,6]
Il Pesce. Modena: Pubblicità Italia.[1]
Rivista italiana di acquacoltura. Verona; formerly *Rivista italiana di piscicoltura e ittiopatologia*. Treviso: Associazine Piscicoltori Italiana.[1]

Japan

Bulletin of National Research Institute of Aquaculture. Nansei, Mie.[2,4,5,9]
Scientific Reports of the Hokkaido Salmon Hatchery. Sapporo.
Technical Reports of the Hokkaido Salmon Hatchery. Sapporo.

Philippines

SEAFDEC Asian Aquaculture and Aqua Farm News. Iloilo: South East Asian Fishery Development Center.
Asian Fisheries Science. Manila: Asian Fishery Society.

Poland

Archives of Polish Fisheries, formerly *Roczniki Nauk Rolniczych—serie H—Rybactwo*. Warsaw.[1,8]
Polish Archives of Hydrobiology. Warsaw: Polska Nauk.[4,8]

Rumania

Analele Universitatii din galati. Fasc. VII. Tehnica piscicola. Galati.[1]
Buletinul de cercetari piscicole. Nucet, Jud. Dimbovita.[1,2,5,8,9]

The Netherlands

Aquaculture. Amsterdam: Elsevier.[1,2,3,4,5,6,7,8,9]

United Kingdom

Aquacultural Engineering. London: Applied Sciences Publishers.[1,2,5]
Aquacultural International. London: Chapman & Hall.
Aquaculture Research, formerly *Aquaculture and Fisheries Management.* Oxford: Blackwell Scientific Publications.[1,8]
Fish Farming International. London: Emap Heighway Publications.[1,3,6]
Journal of Aquaculture in the Tropics. Oxford: IBH Publ.[1]
Review in Fish Biology & Fisheries. London: Chapman & Hall.

United States

Aquaculture Magazine. Little Rock, Arkansas: Briggs Associates.[1,3]
Fisheries. Bethesda: American Fisheries Society.[1,2,6,8]
Marine Fisheries Review. Washington: NOAA, US Department of Commerce.
Progressive Fish Culturist and Transactions of the American Fisheries Society. Bethesda, Maryland: American Fisheries Society.
World Aquaculture. Baton Rouge, Louisiana: World Aquaculture Society.

In France, the following journals can be consulted in the following libraries:

(1) Laboratoire d'ichtyologie générale et appliquée du Muséum National d'Histoire Naturelle. 43, rue Cuvier, 75231 Paris Cedex 05. Tél. 40.79.37.46.
(2) IFREMER—Centre de Nantes. Service de la Documentation et des Publications. Rue de l'ile d'Yeu, B.P. 1049, 44037 Nantes Cedex 01. Tél. 40.37.40.00.
(3) CEMAGREF—Groupement de Bordeaux. 50, avenue de Verdun, Gazinet, B.P. 3, 33610 Cestas Principal. Tél. 56.36.09.40.
(4) Laboratoire Arago. Bibliothèque. 66650 Banyuls-sur-mer. Tél. 68.88.00.40.
(5) IFREMER—Centre de Brest. Service de la Documentation et des Publications. BP. 70. 29263 Plouzané. Tél. 98.22.43.64.
(6) Institut océanographique. Bibliothèque. 195, rue Saint-Jacques, 75005 Paris. Tél. 43.25.63.10.
(7) Musée océanographique. Bibliothèque. Avenue Saint-Martin, Monaco-Ville, MC 98000 Monaco. Tél. 93.15.36.00.

Author index

P. Bergot	INRA, Station d'Hydrobiologie, Unité d'Elevage Larvaire BP 3, 64310 Saint Pée-sur-Nivelle, France.
R. Berka	Institute of Fish, Vodnany, Czech Republic.
R. Billard	Laboratoire d'Ichtyologie Générale et Appliquée, Muséum National d'Histoire Naturelle, 43, rue Cuvier, 75231 Paris, France.
Th. Boujard	INRA Station d'Hydrobiologie BP 3, 64310 St. Pée/Nivelle, France.
A. Demael	Laboratoire de Physiologie Générale et Comparée. Université Claude Bernard Lyon I, 69622 Villeurbanne, France.
B. Fauconneau	INRA, Campus de Baulieu, 35042 Rennes, France.
P. Haffray	SYSAAF, Section Aquacole, ENSAR, 65, rue de St. Brieuc, 35042 Rennes, France.
M.G. Hollebecq	INRA, Laboratoire de Génétique des Poissons, 78350 Jouy en Josas, France.
M. Jacquinot	Directeur Marketing de la Sté GEM, 58A rue du Dessous des Berges, 75013 Paris, France.
P. Kestemont	Facultés Universitaires Notre-Dame de la Paix, Unité d'Ecologie des Eaux Douces, 61, rue de Bruxelles B-5000 Namur, Belgium.
B. Lanoiselée	ACA. Pinchat, 37210 Vouvray, France.
H Le Louarn	Laboratoire d'Ecologie Hydrobiologique, INRA/ENSAR, 65, rue de St. Brieuc, 35042 Rennes, France.
C. Mariojouls	Institut National Agronomique CEREOPA, 16, rue Claude Bernard, 75005 Paris, France.
J. Marcel	ITAVI, Cellule Technique Aquacole, 28 rue du Rocher, 75008 Paris, France.
P. Melotti	Zooculture Institute, Bologna University, Bologna, Italy.
M. Natali	Ichthyological Centre of Lake Trasiment, Sant'Arcangelo of Perugia, Italy.

M. Morand	Laboratoire Vétérinaire Départemental du Jura, BP. 376, 39016 Lons le Saunier, France.
C. Salmomoni	AGROBIOTEC System, Via Bacciliera BP. 376, 40012, Calderara di Reno, Italy.
J. Sevrin-Reyssac	Laboratoire d'Icthyologie Générale et Appliquée, Muséum National d'Histoire Naturelle, 43 rue Cuvier, 75231 Paris, France.
M.A. Szumiec	Science Academy of Poland, Experimental Fish Station, Golysz 43, 422 Chybie, Poland.
L. Varadi	Fish Research Institute, Po. Box 547, Szarvas, Hungary.

Subject index